Protecting-Group-Free
Organic Synthesis

Protecting-Group-Free Organic Synthesis

Improving Economy and Efficiency

Edited by
Rodney A. Fernandes

Indian Institute of Technology Bombay
Mumbai, India

Registered Office(s)
John Wiley & Sons, Inc., 111 River Street, Hoboken, NJ 07030, USA
John Wiley & Sons Ltd, The Atrium, Southern Gate, Chichester, West Sussex, PO19 8SQ, UK

Editorial Office
9600 Garsington Road, Oxford, OX4 2DQ, UK

For details of our global editorial offices, customer services, and more information about Wiley products visit us at www.wiley.com.

Wiley also publishes its books in a variety of electronic formats and by print-on-demand. Some content that appears in standard print versions of this book may not be available in other formats.

Library of Congress Cataloging-in-Publication Data

Names: Fernandes, Rodney A., 1972– editor.
Title: Protecting-group-free organic synthesis : improving economy and efficiency / edited by Rodney A. Fernandes.
Description: First edition. | Hoboken, NJ : John Wiley & Sons, 2018. | Includes bibliographical references and index. |
Identifiers: LCCN 2018002472 (print) | LCCN 2018010625 (ebook) | ISBN 9781119295198 (pdf) | ISBN 9781119295228 (epub) | ISBN 9781119295204 (cloth)
Subjects: LCSH: Organic compounds–Synthesis.
Classification: LCC QD262 (ebook) | LCC QD262 .P77 2018 (print) | DDC 547/.2–dc23
LC record available at https://lccn.loc.gov/2018002472

Cover Design: Wiley
Cover Image: © Aleksandr Simonov/Shutterstock

Set in 10/12pt Warnock by SPi Global, Pondicherry, India
Printed in Singapore by C.O.S. Printers Pte Ltd

10 9 8 7 6 5 4 3 2 1

Editor Details

Rodney A. Fernandes received his PhD from National Chemical Laboratory, Pune, Maharashtra, under the tutelage of Dr. Pradeep Kumar in January 2003. Subsequently he worked with Prof. Yoshinori Yamamoto as a postdoctoral fellow at the Chemistry Department, Tohoku University, Sendai, Japan (January 2003–December 2003). He then moved to Prof. Reinhard Brückner's laboratory at the Institute of Organic Chemistry and Biochemistry, University of Freiburg, Germany (July 2004–March 2006), as an Alexander von Humboldt fellow and then as DFG postdoctoral fellow (April 2006–June 2006). He started his independent research at the Instituto de Quimica, Universidad Nacional Autonoma de Mexico (UNAM), Mexico City (September 2006–July 2007), as assistant professor. He joined the Department of Chemistry of IIT Bombay (Mumbai, India) as assistant professor in August 2007 and became an associate professor in January 2011. He was promoted to full professor in May 2015. He has 89 publications and 8 patents to his credit and has guided 10 PhD's at present. His area of research includes asymmetric synthesis of bioactive natural products, total synthesis, development of synthetic methodologies, and organometallic chemistry, including asymmetric catalysis. He was the recipient of the Indian National Science Academy (INSA) Young Scientist Medal Award in Chemical Sciences in 2004. He is elected fellow of Maharashtra Academy of Sciences (2015). He served as Dean Academic at IIT-Goa on deputation from IIT-Bombay (August 2017–July 2018).

Contents

List of Contributors

Trapti Aggarwal
Department of Chemistry
University of Delhi
India

Rakeshwar Bandichhor
API Research & Development
Integrated Product Development
Dr. Reddy's Laboratories Ltd.,
Hyderabad, India

Alakesh Bisai
Department of Chemistry
Indian Institute of Science Education and
Research Bhopal Bhauri
Bhopal, India

Vishnumaya Bisai
Department of Chemistry
Indian Institute of Science Education and
Research Tirupati
India

Alejandro Cordero-Vargas
Instituto de Química
Universidad Nacional Autónoma de
México
Ciudad de México, México

Rodney A. Fernandes
Department of Chemistry
Indian Institute of Technology Bombay
Mumbai, India

Isao Kadota
Department of Chemistry
Graduate School of Natural Science and
Technology
Okayama University
Japan

Remya Ramesh
Division of Organic Chemistry
CSIR-National Chemical Laboratory
Pune, India

Fernando Sartillo-Piscil
Centro de Investigación
de la Facultad de Ciencias
Químicas and
Centro de Química de la Benemérita,
Universidad Autónoma de Puebla
México

Swapnil Sonawane
API Research & Development
Integrated Product Development
Dr. Reddy's Laboratories Ltd.,
Hyderabad
India

D. Srinivasa Reddy
Division of Organic Chemistry
CSIR-National Chemical Laboratory
Pune, India

Hiroyoshi Takamura
Department of Chemistry
Graduate School of Natural Science and
Technology
Okayama University
Japan

Tomonari Tanaka
Department of Biobased Materials
Science

Graduate School of Science and
Technology
Kyoto Institute of Technology
Japan

Akhilesh K. Verma
Department of Chemistry
University of Delhi
India

Foreword by Prof. W. Hoffmann

Protecting-group-free synthesis has come into focus in the twenty-first century, as current active pharmaceutical ingredients and other targets of organic synthesis have become increasingly more complex, whereby efficiency in synthesis gets a pressing issue. Efficiency in synthesis depends critically on chemoselectivity both in skeleton-building transformations and in refunctionalization reactions. Any lack in chemoselectivity requires protection of the affected functional groups. This and the ultimate deprotection steps decrease the efficiency of a synthesis. Accordingly, the extent of protecting-group use in synthesis is a direct indicator for a lack of chemoselectivity in the transformations applied. While total chemoselectivity in all transformations will remain an utopic goal for long, protecting-group-free synthesis is within closer reach, as it depends not only on functional group-tolerant skeleton-building transformations, such as free radical reaction cascades, transition metal-catalyzed sequences, or biocatalytic events, but protecting-group-free synthesis has in addition a strong component from strategic synthesis planning. The aim is to avoid altogether the incompatibility of vulnerable functional groups with conditions from the necessary skeleton-building reactions. While the latter form the core of a synthesis plan, there is still the option to change the sequence of individual steps in a multistep synthesis to introduce a vulnerable functional group – not before but after the offending skeleton-building step has been executed.

Looking at the targets of organic synthesis, it is trivial to note that protecting-group-free synthesis will be easier to attain with molecules that have a lower degree of functionalization. In turn, protecting-group schemes will prevail for long, when, e.g. the synthesis of polypeptides or polysaccharides is concerned. To render their synthesis protecting-group-free may at present even be counterproductive when aiming for overall synthetic efficiency. That is, protecting-group-free synthesis has no merit in itself when it is judged by the overall economy of a synthesis.

The editor and authors have collated in this volume an impressive number of protecting-group-free syntheses. This number surprises in view of the limited chemoselectivity of present-day synthetic methods. Yet, this number is at the same time encouraging, showing that protecting-group-free synthesis is a valuable goal that can be frequently reached with reasonable effort.

The listed beautiful syntheses in this book have reached the status of being protecting-group-free by significant intellectual input in synthesis planning. To extract this aspect from the individual examples will be the pleasure for the connoisseur reader.

Protecting-group-free synthesis is challenging the present limitations in chemoselectivity of synthetic transformations. In due time chemoselectivity should become increasingly more perfect to the point that protecting-group-free synthesis will in the end become accepted common practice.

Marburg, 28 November 2017

Reinhard W. Hoffmann
Fachbereich Chemie der
Philipps Universität Marburg

Foreword by Prof. G. Mehta

During the advance of synthetic organic chemistry, particularly in the second half of the last century, protecting-group maneuvers emerged as a legitimate and often essential tactic in pursuit of multistep synthesis of complex targets. Indeed, devising new protective groups and deprotection protocols became an active area in itself, and the seminal series of *Greene's Protective Groups in Organic Synthesis* (Volumes 1–4 from 1980 to 2007) with nearly 7000 references bear testimony to the activity and prevailing interest in this area. However, these worthy efforts on protection–deprotection chemistry, unavoidable at the times and contexts, also led to a quest for the avoidance of these "wasteful steps." As green and sustainable chemistry concerns surfaced and drew traction, the assertion that "the best protecting group is one that is not required" gained momentum. In this developing scenario, tactics and strategies deployed in multistep syntheses and total syntheses of natural products that involved circumvention of protecting-group maneuvers came to the fore. Indeed, the past couple of decades has witnessed impressive advances in protecting-group-free (PGF) synthesis, and there is a considerable perceived premium associated with such undertakings.

Thus, the book *Protecting-Group-Free Organic Synthesis: Improving Economy and Efficiency* edited by Rodney A. Fernandes, an active researcher himself, with contributions from many notable practitioners of organic synthesis, is a topical offering and a reminder that the days of "long" and "any how" synthesis are now passé and the need for shorter, efficient syntheses do not permit the luxury of protective group interventions. There is little doubt that in future syntheses that imbed protecting-group operations will be discounted unless a compelling case can be brought out for their use. Such forebodings are already visible in reviewer reports and critical assessment of the quality of a synthetic effort. The strides made in PGF synthesis in recent years are indeed impressive with over 100 PGF syntheses. Many of these PGF syntheses target scarce bioactive natural products that require scale-up and price competitiveness, and such efforts greatly enhance the centrality, utility, and potential of organic synthesis. It is hoped that PGF strategy will find increasing applications in API manufacturing and extend to carbohydrates, peptide, and nucleic acid synthesis where multiple protection–deprotection interventions are generally considered inevitable.

This book provides a diverse coverage of the nascent and emergent field of PGF syntheses of molecules of varying complexity that range from pharmaceuticals to natural products to biopolymers. The ideas based on harnessing cascade/domino processes, deployment of latent functionalities, and exploitation of hidden symmetry have been well articulated in different chapters and should be of interest to the synthetic organic chemistry community in academia as well as industry, and stimulate new directions and tactics in their synthetic efforts. It is also to be expected that PGF synthesis endeavors will lead to newer advances in reagent development and catalyst design for enhanced functional group selectivity, an operational requirement for PGF synthesis. All in all, the book *Protecting-Group-Free Organic Synthesis: Improving Economy and Efficiency* is a welcome contribution that should be of general interest to the synthetic organic chemistry community, and the editor and contributors deserve to be complemented for their efforts.

Hyderabad, 11 December 2017

Goverdhan Mehta
University Distinguished Professor &
Dr. Kallam Anji Reddy Chair
School of Chemistry
University of Hyderabad, India

Preface

Modern organic synthesis has set high standards for its practitioners today. The art of total synthesis has always inspired strong minds who ventured on the tough path of target-oriented synthesis. The last two decades have seen tremendous growth in the complexity of natural products synthesized. I have been always inspired by the mesmerizing total synthesis work by professors – Woodward, Corey, Nicolaou, Kishi, Danheiser, Danishefsky, Paquette, Trauner, Denmark, and many more. Total synthesis has been referred to as the art of building molecules. It poses a myriad of synthetic challenges and requires unabated efforts, overcoming unforeseen hurdles that spring up between putting a proposed strategy on paper and actually executing it in the laboratory. During my days as a research scholar, I was awestruck by the articles from Nicolaou's group on the total synthesis of CP molecules, which is compared to Theseus, a mythical king, who battled and overcame foes (Carl A. P. Ruck and Danny Staples (1994). *The World of Classical Myth*. ch. IX, "Theseus: Making the New Athens", pp 203 – 222. Durham, NC: Carolina Academic Press).

Of the many challenges, chemoselectivity, which includes efficient differentiation of functional groups without masking, has been the toughest challenge imposed upon a total synthesis chemist. A target-oriented synthesis demands that the molecule is synthesized with the correct placement of all its functionality, in addition to the correct stereochemistry in chiral molecules. This helps in validating the proposed structure. Protecting-group-free (PGF) synthesis is one parameter that adds to the overall efficiency and economy of a synthesis, apart from atom and redox economies. In order to achieve this, a chemist needs a clear understanding of functional group reactivity, compatibility of reagents, and reaction conditions. Even though we may be proficient in all the so-called tactics, the ultimate target may be far from reach. Hence it is rightly said that if even the last step fails in a total synthesis, it is enough to jeopardize the whole strategy and hard work put in.

However, with the advent of powerful new reactions and compatibility of reagents and their mechanistic understanding, organic synthesis without protecting groups has now been realized. There have been tireless attempts by many synthetic chemists to design strategies either with minimal or no use of protecting groups, aiming to come closer to achieving an "ideal synthesis." It would be next to impossible to condemn the use of protecting groups, but a sound knowledge of new chemistry, known PGF syntheses, and a desire to practice the latter will go a long way in organic synthesis. While there are a few books available on the development and use of protecting groups in detail, there are currently no books available to the best of my knowledge on practicing/practiced PGF syntheses. Details of the latter may be found as scattered occasional reviews in some forefront journals of organic chemistry. I believe this compilation is the first of its kind based on the syntheses practiced with no use of protecting groups, contributing directly to step economy and hence to the efficiency and economy of the syntheses.

This book intends to give a comprehensive account of practiced, known PGF syntheses of molecules of medium to high complexity. The introductory chapter gives a concise review of historical developments, need, the concepts, and future prospects of PGF syntheses. The next three chapters cover extensive literature on total syntheses of many molecules without protecting groups. This book includes over 100 syntheses that have been achieved without protecting groups. Some are beautifully crafted based on cascade/domino reactions, while others involve rearrangements. PGF syntheses of drugs and related pharmaceuticals with excellent examples of several molecules are discussed as a separate chapter. Synthesis of various heterocycles and carbohydrate-based PGF syntheses will enlighten heterocyclic chemists who are majorly engaged in drug discovery. Moving ahead, more details of practicing PGF synthesis of glycoconjugates, peptides, and biopolymers constitute another relevant chapter. The book winds up with the use of latent functionality-based approaches to target molecules and the beautiful exploration of hidden symmetry (latent symmetry considerations) to achieve the synthesis of nonsymmetric molecules.

I would like to acknowledge all the students from my research group for help in creating some of the schemes in ChemDraw. My family spared me time to work on this manuscript, and my wife, Moneesha, also a chemist, is thanked for proofreading the chapters. I am also sincerely grateful to Prof. Mahesh Lakshman for his suggestions during the proposal stage of this book. I thank my parent institute, Indian Institute of Technology Bombay, for excellent SciFinder search facility and access to other online literature. The final stages post manuscript submission including proof reading were completed at IIT-Goa, while on deputation as Dean Academic Programme and I would like to thank the Director, IIT-Goa for encouragement and support. I also express my gratitude to all the authors for agreeing to contribute to this book without any reservations or demands. I also thank Professor Goverdhan Mehta and Professor Reinhard W. Hoffmann for contributing the Forewords. My apologies if any known PGF synthesis was unintentionally not covered in this book by myself or any contributing author.

Rodney A. Fernandes

1

Introduction: Concepts, History, Need, and Future Prospects of Protecting-Group-Free Synthesis

Rodney A. Fernandes

Department of Chemistry, Indian Institute of Technology Bombay, Mumbai, India

"There is excitement, adventure and challenge, and there can be great art in organic synthesis. These alone should be enough, and organic chemistry will be sadder when none of its practitioners are responsive to these stimuli."

— **R. B. Woodward, 1956**

For "ideal synthesis" "– a sequence of only construction reactions involving no intermediary refunctionalizations, and leading directly to the target, not only its skeleton but also its correctly placed functionality."

— **Hendrickson, 1975**

1.1   Introduction, Concepts, and Brief History

Nature, an architect par excellence, produces hundreds of compounds beautifully crafted and is the master chemist of all. These intriguing molecules have challenged many practitioners of organic synthesis as how to achieve an ideal synthesis that closely resembles nature's creation. The design of a synthetic strategy for a complex molecule from simple synthons and achieving it is an amalgamation of ingenuity, creativity, and determination. Organic synthesis has evolved from the beginning of this century, and chemists have mastered the art of building molecules using the arsenal of reactions, reagents, and analytical methods. The astonishing progress in the last few decades in new methods development, availability of new reagents, and powerful techniques for reaction analysis have changed the dimension and image of the art of organic synthesis. Hence it is rightly said today that with reasonable effort and time, any isolated compound from natural sources with any level of complexity can be synthesized. The remarkable synthetic accomplishments over the years should be considered among the top achievements of human genius.

Protecting-Group-Free Organic Synthesis: Improving Economy and Efficiency, First Edition.
Edited by Rodney A. Fernandes.
© 2018 John Wiley & Sons Ltd. Published 2018 by John Wiley & Sons Ltd.

In organic synthesis, of the three challenges – chemoselectivity, regioselectivity, and stereoselectivity – the most demanding and strenuous to achieve is chemoselectivity [1]. How to differentiate functional groups without selective masking (chemoselectivity) has always been a concern while designing synthetic strategies. A target-oriented synthesis often demands completion of synthesis with many closely placed similar functional groups involving a high level of selectivity, and hence, synthetic strategies, though not desirable, inevitably need to use protecting groups. Hence, most total synthesis chemists invariably follow the commonly available books on various protecting groups and the ways to introduce and also to remove them [2]. A given molecule can be synthesized in many ways by strategic deconstruction reactions or retrosynthesis, which allows many possible options to build the molecule [3]. It is this scope that results in different ways of functional group modifications, some of which may be far from ideal construction reactions, straying away from an ideal synthesis. Hendrickson developed a rigorous system of codification of construction reactions to build a complex molecule [4]. It can be inferred from his paper that an "ideal synthesis" would require "– a sequence of only construction reactions involving no intermediary refunctionalizations, and leading directly to the target, not only its skeleton but also its correctly placed functionality." Thus there exists a need for truly constructive or skeleton-building reactions in total synthesis. Although this concept has inspired many minds to design efficient strategies, the practice of total synthesis may need a long way to go to achieve an ideal protecting-group-free (PGF) synthesis, the nature's way [5]. There are many complex molecules with multiple functionalities, and their synthesis inevitably necessitates protecting groups due to the close similarity of functional groups reactivity. In many cases, cascade reactions and rearrangements are sought after to achieve a PGF-based close to an ideal synthesis. Many syntheses are biomimetic and therefore based closer to the biosynthesis pathway and use the natural reactivity of functional groups. This sounds good when complex molecules have an all-carbon framework and/or minimal functional groups. This can be exemplified by Anderson's synthesis of α-cedrene (**5**; Scheme 1.1) [6]. A pentane solution of nerolidol (**1**) was treated with formic acid and then with trifluoroacetic acid (TFA) for 2 h to obtain α-cedrene (**5**) in about 20% yield. This synthesis involving the bisabolene to spirane intermediates (type **2** and **3**, respectively) closely mimics the parallel biogenetic pathway.

Another closely related synthesis by Corey and Balanson [7] involved the ring opening of cyclopropane **12** generating a carbonium ion and subsequent incipient carbanion **13**, which triggers two ring closures giving cedrone **14** (Scheme 1.2), from which the synthesis of α-cedrene (**5**) is known [8]. Addition of lithiated compound **7** to enol ether enone **6** gave compound **8**. This on DIBAL-H reduction to **9** and regioselective cyclopropanation provided **10**. Further Collins oxidation gave ketone **11**, which was then subjected to rearrangement to deliver α-cedrene (**5**).

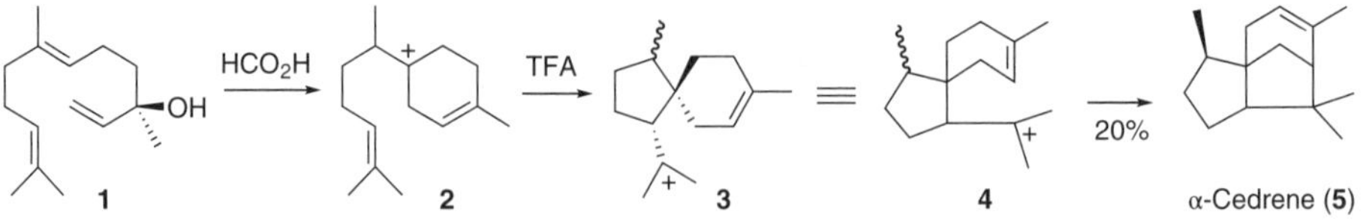

Scheme 1.1 Anderson's synthesis of α-cedrene (**5**).

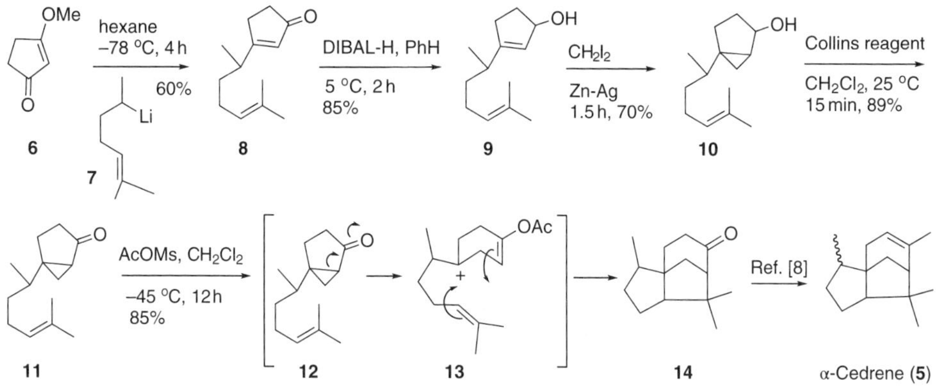

Scheme 1.2 Corey's synthesis of α-cedrene (**5**).

Historically, many early syntheses were reported without employment of protecting groups. The targets were simple at that time and had limited functionality, and masking groups was not a necessity. Thus it is rightly said that practicing PGF synthesis is not by synthetic planning but out of choice or necessity. Hence many a time the first synthesis of a newly isolated natural product of reasonable complexity is well praised and has its own charm, even though the second synthesis could be shorter, PGF, and much more efficient. The syntheses of early times could be evaluated for efficiency even though feasibility was what counted the most. The concept of PGF synthesis was not as developed and sought after as it is today. For example, the synthesis of tropinone (**21**) by Robinson in 1917 is considered as one of the greatest achievements in organic synthesis as it was PGF, and the choice of materials used for its preparation had a natural reactivity that followed a distinct pathway with minimal side reactions (Scheme 1.3) [9]. The synthesis illustrates the genius of Robinson, and it could partly be attributed to the inherent symmetry of the natural product and his knowledge of alkaloid biogenesis. The materials used are succinaldehyde (**15**), methylamine, and acetonedicarboxylic acid (ADCA, **17**) in water as a medium, reacted by a distinct cascade reaction path involving imine formation, Mannich reaction, and, lastly, double decarboxylation during acidic work-up, to provide tropinone in moderate 42% yield. This synthesis has entered in every account reported thereafter based on the concepts, be it PGF

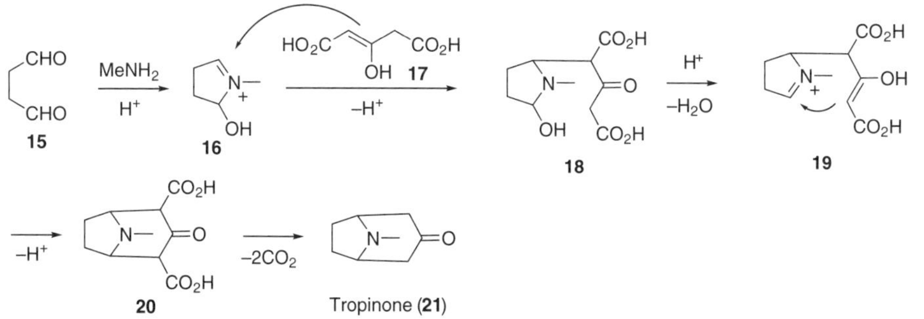

Scheme 1.3 Robinson's synthesis of tropinone (**21**) in 1917.

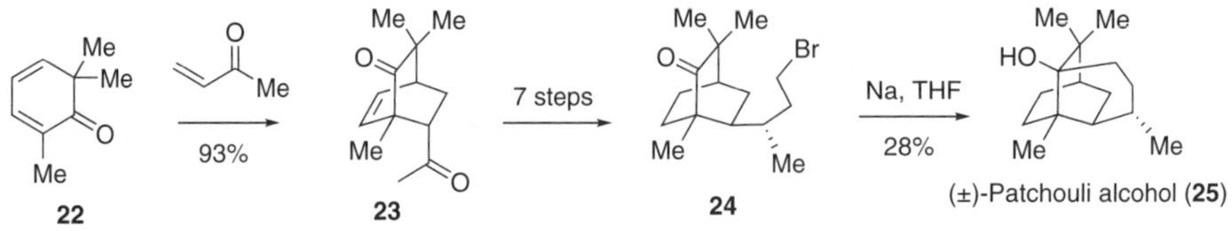

Scheme 1.4 Danishefsky's synthesis of (±)-patchouli alcohol.

syntheses, total syntheses, ideal synthesis, green chemistry, or modern organic synthesis. This synthetic strategy conceptually still poses a challenge to future chemists to find a catalytic system that could make acetone to successfully replace acetonedicarboxylic acid (it is known that this gives very low yields in comparison with ADCA). This would then qualify for a truly ideal synthesis or closer to atom-economic synthesis.

Danishefsky's synthesis of (±)-patchouli alcohol (**25**) in 1968 represents another early example of an efficient PGF synthesis (Scheme 1.4) [10]. The natural product had limited functional groups (only one OH group), which made the design of synthetic strategy simpler. The strategy was based on skeleton-building steps with minimum side reactions. The initial Diels–Alder reaction of **22** with methyl vinyl ketone set the [2.2.2] bicyclic system **23** in place. The remaining steps were toward the construction of the third ring.

Greene and coworkers in 1978 reported an efficient conversion of α-santonin (**26**) to (−)-estafiatin (**30**; Scheme 1.5) [11]. (−)-Estafiatin was isolated from the bitter herb *Artemisia mexicana* in 1963 by Sanchez-Viesca and Romo [12]. α-Santonin (**26**) was converted in three steps to produce compound **27**. Further reduction of the enone with NaBH$_4$ in pyridine and elimination of the alcohol in HMPA at 250 °C gave a mixture of di- and trisubstituted olefins from which the latter diene **28** was separated. Further addition of α-selenide to the lactone **28** and elimination gave the *exo*-methylene compound **29**. Selective epoxidation of the triene from the less hindered α-face produced (−)-estafiatin (**30**) as the major product. The synthesis represented an efficient conversion of one natural product to the other.

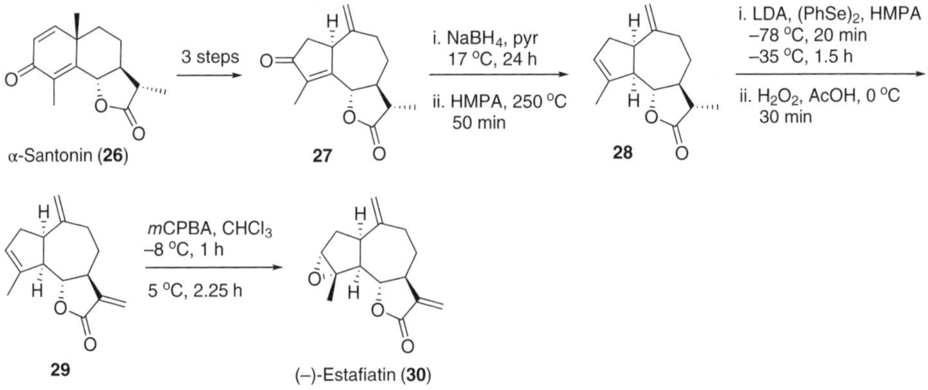

Scheme 1.5 Total synthesis of (−)-estafiatin (**30**).

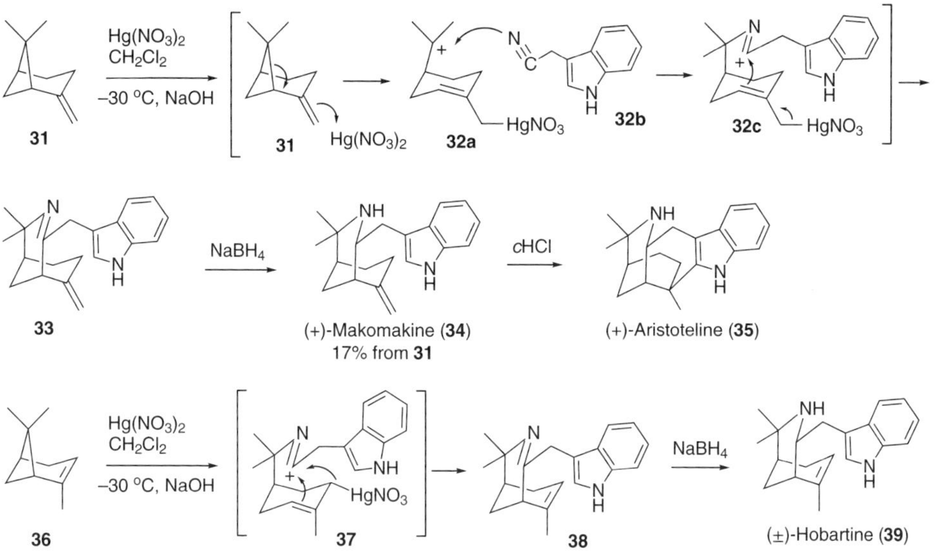

Scheme 1.6 Kenney's synthesis of (+)-makomakine, (+)-aristoteline, and (±)-hobartine.

While a few reports on PGF synthesis of natural products appeared during the 1970s, Stevens and Kenney in 1983 executed an expeditious synthesis of (+)-makomakine (**34**), (+)-aristoteline (**35**), and (±)-hobartine (**39**) (Scheme 1.6) [13]. The reaction of (−)-β-pinene (**31**) with indol-3-ylacetonitrile **32b** in the presence of Hg(NO₃)₂ led to [3.3.1] bicyclic structure **33** as a single diastereomer. The most plausible pathway involves the opening of cyclobutane **31**, giving cation **32a**. Addition of nitrile nitrogen generates iminium cation **32c**, which stereoselectively closes to give imine **33**. Reduction of the imine **33** gave (+)-makomakine (**34**), isolated in 17% overall yield from **31**. Treatment of **34** with conc. HCl induced C-2 indole cyclization, producing (+)-aristoteline (**35**). A similar sequence from (+)-α-pinene (**36**) gave **38** that was reduced to (±)-hobartine (**39**). This was obtained as a racemate as the intermediate allylic mercurial compound **37** can cyclize on both faces of the iminium cation.

At the same time Weinreb and coworkers reported the PGF synthesis of (±)-cryptopleurine (**48**) as shown in Scheme 1.7 [14]. Homoveratric acid **40** was condensed with *p*-anisaldehyde to afford the olefin **41**, which, over four steps including esterification, cyclization, ester reduction, and oxidation, was converted into aldehyde **42**. Subsequent Wittig olefination with **43** resulted in 3 : 1 (*E/Z*)-olefin mixture; the acid obtained was converted into amine **44**. The latter was converted into methylol acetate **45** and then heated in a sealed tube to obtain the cyclized product **47** via intermediate **46**. It was observed that the (*Z*)-isomer did not react and was recovered as the amide (*Z*)-**44**. Subjecting the methylol acetate prepared from this amide (*Z*)-**44** gave no lactam on heating, possibly due to nonavailability of required conformation for cyclization. Reduction of lactam **47** with LiAlH₄ gave (±)-cryptopleurine (**48**).

In 1989, Heathcock et al. carried out extensive work resulting in the PGF total synthesis of racemic fawcettimine (**58**) [15]. Reaction of cyano enone **49** with bis-silane **50** under TiCl₄ conditions gave **51** quantitatively (Scheme 1.8). Further oxidation provided

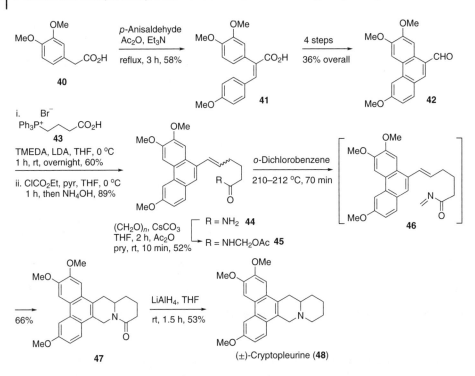

Scheme 1.7 Total synthesis of (±)-cryptopleurine (**48**) by Weinreb et al. in 1983.

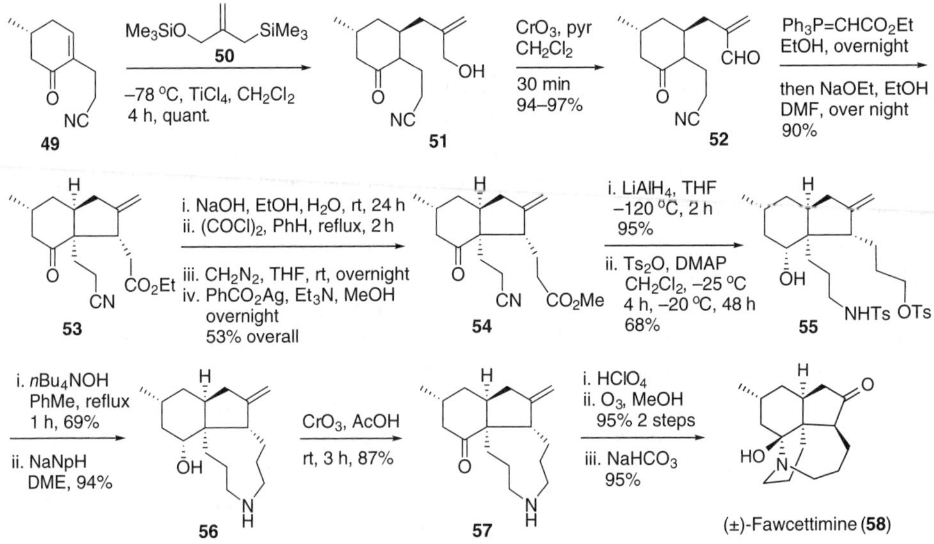

Scheme 1.8 Total synthesis of (±)-fawcettimine (**58**) by Heathcock in 1989.

the aldehyde **52**, which upon Wittig olefination and intramolecular cyclization gave **53**. The reaction involved a selective 1,4-addition over the other possible 1,6-addition, following Baldwin's rules for 5-*exo-trig* cyclization. The reaction was also diastereoselective, giving **53** as a single diastereomer. Further the Arndt–Eistert homologation of **53** provided **54** in overall 53% yield. The next reduction was dependent on temperature, giving the α-hydroxy group in 10 : 1 ratio at –110 to –120 °C. Both the amino and hydroxyl groups were tosylated to give **55**. Further intramolecular cyclization and tosyl removal gave amine **56**. The chromic acid oxidation of **56** provided **57**, which was converted into perchlorate salt and subjected to ozonolysis followed by quenching with NaHCO$_3$ to give fawcettimine (**58**), isolated as hydrobromide salt on treatment with aqueous HBr.

In 1993, Heathcock reported an elegant synthesis of (–)-alloaristoteline, (–)-serratoline, and (+)-aristotelone [16]. Details of this are presented in Chapter 4. In the 1990s, the total syntheses of many natural products were reported without using protecting groups. Thus this marked the beginning where efficiency replaced feasibility in total synthesis. The subsequent chapters describe many such PGF syntheses.

1.2 Need and Future Prospects of Protecting-Group-Free Synthesis

Understanding the redox economy of a synthetic design is very important. Many syntheses use several functional group manipulations, resulting in unwarranted loss in atom economy and lengthy sequences. Baran demonstrated a method for determining the % ideality in a synthesis [17]. This is given as follows:

$$\%ideality = \frac{number\ of\ construction\ reactions + number\ of\ startegic\ redox\ reactions}{total\ number\ of\ steps} \times 100$$

It would be wise that most syntheses for a given target molecule are evaluated for their % ideality and compared. The higher the number of construction and strategic redox reactions in a synthesis, the higher is the ideality. Nature's perfection lies in all the constructive reactions being involved in the biogenetic path. However, % ideality is not just one parameter for efficiency. Many other parameters need to be optimized for an efficient synthesis, e.g. easier purifications, chemical costs and availability, number of steps, reaction yields, etc. Thus, overall yield, number of steps, and percent ideality would primarily govern the efficiency of synthesis. PGF synthesis is one criterion that would control the step economy and add to overall efficiency.

With growing expenses and environmental concerns with respect to use of hazardous chemicals, wastage, and disposal as big concerns, an efficient synthesis meeting atom economy, step economy, and redox economy is highly desirable. It should be possible to provide large quantities of complex natural product with minimum efforts and expense. Practicing close to an "ideal synthesis" is therefore a daunting task. It highly demands strategies that meet Hendrickson's criteria of ideal synthesis. While designing strategies, one needs to minimize nonstrategic redox manipulations, indirect functional group interconversions, and protecting-group manipulations [4, 17]. All these concession steps may be unavoidable but add to the step count in total synthesis, resulting in wastage of material, especially in low-yielding steps. A nonstrategic redox manipulation can be exemplified by an ester group reduction to an alcohol and reoxidation to an aldehyde. Sometimes it is possible for an ester group to be reduced partially to an aldehyde, but this depends on the substrate, and in most cases it results in overreduction or mixtures. Hence as a precaution, one would first reduce the ester to an alcohol and then reoxidize it to the aldehyde, without a concern for the additional step and an added nonstrategic redox count. It is also possible that the sensitive aldehyde is brought in as an acetal, but this too adds to protecting-group manipulation. Amine functionality is many times derived through an azide. While the azide is quite a stable precursor and can be brought into the substrate, perhaps, from the beginning of a synthesis, and then reduced at the correct time, it is sometimes derived from the alcohol by first conversion to a halide or mesyl/tosyl derivative that is then displaced by azide. All these add to indirect functional group interconversions, and though do not count toward protecting groups, they add to the step count. Thus a PGF synthesis should also cater to atom economy and redox economy.

However, with the advent of powerful new reactions and their mechanistic understanding and compatibility of modern reagents, PGF organic synthesis has now been realized. Tireless attempts have been made by many synthetic chemists to design strategies avoiding or with minimal use of protecting groups so as to come closer to achieving an "ideal synthesis." Cascade or domino reactions and rearrangements to bring quick molecular complexity with skeleton-building reactions are present-day strategies. It would be next to impossible to condemn the use of protecting groups, but a sound knowledge of new chemistry, known PGF syntheses, and a desire to practice the latter will go a long way in organic synthesis.

References

1 (a) Trost, B.M. (1983). *Science* 219: 245. (b) Shenvi, R.A., O'Malley, D.P., and Baran, P.S. (2009). *Acc. Chem. Res.* 42: 530.

2 (a) Kocienski, P.J. (2005). *Protecting Groups*, 3e. New York: Theme. (b) Greene, T.W. and Wuts, P.G. (1999). *Protecting Groups in Organic Syntheses*, 3e. Hoboken: Wiley.

3 (a) Sierra, M.A. and de la Torre, M.C. (2004). *Dead Ends and Detours, Direct Ways to Successful Total Synthesis*. Weinheim: Wiley-VCH. (b) Nicolaou, K.C. and Sorensen, E.J. (1996). *Classics in Total Synthesis*. New York: VCH. (c) Nicolaou, K.C. and Snyder, S.A. (2003). *Classics in Total Synthesis II*. Weinheim: Wiley-VCH.

4 Hendrickson, J.B. (1975). *J. Am. Chem. Soc.* 97: 5784.

5 For reviews on protecting-group-free syntheses see:(a) Hoffmann, R.W. (2006). *Synthesis* 3531. (b) Young, I.S. and Baran, P.S. (2009). *Nat. Chem.* 1: 193. (c) Saicic, R.N. (2014). *Tetrahedron* 70: 8183.

6 Andersen, N.H. and Syrdal, D.D. (1972). *Tetrahedron Lett.* 13: 2455.

7 Corey, E.J. and Balanson, R.D. (1973). *Tetrahedron Lett.* 14: 3153.

8 Stork, G. and Clarke, F.H. Jr. (1955). *J. Am. Chem. Soc.* 77: 1072.

9 Birch, A.J. (1993, 1938–1996). *Notes Records R. Soc. Lond.* 47 (2): 277.

10 Danishefsky, S. and Dumas, D. (1968). *Chem. Commun.* 1287.

11 Edgar, M.T., Greene, A.E., and Crabbé, P. (1979). *J. Org. Chem.* 44: 159.

12 Sanchez-Viesca, F. and Romo, J. (1963). *Tetrahedron* 19: 1285.

13 Stevens, R.V. and Kenney, P.M. (1983). *J. Chem. Soc., Chem. Commun.* 384.

14 Bremmer, M.L., Khatri, N.A., and Weinreb, S.M. (1983). *J. Org. Chem.* 48: 3661.

15 Heathcock, C.H., Blumenkopf, T.A., and Smith, K.M. (1989). *J. Org. Chem.* 54: 1548.

16 Stoermer, D. and Heathcock, C.H. (1993). *J. Org. Chem.* 58: 564.

17 Gaich, T. and Baran, P.S. (2010). *J. Org. Chem.* 75: 4657.

2

Protecting-Group-Free Synthesis of Natural Products and Analogs, Part I

Rodney A. Fernandes

Department of Chemistry, Indian Institute of Technology Bombay, Mumbai, India

2.1 Introduction

Synthetic strategies efficient in step economy apart from atom economy are in need today while performing modern organic synthesis. Practitioners of organic synthesis today are aware of the consequences of concession steps in synthesis design that lower overall efficiency [1]. Achieving atom economy in most reaction steps is very difficult due to the condensation nature of many reactions, side product formation, low-yielding steps, product isolation problems, use of sophisticated and/or stoichiometric reagents, etc. This greatly affects the overall efficiency of a synthesis. It is quite possible to achieve step economy by reducing nonstrategic steps like functional group interconversions and designing protecting-group-free (PGF) synthesis [2]. These also add to the overall efficiency of a synthesis apart from atom economy. PGF synthesis is not easy and is highly target dependent. It is next to impossible practicing PGF syntheses, say, in polypeptides and polysaccharides, due to the involvement of several functional groups of similar reactivity. On the other hand, natural products with an all-carbon framework or with minimal functional groups could be synthesized easily by PGF synthesis. Many times, cascade or domino reactions or rearrangements are sought after for generating molecular complexity as these involve multiple bond formations that can add value, if these are skeleton constructive steps [3]. A sound knowledge of the reactivity pattern of the chosen synthons and modern reagents that are compatible with multiple reaction sites, and/or further exploiting certain features in the target molecule like symmetry, is desirable for practicing PGF synthesis.

Embarking on the design of a PGF synthesis is usually not by synthetic planning, but out of choice or necessity. When a new natural product is isolated and shows potent bioactivity, most practitioners of modern organic synthesis would be in a race to be the first to achieve its synthesis. Here, the efficiency of the synthesis takes second place, while time is the prime ruler. It is the first synthesis that is well acclaimed and celebrated for a reasonably complex molecule, no matter how little efficient it is. The first synthesis indeed deserves the prize as it authenticates the proposed structure for a newly isolated molecule. It may also ascertain the absolute stereochemistry and hence the absolute

Protecting-Group-Free Organic Synthesis: Improving Economy and Efficiency, First Edition.
Edited by Rodney A. Fernandes.
© 2018 John Wiley & Sons Ltd. Published 2018 by John Wiley & Sons Ltd.

structure for chiral molecules. Many times, the first synthesis has helped in correcting erroneously assigned structures, indicating revision of the full structure or the proposed stereochemistry, in case of chiral molecules.

There is uttermost need today to develop powerful reactions or reagents that can also be of general nature and adoptable for complex molecule synthesis. Several new methods or reagents are reported every day; however many suffer in generality. The much sought-after reactions of the decade, C–H activations [4], have been looked at with high expectations to enable total synthesis [5]. Similar to auxiliary-based synthesis, the use of directing groups is not atom economical, requiring the attachment and detachment of the auxiliary or directing group. This again necessitates the design of directing groups that are easily attachable and also detachable. The case of using protecting groups in stoichiometric amounts is also similar. There have been a few reports of directing groups that are of a transient nature, circumventing some of the issues of auxiliary attachment methods [6]. These self-attaching groups, when used in catalytic amount, hold promise for efficiency in organic synthesis.

Nevertheless, literature is bountiful with examples of total syntheses of natural products ranging from simple structure to those of complex nature, being synthesized devoid of protecting groups. This chapter discloses several such syntheses arranged in order of presence of no rings (acyclic) to monocyclic, bicyclic, tricyclic, polycyclic, and macrocyclic type. The chapter is not comprehensive, and subsequent chapters will cover many other PGF-based syntheses.

2.2 Mytilipin A

Mytilipin A (**9**) is a hexachlorosulfolipid and a member of a growing family of chlorosulfolipids [7]. Vanderwal et al. [8] reported a short synthesis of **9** as shown in Scheme 2.1. *anti*-Dichlorination of crotyl alcohol (**1**) with Cl_2 in the presence of Et_4NCl provided **2**. Oxidation of **2** with Dess–Martin periodinane (DMP) followed by careful work-up gave the sensitive and volatile aldehyde **3**, which was immediately subjected to allylation with bromoallylaluminum reagent. This resulted in the intermediate bromohydrin in high diastereocontrol (Felkin–Anh and Cornforth models), which with aqueous base was converted into allyl epoxide (±)-**4**. The alkene partner **6** was prepared from 8-bromo-1-octene **5** by converting into the homolog aldehyde and then Takai chloro-olefination. The metathesis of **6** with **4** using Grubbs' cycloadamantyl catalyst provided the Z-alkene **7** with complete control of alkene geometry ($Z/E = 20 : 1$). Chlorinolysis of vinyl epoxide **7** proceeded smoothly with inversion of configuration to form **8**. Further chlorination of **8** with Et_4NCl with complete chemoselectivity and high diastereoselectivity led to the hexachlorinated product, which on sulfation of the secondary alcohol (Carreira's conditions) [9] provided efficiently mytilipin A (**9**).

They also considered the resolution of vinyl epoxide (±)-**4** adopting the *meso*-epoxide desymmetrizing chlorinolysis of Denmark and coworkers [10] (Scheme 2.2). This resulted in (+)-**4**, obtained in 93.5 : 6.5 er and 43% yield (at 57% conversion). The stereochemical outcome of this resolution resulted in opposite stereochemistry in (+)-**4**, leading to the synthesis of the unnatural enantiomer of mytilipin A following the same steps as in Scheme 2.1.

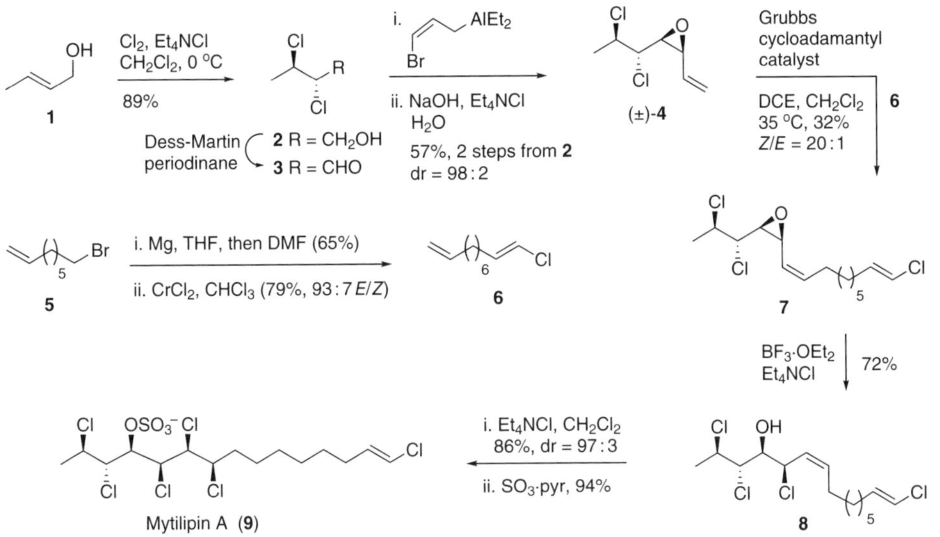

Scheme 2.1 Total synthesis of mytilipin A (**9**) by Vanderwal et al.

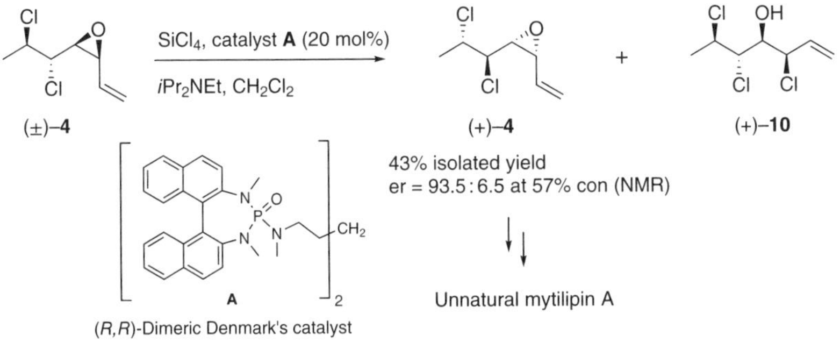

Scheme 2.2 Kinetic resolution of (±)-**4** to enantio-enriched vinyl epoxide (+)-**4**.

2.3 Chokols

The chokols having antifungal activity are 2,6-cyclofarnesanes, isolated from the stroma of the timothy-grass *Phleum pratense* infected by the endophytic fungus *Epichloe typhina* [11]. Barrero and coworkers reported a short protocol for the synthesis of chokols K, E, and B, which includes Ti(III)-mediated diastereoselective radical cycliza-tion with the hydroxyl group directing the stereochemical outcome [12]. The synthesis of chokols started from naturally occurring (+)-nerolidol (**11**) as shown in Scheme 2.3. The epoxidation of **11** with VO(acac)$_2$ delivered 1,2-monoepoxy derivative **12** in 72% yield. The latter on Ti(III)-mediated diastereoselective cyclization afforded **13** with cyclopentane unit. Further, deoxygenation of primary alcohol was carried out chemoselectively via the xanthate and reduction with Bu$_3$SnH, AIBN to furnish the chokol K (**14**). This on treatment with AD-mix-α delivered a 1 : 1 mixture of chokol E

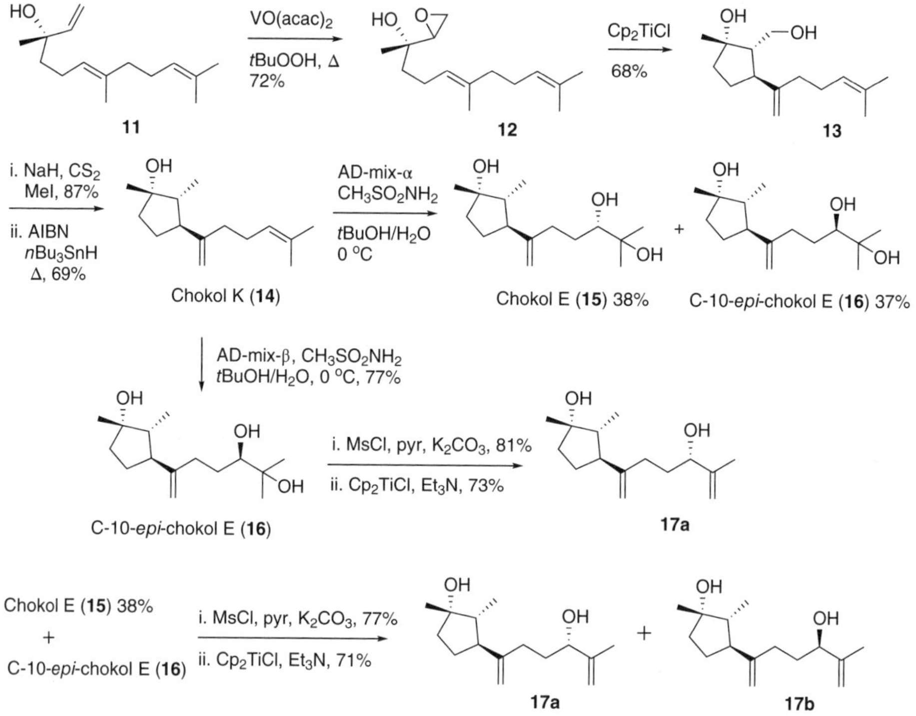

Scheme 2.3 Synthesis of chokols K (**14**), E (**15**), and B (**17a** or **17b**).

(**15**) and C-10-*epi*-chokol E (**16**). However, the reaction of chokol K (**14**) with AD-mix-β provided **16** as a single diastereomer. The structure of chokol B is either **17a** or **17b**. Thus the conversion of **16** into the epoxide and titanocene(III)-mediated [13] opening of the oxirane gave **17a**. Similarly the mixture of **15** and **16** delivered the mixture of **17a** and **17b**. Since the NMR spectra of these have overlapping signals and the optical rotation of natural chokol B is unknown, its structure could not be assigned.

2.4 (±)-Diospongin A

Tandem reactions have significance in the generation of molecular complexity and are step economic [3]. Hong et al. [14] developed tandem cross-metathesis/thermal S_N2' methodology to construct *O*-heterocycles (Scheme 2.4). The strategy was utilized in the PGF synthesis of (±)-diospongin A (**23**). Addition of allylmagnesium bromide to benzaldehyde (**18**) gave **19**. Terminal olefin cleavage and allylation provided the diol **20** (*syn/anti* = 5 : 1) with readily separable diastereomers. The tandem cross-metathesis of *syn*-diol **20** with allyl bromide and heating the mixture in toluene resulted in S_N2' displacement of bromide, giving the desired 4-hydroxy-2,6-*cis*-tetrahydropyran **21b** (**21a** : **21b** = 1 : 5, 83%). The subsequent cross-metathesis of **21b** with styrene led to **22**. The latter on regioselective Wacker-type oxidation gave (±)-diospongin A (**23**).

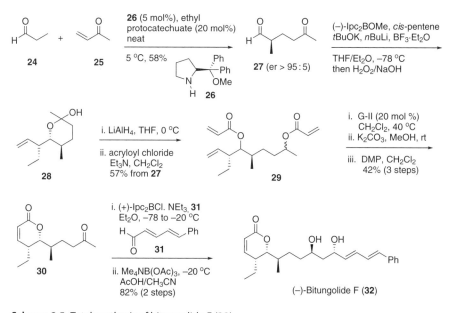

Scheme 2.4 Synthesis of (±)-diospongin A (**23**).

2.5 (–)-Bitungolide F

Bitungolide F (**32**) containing γ-ethyl-substituted α,β-unsaturated 5,6-dihydropyran-2-one framework was isolated by Tanaka et al. [15] from Indonesian sponge *Theonella cf. Swinhoei*. The compound displayed cytotoxic effect against 3Y rat normal fibroblast cells and inhibition toward VH1-related dual-specificity phosphatase. Cossy et al. developed a PGF synthesis of bitungolide F as shown in Scheme 2.5 [16]. The newly

Scheme 2.5 Total synthesis of bitungolide F (**32**).

developed, highly enantioselective Chi and Gellman's [17] organocatalytic Michael addition between propanal **24** and methyl vinyl ketone **25** offered α-alkylated dicabonyl **27** in fair yield and good enantioselectivity (58%, er > 95 : 5). The compound **27** on boron-mediated asymmetric pentenylation formed homoallylic alcohol, which hemiketalized simultaneously to give **28**. The latter was reduced with LiAlH$_4$ to offer the diol, which was acylated using acryloyl chloride to provide **29**. This on ring-closing metathesis with Grubbs II catalyst led to the six-membered lactone. Further saponification using 5% KOH gave the free alcohol, which on oxidation with DMP provided ketone **30**. The latter was treated with (+)-Ipc$_2$BCl in chiral aldol reaction with dienal **31** to give aldol product (dr = 5 : 1), which was later reduced to 1,3-*anti*-diols using Me$_4$NB(OAc)$_3$, resulting in (−)-bitungolide F (**32**).

2.6 (+)-Brevisamide

In 2011, Zakarian and coworkers [18] reported a concise, enantioselective PGF total synthesis of diversely functionalized cyclic ether natural product (+)-brevisamide (**43**), a metabolite isolated from dinoflagellate *Karenia brevis* [19]. The synthetic plan was based on a catalytic asymmetric Henry reaction and Achmatowicz rearrangement as key steps. The end game involved a Stille cross-coupling reaction to install the conjugated (*E,E*)-dienal (Scheme 2.6).

The commercially available ethyl 3-(furan-2-yl)propionate **33** was converted into **34** over 5 conventional steps involving ester reduction to alcohol, oxidation to aldehyde, terminal alkyne generation, methylation, and furan formylation. The key Henry

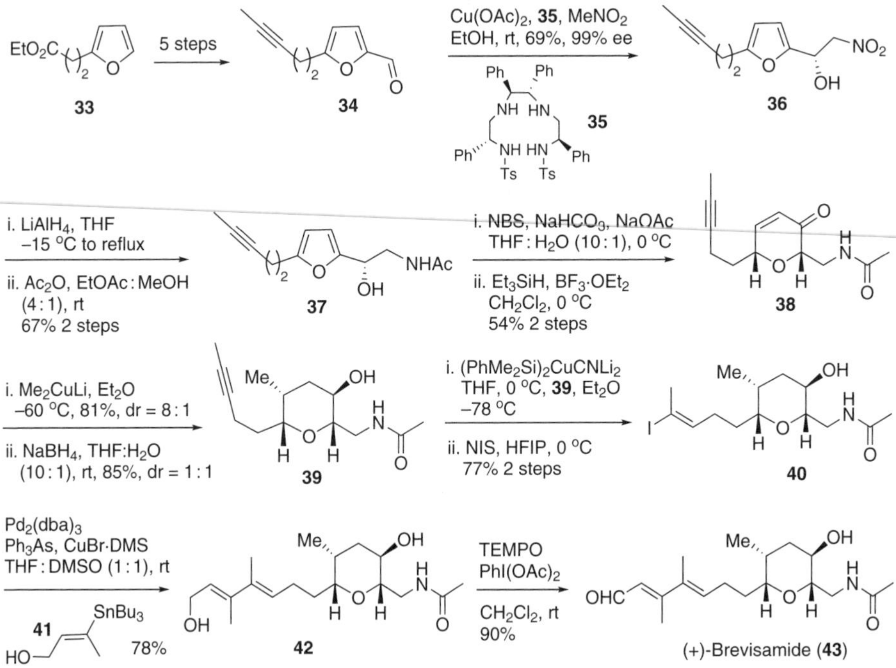

Scheme 2.6 Zakarian's total synthesis of (+)-brevisamide (**43**).

reaction was attempted on **34** using the conditions developed by Wan and coworkers [20], giving **36** in 69% yield and 99% ee. LiAlH$_4$ reduction to amine and selective acylation provided the amide **37**. The cyclic hemiacetal was obtained by Achmatowicz rearrangement [21], and further reduction with BF$_3$.OEt$_2$ and Et$_3$SiH produced **38** in good yield. Addition of methyl group (dr = 8 : 1) and reduction of keto group afforded the more stable diastereomer **39** in 1 : 1 ratio (separable). The undesired epimer was recycled by oxidation–reduction sequence. Further installation of vinyl iodide with silylcupration followed by iododesilylation with *N*-iodosuccinimide in hexafluoroiso-propanol (HFIP) provided **40** in complete stereoselectivity and high regioselectivity. The end-game Stille cross-coupling with vinyltin reagent **41** to **42** (78%) followed by chemoselective oxidation of the allylic alcohol using TEMPO in the presence of PhI(OAc)$_2$ yielded (+)-brevisamide (**43**) in 90% yield.

2.7 21,22-*Diepi*-membrarolin

Metal-oxo-mediated oxidative cyclization of 1,5-diene leads to substituted tetrahydro-furan (THF). This strategy was employed by Hu and Brown in the synthesis of 21,22-*diepi*-membrarollin (**53**; Scheme 2.7) [22]. The KMnO$_4$-mediated oxidative

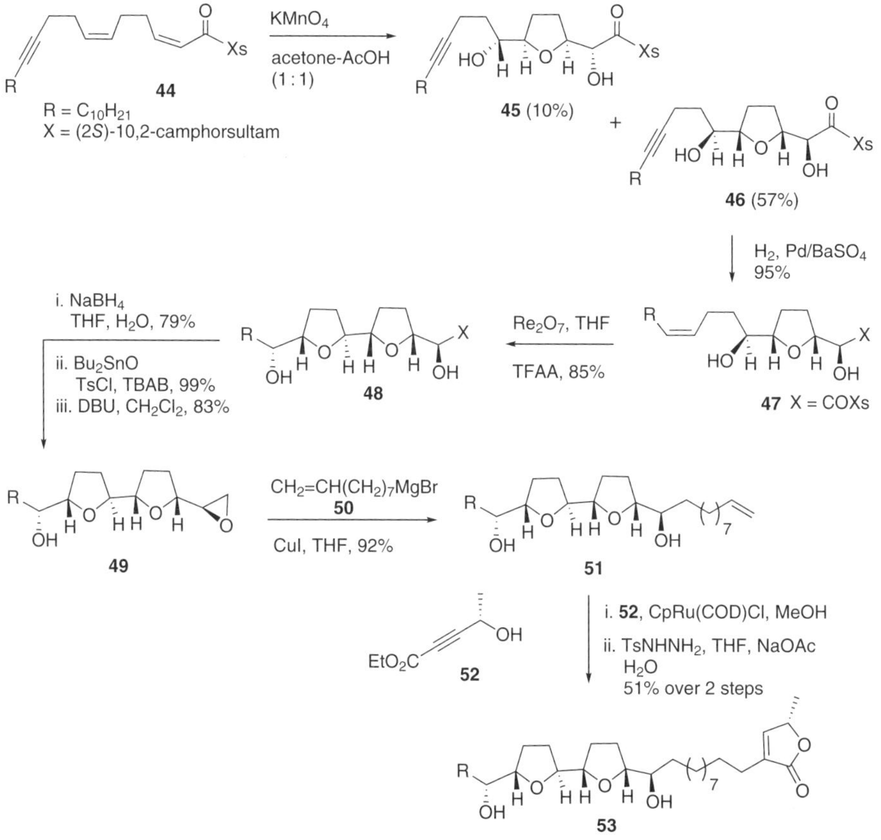

Scheme 2.7 Synthesis of 21,22-*diepi*-membrarollin (**53**).

cyclization of dienyne **44** afforded a separable mixture of THF diols **45** and **46** (1 : 6 ratio). Partial reduction of alkyne **46** gave the (*Z*)-olefin **47**, which was subjected to Re_2O_7-promoted oxidative cyclization, affording the bis-THF **48** in excellent yield as a single isolated diastereomer. Reductive removal of the chiral auxiliary, tosylation of the primary hydroxyl, and displacement of the tosyl group gave the epoxide **49**. Epoxide opening with Grignard reagent **50** gave the olefin **51**. From here, known established steps were employed to complete the synthesis of 21,22-*diepi*-membrarollin **53** [23]. This involved the formal Alder-ene reaction of **51** with alkyne **52** to give butenolide, which upon reduction of the nonbutenolide double bond by diimide gave **53** in 51% yield over two steps.

2.8 (±)-Pogostol and (±)-Kessane

Pogostol, isolated from the patchouli plant, has been used in Chinese medicine and has antiemetic properties [24]. Kessane, a structurally related molecule, was isolated from *Valeriana officinalis* [25]. A common synthesis of pogostol (**59**) and kessane (**62**) by Booker-Milburn and coworkers [26] is shown in Scheme 2.8. Addition of 3-butenylmagnesium bromide to **54**, followed by TMSCl and Et_3N, afforded the enol ether **55** as a single diastereomer, which was then subjected to cyclopropanation using Et_2Zn/CH_2I_2 to give **56**. Further the Fe(III)-mediated ring expansion cyclization reaction with $Fe(NO_3)_3$ and 1,4-cyclohexadiene (as hydrogen atom donor) resulted in the bicyclic compound **57**. Remarkably, the tertiary hydroxyl group remained intact without elimination. Compound **57** on Wittig olefination and hydroboration–oxidation provided the ketone **58** that on Tebbe methylenation afforded (±)-pogostol (**59**) and (±)-*epi*-pogostol (**60**). The latter on reaction with *N*-iodosuccinimide gave a mixture of iodo-ethers **61a** and **61b**, which on hydrogenation provided (±)-kessane (**62**).

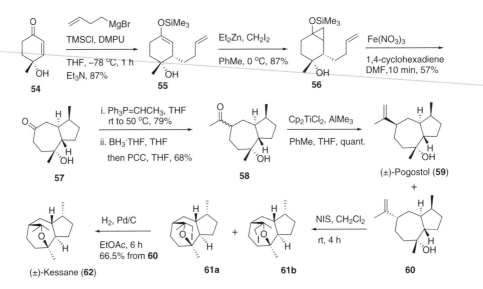

Scheme 2.8 Total synthesis of (±)-pogostol (**59**) and (±)-kessane (**62**).

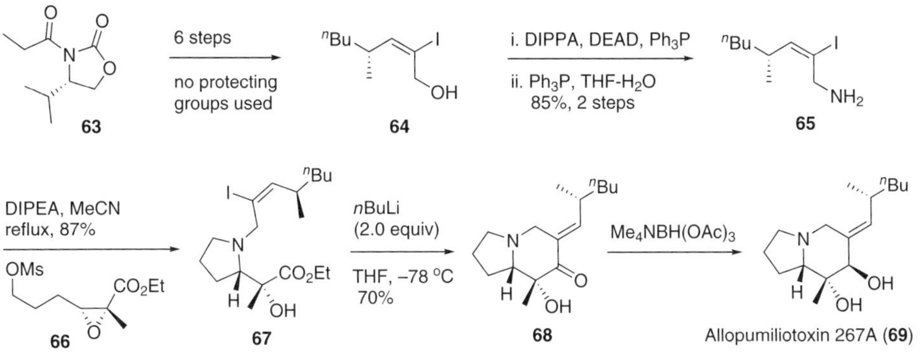

Scheme 2.9 Total synthesis of (+)-allopumiliotoxin 267A (**69**).

2.9 (+)-Allopumiliotoxin 267A

Allopumiliotoxin 267A (**69**) is one of the members of amphibian alkaloids. A short and efficient total synthesis of **69** was achieved by Wang and coworkers (Scheme 2.9) [27]. A reported route was employed from the Evans auxiliary **63** to get **64** over six steps [28]. Azide displacement by using Mitsunobu conditions followed by reduction gave amine **65**. A one-pot epoxide ring opening of **66** by **65** followed by intramolecular cyclization afforded the pyrrolidine ring **67**. The intramolecular nucleophilic acyl substitution reaction gave the desired bicyclic enone **68** in good yield. Finally substrate-directed reduction of **68** afforded the natural product (+)-allopumiliotoxin 267A (**69**).

2.10 (−)-Hortonones A–C

In 2011, Andersen et al. isolated the hexahydroazulenones hortonones A–C, which are rearranged sesquiterpenoids from the leaves of Sri Lankan *Hortonia* [29]. Hortonone C showed *in vitro* cytotoxicity against human breast cancer MCF-7 cells at 5 μg ml^{-1}. Stambulyan and Minehan reported the synthesis of hortonones A–C (**79–81**) from vitamin D$_2$ via the Inhoffen-Lythgoe diol **71** without any protecting groups (Scheme 2.10) [30]. Oxidative cleavage of vitamin D$_2$ (**70**) with ozone followed by NaBH$_4$ reduction gave low overall yields (~40%) of Inhoffen-Lythgoe diol **71**. Further the catalytic dihydroxylation of the crude ozonolysis product, oxidative cleavage (KIO$_4$, dioxane/H$_2$O), and reduction (NaBH$_4$/MeOH) gave the desired diol **71** in 75% overall yield. Selective tosylation of the primary alcohol, reduction of the tosylate with LiAlH$_4$ to **72**, and further oxidation of the secondary alcohol with DMP provided **73**. The *trans*-fused 5,6-membered ketone **73** was isomerized under basic conditions (NaH, THF) to provide the corresponding *cis*-ketone **74** in 72% yield. The ring expansion was achieved using TMSCHN$_2$ and BF$_3$·OEt$_2$ in CH$_2$Cl$_2$ at −40 °C followed by warming to room temperature to give ketone **75b** in 74% yield and high regioselectivity (1 : 10, **75a** : **75b**). Dehydrogenation of **75b** under Saegusa protocol [31] afforded enone **76** in 94% yield. Further reduction of **76** to **77**, epoxidation, and

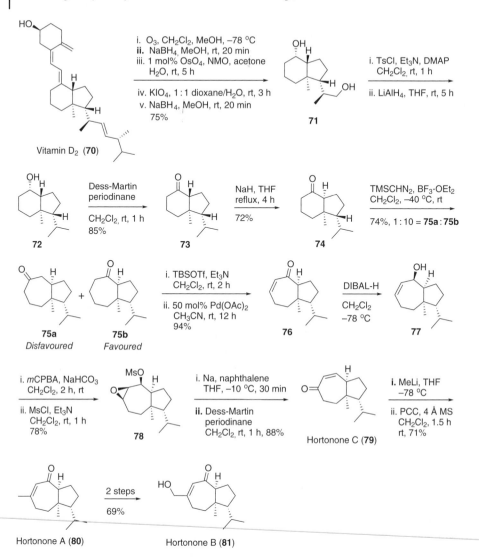

Scheme 2.10 Total synthesis of (−)-hortonones A-C (**79–81**).

mesylation led to **78**. Reduction of the latter with sodium naphthalenide in THF at −10 °C gave the corresponding allylic alcohol, which was then oxidized with the DMP to provide hortonone C (**79**) in 88% yield. Further addition of methyl lithium to hortonone C and transposed oxidation of the tertiary allylic alcohol with PCC gave hortonone A (**80**) in 71% yield. Oxidation of hortonone A to hortonone B (**81**) is completed via enolization (TBSOTf, DIPEA, −78 °C), regioselective epoxidation (1.1 equiv *m*CPBA, CH₂Cl₂, NaHCO₃, −20 °C), and aqueous hydrolysis in overall 69% yield.

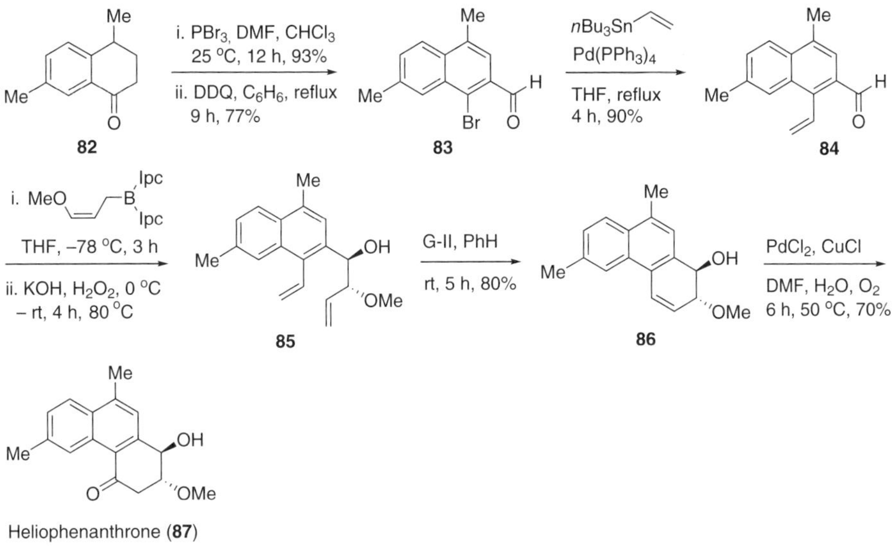

Heliophenanthrone (**87**)

Scheme 2.11 Synthesis of (−)-heliophenanthrone (**87**).

2.11 (−)-Heliophenanthrone

Heliophenanthrone (**87**) was isolated in 2003 from the aerial part of *Heliotropium ovalifolium* [32]. A short synthesis of **87** was achieved by Mukherjee and Sarkar (Scheme 2.11) [33]. The synthesis began with Vilsmeier–Haack-type formylation of easily available α-tetralone (**82**), followed by aromatization with DDQ to form **83**. The vinyl group was introduced using well-explored Stille-coupling reaction to get **84**. The aldehyde **84** underwent reaction with [(Z)-γ-4-methoxy-allyl]diisopinocamphenylborane to afford the β-methoxy homoallyl alcohol **85** in 80% yield and 90% ee. Ring-closing metathesis subsequently provided **86**. The latter on regioselective Wacker oxidation at the bay region furnished efficiently heliophenanthrone (**87**) in 70% yield.

2.12 (−)-Pycnanthuquinone C

The natural product pycnanthuquinone C was isolated from the brown alga *Cystophora harveyi* [34]. Its biological activity has not been reported, probably due to paucity of material. Trauner and coworkers [35] reported a biomimetic synthesis of (−)-pycnanthuquinone C (**91**) through the vinyl quinone Diels–Alder (VQDA) reaction as a key step (Scheme 2.12). The total synthesis started with a Heck cross-coupling reaction of bromohydroquinone **88** and monoterpene alcohol (−)-linalool **89** followed by MnO_2 oxidation to give vinyl quinone **90**. Heating the solution of **90** in a biphasic toluene/water mixture at 60 °C gave pycnanthuquinone C (**91**) and its epimer **92** as a 5 : 4 mixture in 37% total yield. This reaction involves the diastereoselective intramolecular VQDA reaction followed by nonselective H_2O addition to the isoquinone methide intermediate, followed by oxidation to deliver (−)-pycnanthuquinone C (**91**) and its

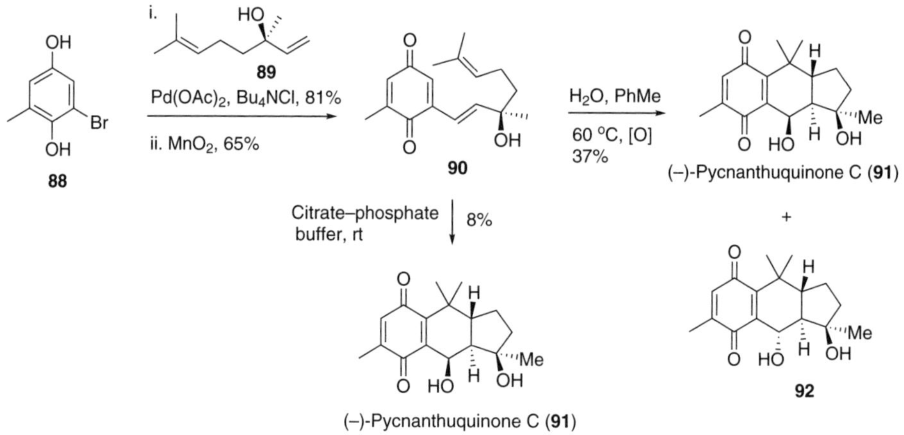

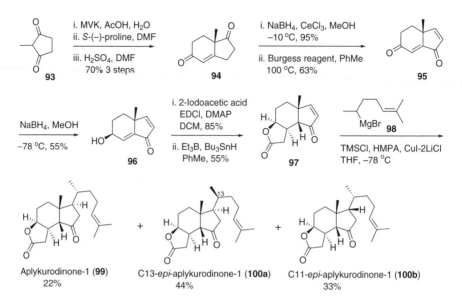

Scheme 2.12 Total synthesis of (−)-pycnanthuquinone C (**91**).

diastereomer **92**. When the VQDA reaction of **90** was carried out utilizing biomimetic conditions in citrate–phosphate buffer at pH 5 and room temperature, it gave only (−)-pycnanthuquinone C (**91**), but in very low yield.

2.13   (+)-Aplykurodinone-1

Aplykurodinone-1 (**99**) is a degraded steroid isolated in 2005 from the sea hare *Syphonota geographica* [36]. Tang et al. [37] reported the synthesis of aplykurodinone-1 (**99**) by utilizing Hajos–Parrish ketone **94** [38] as shown in Scheme 2.13. The synthesis

Scheme 2.13 Tang's total synthesis of (+)-aplykurodinone-1 (**99**).

of **94** started from 2-methyl-1,3-cyclopentanedione **93** using the Robinson annulation protocol, which gave 70% yield and 99% ee. This on stereo- and regioselective reduction using sodium borohydride in the presence of cerous chloride afforded the alcohol in high yield. Elimination of the resultant alcohol with the well-known Burgess reagent [39] surprisingly gave the ketone **95**. The mechanism was later demonstrated as one-step aerobic oxidation/elimination sequence. Further stereo- and regioselective reduction of ketone with sodium borohydride in methanol gave moderate yield of desired alcohol **96**. Esterification of alcohol **96** with 2-iodoacetic acid and then triethyl boron and tributyltin hydride-promoted free radical Michael addition gave the intermediate **97**. This on Michael addition of Grignard reagent **98** provided (+)-aplykurodinone-1 (**99**) in 22% yield along with C13-*epi*-aplykurodinone-1 (**100a**) and C11-*epi*-aplykurodinone-1 (**100b**) in 44 and 33% yields, respectively. The synthesis of aplykurodinone-1 (**99**) was completed in nine steps with 2.4% overall yield.

2.14 (±)-Hippolachnin A

Lin et al. isolated hippolachnin A (**109**) recently from the marine sponge *Hippospongia lachne* from the South China Sea [40]. It showed potent antifungal activity against *Cryptococcus neoformans*, *Trichophyton rubrum*, and *Microsporum gypseum*. Hippolachnin A has a unique structure with six contiguous stereocenters. The Brown and Wood groups independently developed two complementary routes to the total synthesis of hippolachnin A (Schemes 2.14 and 2.15, respectively) [41]. From these efforts, a combined synthesis with featured steps from both the earlier syntheses was developed in a collaborative research [41]. The Brown's route is depicted in Scheme 2.14. Cycloaddition of quadricyclane **101** with olefin **102** under microwave conditions and aqueous NaOH quenching led to compound **103** in 50% yield

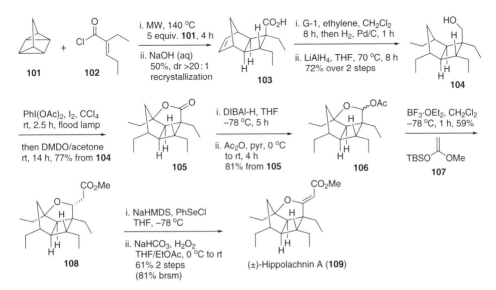

Scheme 2.14 Brown's independent synthesis of (±)-hippolachnin A (**109**).

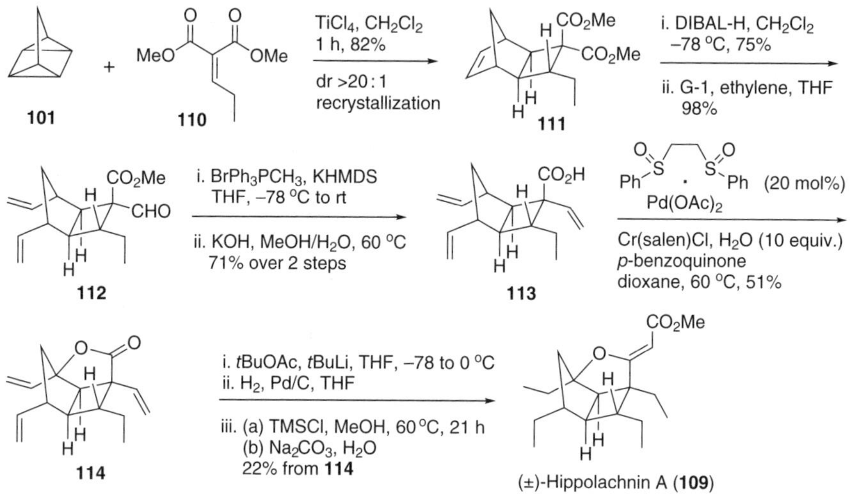

Scheme 2.15 Wood's independent synthesis of (±)-hippolachnin A (**109**).

(dr > 20 : 1) after recrystallization. The ring-opening cross-metathesis/hydrogenation and further reduction of acid with LiAlH$_4$ gave alcohol **104**. The subsequent Csp3-H oxidation under flood lamp irradiation using PhI(OAc)$_2$ and I$_2$ led to THF, which was oxidized with DMDO to get the lactone **105**. The latter was reduced to lactol and acetylated to compound **106**. Addition of silyl ketene acetal **107** to the acetate **106** gave the ester **108** in 59% yield. The final steps were same as Carreira's synthesis [42]. The α-phenylselenylation followed by oxidative elimination resulted in hippolachnin A (**109**) in 61% yield.

Wood's synthesis is shown in Scheme 2.15. Lewis acid-promoted cycloaddition of quadricyclane **101** with olefin **110** led to compound **111** in 82% yield and dr >20 : 1 after recrystallization. Further reduction of ester to aldehyde and ring-opening cross-metathesis gave compound **112**. This on Wittig olefination and ester hydrolysis gave the acid **113**. The latter was subjected to White's allylic oxidation protocol [43], wherein addition of excess water was found to give good yields of the lactone **114** (51%). Further addition of *tert*-butyl acetate gave the vinylogous carbonate as a mixture of diastereomers. This on reduction of terminal olefins and transesterification with MeOH provided (±)-hippolachnin A (**109**) in overall 22% yield from **114**.

Both Brown's and Wood's work on (±)-hippolachnin A were displayed at a National Organic Symposium [44], where they became aware of each other's work. Considering the similarities and also complementary nature of strategies, both groups embarked on a collaborative approach to (±)-hippolachnin A (Scheme 2.16) [41]. Brown's approach had an efficient beginning for synthesis of compound **103**, while the end game was efficient in Wood's synthesis. The compound **103** was obtained as in Scheme 2.14. Ring-opening cross-metathesis gave compound **115**. The Pd-catalyzed allylic oxidation (of Wood's synthesis) on compound **115** led to lactone **116** in 70% yield. Further similar steps of Wood's approach of installing unsaturated methyl ester provided (±)-hippolachnin A (**109**) in 15% yield (26% brsm). Thus the combined synthesis (Scheme 2.16) was achieved efficiently and required only seven linear steps.

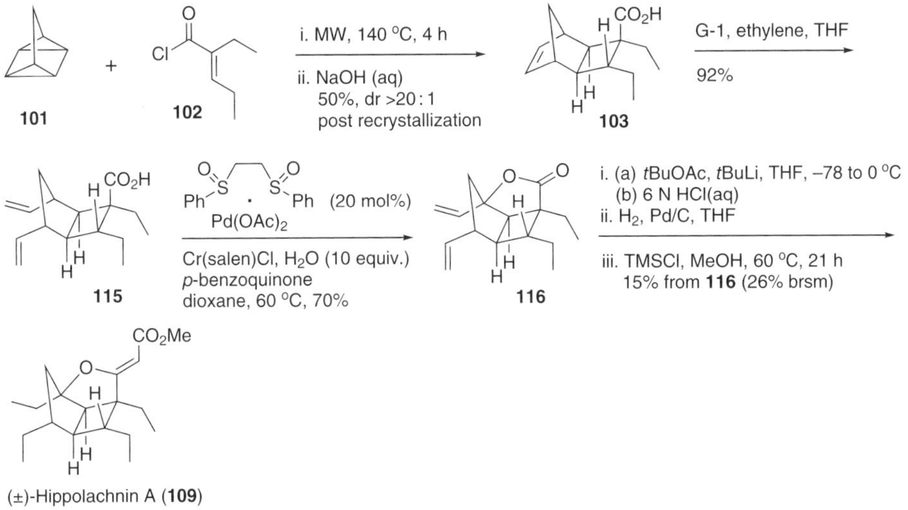

Scheme 2.16 Collaborative total synthesis of (±)-hippolachnin A (109).

2.15 (+)-Linoxepin

Linoxepin (125) was isolated by Schmidt et al. in 2007 from *Linum perenne* [45]. It belongs to the group of aryl dihydronaphthalene lignans and has benzonaphtho[1,8-*bc*]oxepine moiety. Tietze and coworkers reported the first total synthesis of (±)-linoxepin in 10 steps without any protecting groups (Scheme 2.17) [46]. Compound 117 on aldehyde homologation gave 118. Wittig reaction of 118 with methyltriphenylphosphonium bromide and (iodomethyl)trimethylsilane yielded 68% of allylsilane 119. This was then treated with benzyl bromide 120 to give aryl ether 121. Further Sonogashira coupling of propargylic alcohol in the presence of [Pd(PPh$_3$)$_4$], CuI, and Bu$_4$NOAc as base provided alkyne 122 in 98% yield. This alkyne underwent domino carbopalladation/Heck reaction to provide the desired terminal alkene 123a in 76% yield and 123b in 13% yield. The alcohol 123a was oxidized to the corresponding acid, which on esterification gave ester 124 quantitatively. The methyl ester 124 upon selective ozonolysis of the vinyl group in the presence of the dye solvent red 19 [47] followed by reductive work-up directly afforded racemic linoxepin (125) in 80% yield. For the synthesis of both enantiomers (+)-(*R*)-and (−)-(*S*)-linoxepin, the compound 123a was resolved by chiral chromatography (IA column) and then carried forward through the same steps as in Scheme 2.14. From the crystal structure analysis of unnatural (−)-linoxepin, it was concluded to have (*S*)-configuration and the natural product to have (*R*)-configuration.

At the same time Lautens and coworkers [48] synthesized (+)-linoxepin (125) in eight steps by using the modified version of the Catellani reaction [49]. Formylation of 126 and reduction gave benzyl alcohol 127 (Scheme 2.18). Conversion of benzyl alcohol to iodide and ether preparation with phenol 128 gave 129. The Catellani reaction of 129 with 130 (prepared following Zutter procedure [50]) and 131 gave the compound 132 in 92% yield. This intermediate was oxidatively cleaved to aldehyde (98%) under

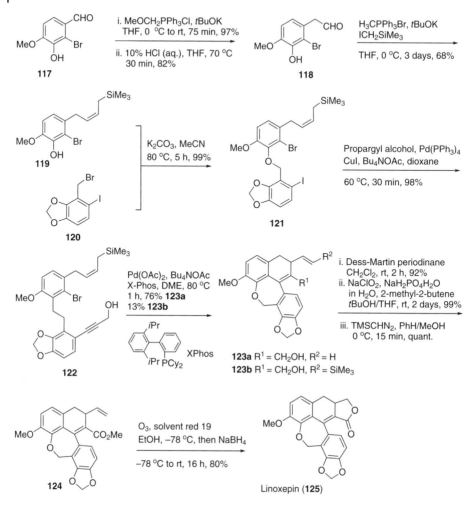

Scheme 2.17 Total synthesis of (+)-linoxepin (**125**).

Lemieux–Johnson conditions followed by TiCl$_4$-mediated condensation to provide **133** in 49% yield. The latter was cyclized by Mizoroki–Heck conditions using cesium acetate as base to afford (+)-linoxepin (**125**) in 76% yield.

2.16 (+)-Antofine and (−)-Cryptopleurine

Antofine and cryptopleurine are tylophorine class of alkaloids having potent anticancer properties and also exhibit antiviral, antibacterial, and anti-inflammatory activities [51]. In 2013, Ying and Herndon [52] reported the synthesis of the unnatural enantiomer of antofine and natural (−)-cryptopleurine (Schemes 2.19 and 2.20). Alkynylproline **135a** was synthesized from the natural enantiomer of proline [53]. The alkynylpiperidines **135b** and **135c** were prepared by resolution of pipecolic acid [54]. The first step of the synthesis involved Sonogashira coupling between 4-bromo-5-iodoveratrole **134** and

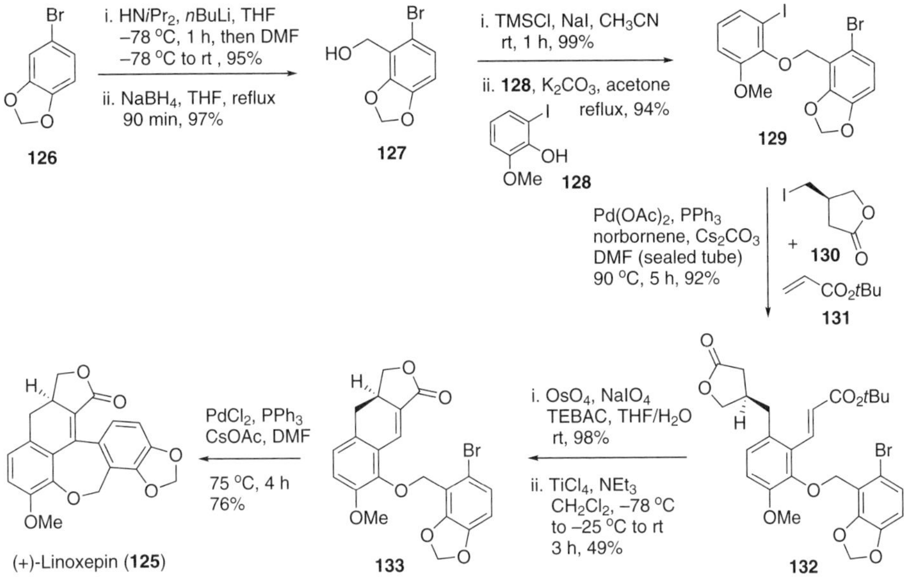

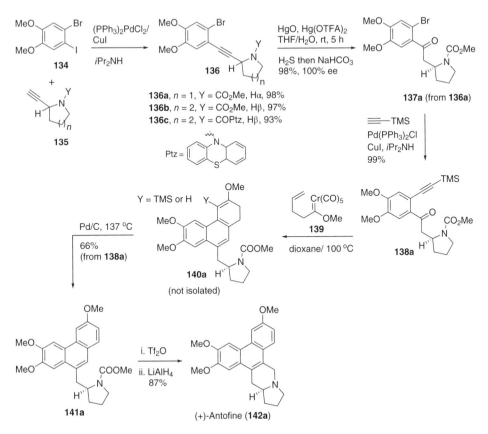

Scheme 2.18 Total synthesis of (+)-linoxepin (**125**).

Scheme 2.19 Total synthesis of (+)-antofine (**142a**).

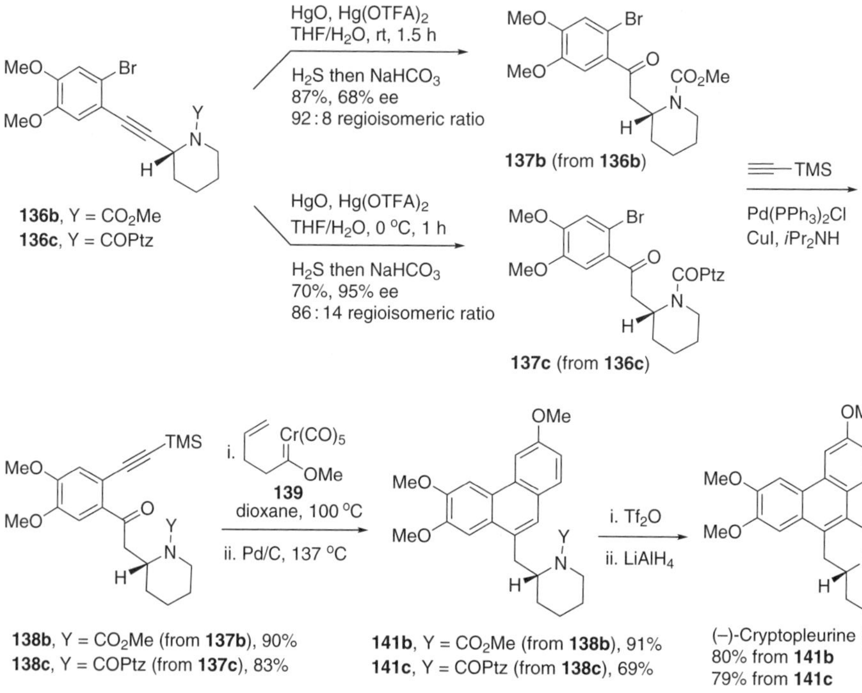

Scheme 2.20 Total synthesis of (−)-cryptopleurine (**142b**).

each of the 2-ethnynylheterocycles **135** (Scheme 2.19) to give **136a–c** in 93–98% yields. The critical alkyne hydration was optimized to avoid racemization of the chiral center and unexpected poor regioselectivity. Thus alkyne **136a** gave **137a** in 98% yields and 100% ee as a single regioisomer. A second Sonogashira coupling of **137a** with TMS-acetylene produced **138a**. This was treated with the carbene **139**, resulting in cyclized product **140a** as a mixture of silylated and desilylated compounds. The mixture on dehydrogenation gave **141a** as a single product. Further the Bischler–Napieralski cyclization followed by reduction of amide produced (+)-antofine (**142a**).

For synthesis of cryptopleurine (**142b**), the hydration of **136b** or **136c** was more prone to racemization and regioisomer formation. Thus, **136b** at best gave **137b** as 92 : 8 regioisomeric mixture in 87% yield and 68% ee (Scheme 2.20) [52]. Similarly, **136c** produced **137c** in 70% yield, 95% ee, and 86 : 14 regioisomeric mixture. These on Sonogashira coupling with TMS-acetylene produced **138b** and **138c**, respectively. Further cyclization with **139** and dehydrogenation gave **141b** and **141c**. These were further cyclized and reduced to furnish (−)-cryptopleurine (**142b**).

2.17 (+)-Tylophorine

A related natural product tylophorine is known to arrest cancer cells in the G1 phase of growth by downregulation of cyclin A2 expression, while (R)-antofine causes cell cycle arrest in the G2/M phase [55]. The unnatural (S)-tylophorine (**151**) was found to be

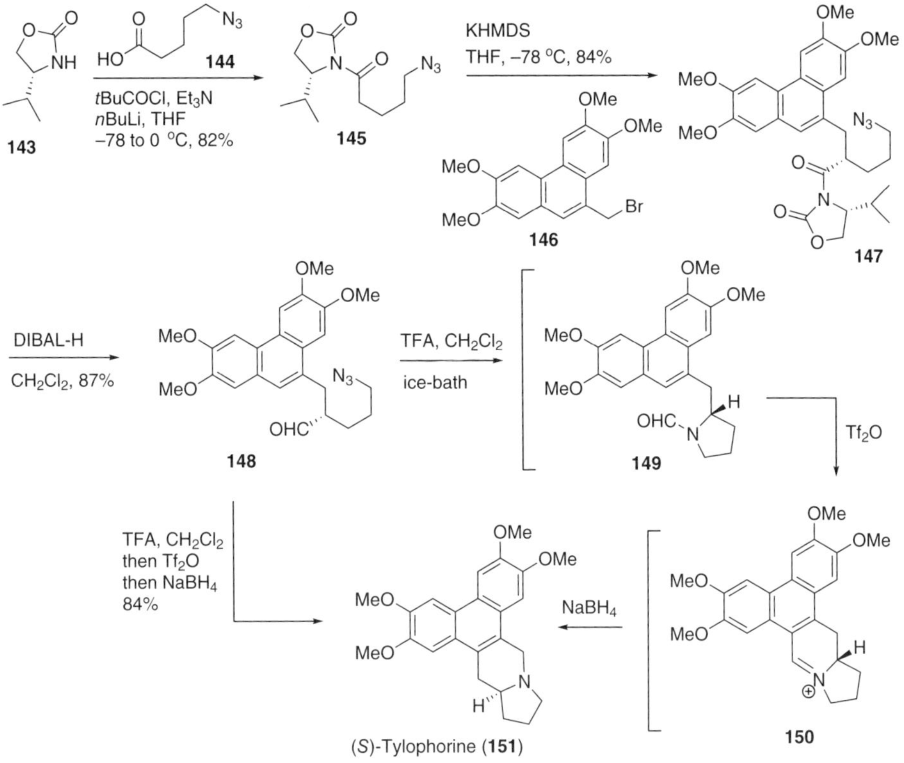

Scheme 2.21 Total synthesis of (*S*)-tylophorine (**151**).

much more potent for cancer cell growth inhibition [56]. Wang et al. [57] developed a one-pot intramolecular Schmidt/Bischler–Napieralski/imine-reduction cascade in the total synthesis of (*S*)-tylophorine (Scheme 2.21). The synthesis commenced with coupling of auxiliary **143** with azido acid **144** to give amide **145**. Stereoselective alkylation of **145** with phenanthryl bromide **146** produced compound **147** as a single diastereomer. DIBAL-H reduction of **147** furnished the azido aldehyde **148**. This on treatment with TFA produced the formamide **149**, which further underwent Bischler–Napieralski reaction upon addition of Tf$_2$O to give **150**, and further addition of NaBH$_4$ furnished (*S*)-tylophorine (**151**).

Opatz and coworkers [58] achieved a concise and modular synthesis of tylophorine, antofine, and deoxypergularinine (Scheme 2.22). Condensation of veratraldehyde (**152**) with homoveratric acid (**153**) gave **154**, which underwent cyclization to furnish phenanthrene-9-carboxylic acid (**155**). Amide formation with pent-4-en-1-amine gave **156**. This on iodocyclization led to **157**, which upon radical cyclization gave **158** that was reduced to racemic tylophorine (**151**). A chiral synthesis was also achieved from **155** by coupling with L-prolinol to give the amide. The hydroxyl group was converted to iodide (*S*)-**157** and cyclized by similar radical conditions to give the amide, which was reduced to (*S*)-tylophorine (**151**). A similar strategy was utilized toward the synthesis of racemic antofine and deoxypergularinine [58].

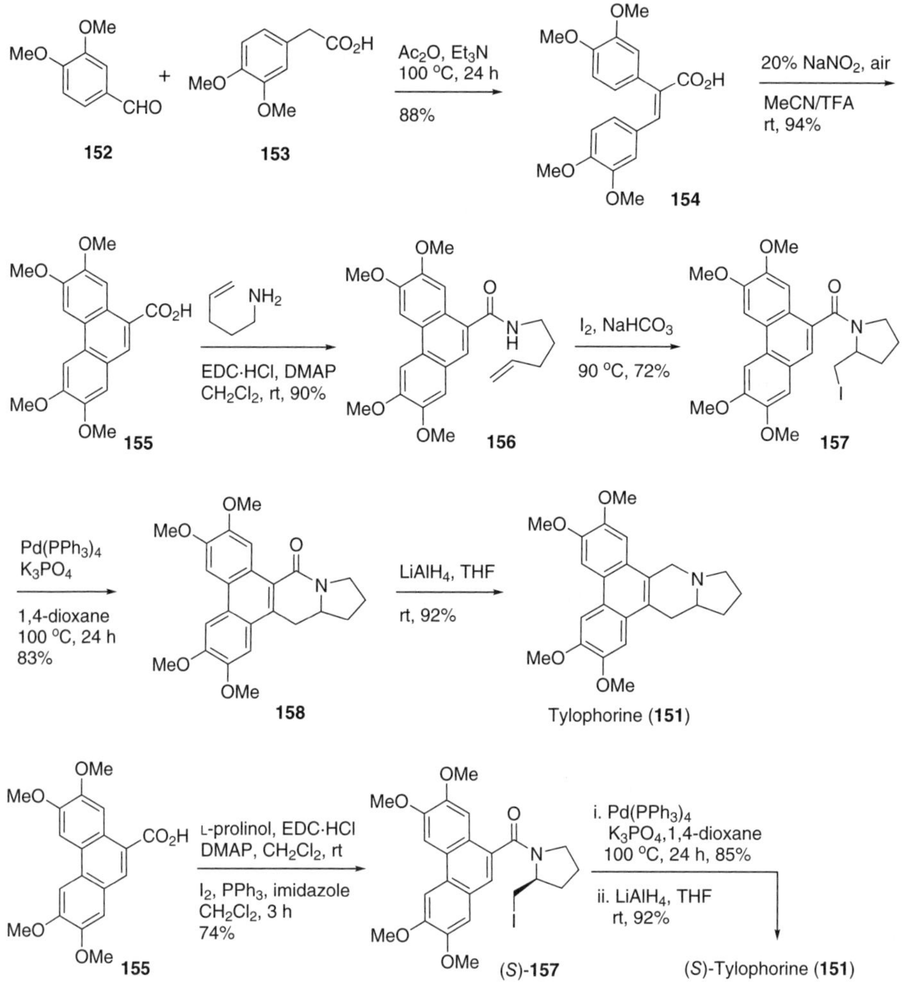

Scheme 2.22 Synthesis of racemic and chiral (*S*)-tylophorine (**151**) by Opatz.

2.18 (±)-Cruciferane

Argade and coworkers developed an efficient aryne insertion strategy to the total synthesis of fused quinazoline alkaloids, such as tryptanthrin, (±)-phaitanthrins A and B, and (±)-cruciferane [59]. These were isolated from *Phaius mishmensis* and *Isatis tinctoria* (*Isatis indigotica* Fortune) and have antiviral and anticancer properties [60]. The insertion reaction of an *in situ* generated benzyne from the precursor **159** with quinazolinone **160** in CH₃CN at 25 °C gave the natural product tryptanthrin (**161**) through N-arylation followed by intramolecular cyclization (Scheme 2.23) [59]. A base-medicated reaction of methyl acetate with tryptanthrin (**161**) delivered phaitanthrin B (**162**). This on NaBH₄ reduction gave (±)-cruciferane (**163**). Similarly, a base-mediated chemoselective aldol reaction of acetone with tryptanthrin (**161**) furnished phaitanthrin A (**164**).

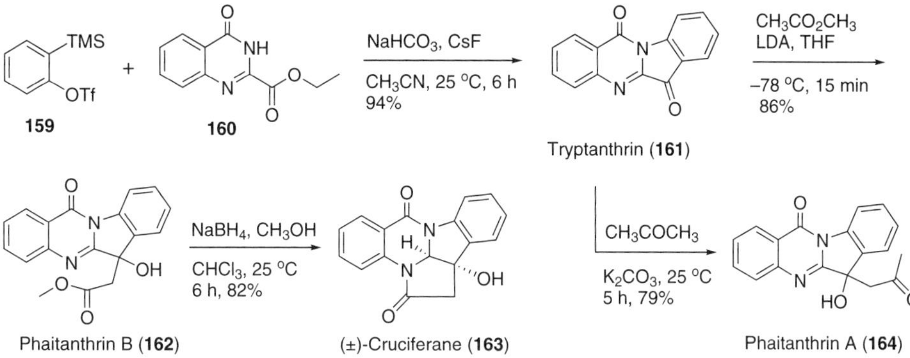

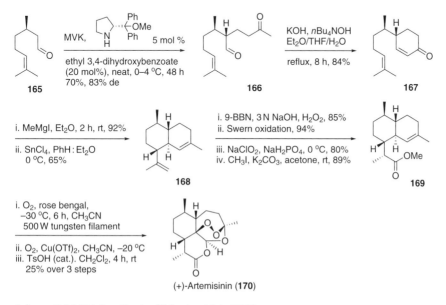

Scheme 2.23 The synthesis of tryptanthrin (**161**), (±)-cruciferane (**163**), and phaitanthrins A (**164**) and B (**162**).

2.19 (+)-Artemisinin

(+)-Artemisinin (**170**) has an unusual trioxane structure with a tetracyclic framework and is a potent therapeutic for malaria [61]. Yadav et al. achieved an eight-step total synthesis of **170** from (R)-citronellal as shown in Scheme 2.24 [62]. Conjugate addition of (R)-citronellal (**165**) to methyl vinyl ketone in the presence of a proline-derived catalyst and ethyl 3,4-dihydroxybenzoate as a cocatalyst following Nicolaou's procedure [63] afforded **166** with 83% de. Ketone **166** was subjected to aldol condensation to produce enone **167**, which on subsequent addition of methylmagnesium iodide and stannous chloride delivered the cyclized ene product **168**. Regioselective

Scheme 2.24 Total synthesis of (+)-artemisinin (**170**).

hydroboration, followed by oxidation to acid and subsequent esterification, then led to ester **169**. Photooxidation of ester **169** following Haynes' protocol [64] produced (+)-artemisinin (**170**).

2.20 (±)-Dievodiamine

Dievodiamine was isolated from *Evodia rutaecarpa* whose fruits were used in traditional Chinese remedies to treat headaches, abdominal pain, inflammation, migraine, chill limbs, nausea, and cancer [65]. However no activity of isolated dievodiamine has been reported. Taylor and coworkers disclosed the first PGF total synthesis of dievodiamine (**182**) in nine steps (Scheme 2.25) [66]. Commercially available indole (**171**) was converted into lactam **172** via Curtius rearrangement followed by electrophilic aromatic substitution reaction. The lactam was subjected to Decker's process [67] with dimethyl anthranilate **173** and POCl$_3$ to deliver dehydroevodiamine hydrochloride (DHED·HCl) **174**. This was then treated with ethynylmagnesium chloride and lithium chloride in toluene to give alkyne **175** in 74% yield. The regioselective transformation of alkyne **175** into the alkenyl stannane **176** was achieved with tributyltin hydride and AIBN in 54% yield. The iodide coupling partner was synthesized from commercially available indole-2-carboxylic acid (**177a**). This was converted into acid chloride **177b**

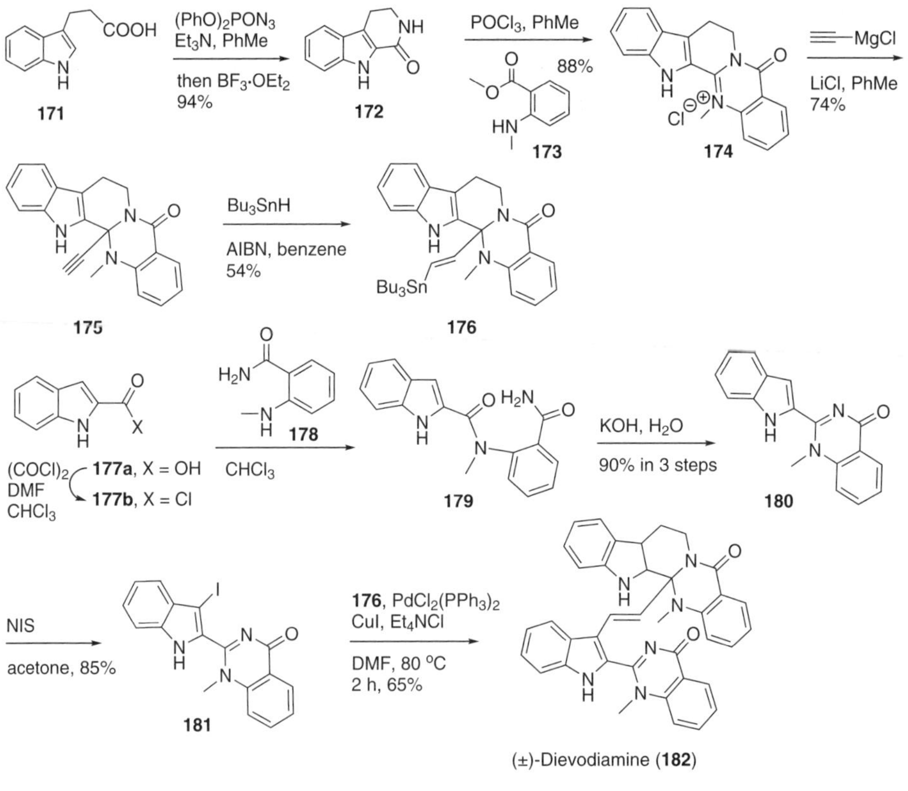

Scheme 2.25 Total synthesis of (±)-dievodiamine (**182**).

and reacted with aniline **178** to form amide **179**, which cyclized under aqueous KOH reflux to afford quinazolinone **180** in excellent yield. It was then treated with *N*-iodo-succinimide in acetone to yield the desired iodo indole **181**. The cross-coupling of iodide **181** and stannane **176** was carried out under palladium catalysis [PdCl$_2$(PPh$_3$)$_2$, Et$_4$NCl] to deliver (±)-dievodiamine (**182**) in 65% yield.

2.21 (−)-Chaetominine

In 2006, the alkaloid (−)-chaetominine (**189**) was isolated by Tan and coworkers [68] from the solid-substrate culture of *Chaetomium* sp. IFB-E015. Owing to its intriguing structural features and the potency against human leukemia K562 and colon cancer SW1116 cell lines [68], (−)-chaetominine has been an attractive synthetic target [69]. A very concise PGF synthesis of (−)-chaetominine (**189**) was accomplished by Huang and coworkers [70] in four steps starting from D-tryptophan (**183**) with a 33.4% overall yield (Scheme 2.26). Aroylation of **183** to **184** followed by dipeptide preparation gave compound **185**. The quinazolinone ring system **186** was prepared by using HC(OMe)$_3$ and reduction of nitro group with low valent titanium. For the epoxide-initiated cascade reaction, treatment of dipeptide **186** with a solution of DMDO in acetone

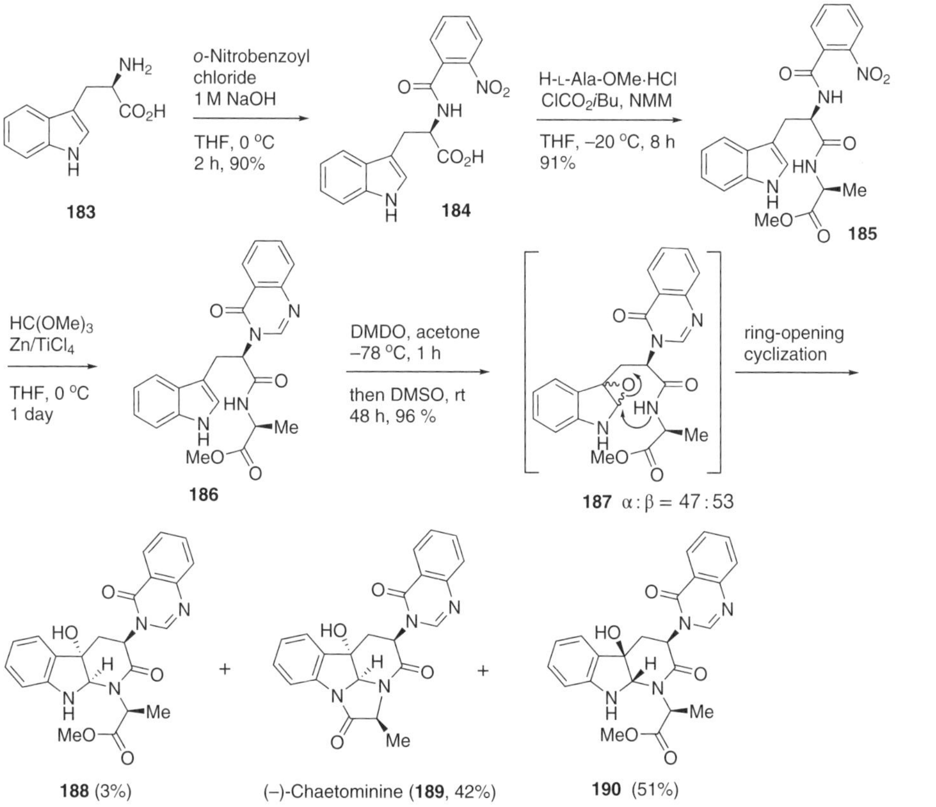

Scheme 2.26 Total synthesis of (−)-chaetominine (**189**) in a four-step process.

at −78 °C for 1 h led to complex mixtures, which implied that the unstable epoxide intermediate failed to cyclize to the desired product **189**. On optimizing the reaction conditions, dipeptide **186** was treated with the combination of DMDO with DMSO under the same reaction conditions that finally produced the desired (−)-chaetominine (**189**) in 42% yield along with a small amount of its lactamization precursor **188** (3%) and the epimer **190** (51%).

2.22 Rubicordifolin

Rubicordifolin, a cytotoxic natural product, was isolated from the Chinese medicinal plant *Rubia cordifolia* [71]. Lumb and Trauner reported the biomimetic synthesis and structure elucidation of rubicordifolin (**196**; Scheme 2.27) [72]. Conjugate addition of a vinyl cuprate derived from vinyl stannane on 2-carbomethoxy naphthohydroquinone **191** followed by oxidation and silyl ether deprotection gave compound **192**. At room

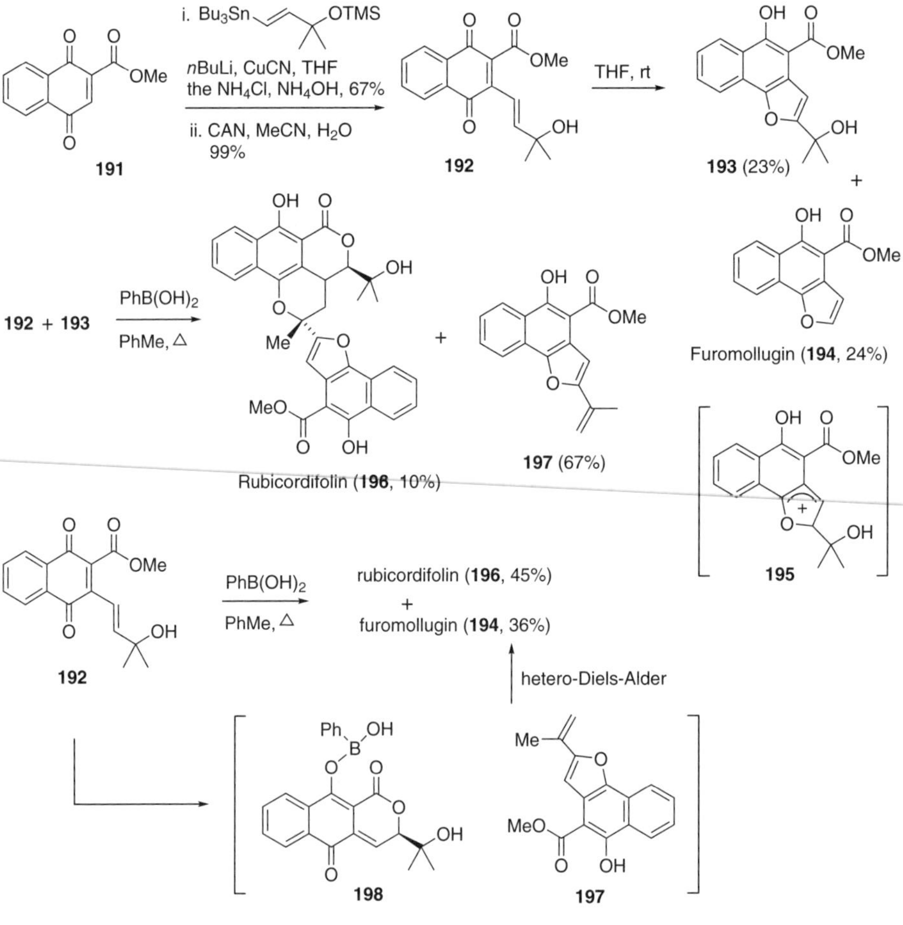

Scheme 2.27 Synthesis of rubicordifolin (**196**).

temperature unstable vinyl quinone **192** underwent cyclization to naphthofuran **193** and furomollugin **194** in 23 and 24% yields, respectively. Compound **193** arises through the cation **195** by deprotonation; if it undergoes *retro*-Friedel–Crafts hydroxyalkyla-tion, it gives **194**. The dimerization of compounds **192** and **193** in the presence of PhB(OH)$_2$ gave rubicordifolin (**196**) in only 10% yield with the major product as **197**. After careful mechanistic interpretation, they found that the compound **192** mainly plays a role in the cyclization and dimerization in the synthesis of rubicordifolin (**196**). Hence they carried out direct cyclization and dimerization of compound **192** in the presence of PhB(OH)$_2$ to afford rubicordifolin (**196**) in 45% yield, and furomollugin (**194**) was isolated in 36% yield. The reaction proceeds through the intermediates **197** and **198** by hetero-Diels–Alder reaction.

2.23 (+)-Caribenol A

Tuberculosis is responsible for over three million deaths worldwide [73]. Caribenol A shows inhibitory activity against *Mycobacterium tuberculosis* (H37Rv), which causes tuberculosis. Yang and coworkers achieved the total synthesis of caribenol A (**207**) as shown in Scheme 2.28 [74]. The halogen–lithium exchange reaction of iodide **199** and then reaction with chiral enone **200** produced allylic alcohol **201**. This on DIBAL-H reduction to the diol and PCC oxidation gave the aldehyde and the transposed ketone from tertiary alcohol. Addition of lithiated ethyl propiolate to the aldehyde and DMP oxidation gave the ketone **202**. The latter on intramolecular Diels–Alder reaction pro-vided the tricyclic compound **203**, which on regioselective hydrogenation followed by treatment with methyl lithium afforded tertiary alcohol **204**. The tertiary hydroxyl was eliminated using Martin's sulfurane to provide a 2 : 1 mixture of **205** and **206**. Compound **205** on ester hydrolysis gave cyclized product caribenol A (**207**). Alternatively, com-pound **203** on selective reduction of the less substituted double bond using Wilkinson's catalyst followed by ester hydrolysis, cyclization, and reduction of the other double bond gave **208**. The latter on hydrazone formation and reaction with iodine provided a 1 : 3 mixture of vinyl iodides **209** and **210**. This mixture on Pd-catalyzed Negishi coupling gave the olefin isomeric caribenol A (17%, not shown) and caribenol A (**207**) in 51% yield.

2.24 Camptothecin and 10-Hydroxycamptothecin

In 1966, the pentacyclic quinoline alkaloid camptothecin (CPT) (**219a**) was discovered from *Camptotheca acuminate* by Wall and coworkers [75]. Since 1970, it has attracted great interest to medicinal and synthetic chemists owing to its cytotoxic potency and anticancer property in drug development [76]. A concise PGF convergent synthetic route to CPT and 10-hydroxycamptothecin was developed by Yao et al. as shown in Scheme 2.29 [77]. The total synthesis commenced with the precursor **214** obtained via intermolecular oxa-Diels–Alder reaction of enol ether **212** and *in situ* generated α,β-unsaturated aldehyde **213** from the corresponding lactone **211**. Compound **214** was coupled with ABC-ring fragment **215** (obtained in three steps from the corresponding aniline-2-aldehyde) using trimethylaluminum to give **216a** and **216b**.

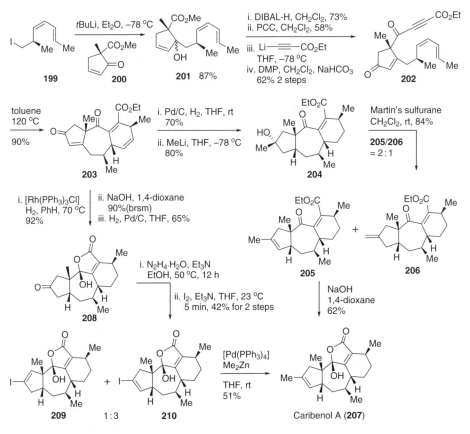

Scheme 2.28 Total Synthesis of (+)-caribenol A (**207**).

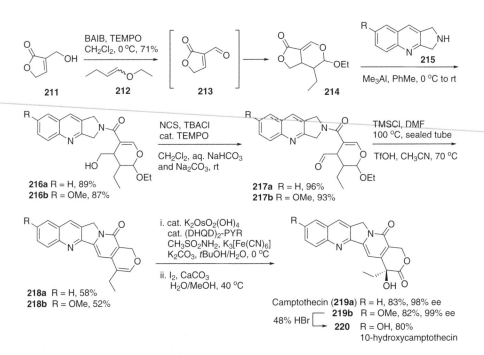

Scheme 2.29 Protecting-group-free total synthesis of camptothecin (**219a**) and 10-hydroxycamptothecin (**220**).

These were then oxidized to aldehydes **217a** and **217b**, and further intramolecular condensation with TMSCl/DMF in the presence of TfOH produced the pentacyclic intermediates **218a** (58%) and **218b** (52%), respectively. These were then subjected to Sharpless asymmetric dihydroxylation and iodine-based hemiacetal oxidation to accomplish the total synthesis of (*S*)-camptothecin (**219a**) and (*S*)-10-methoxycampto-thecin (**219b**), respectively, with excellent yields and enantiopurities. Demethylation of **219b** gave (*S*)-10-hydroxycamptothecin (**220**) in 80% yield.

2.25   (+)-Ainsliadimer A

(+)-Ainsliadimer A (**231**) was discovered by Zhang and coworkers from *Ainsliaea mac-rocephala*, which has been used for the treatment of angina and rheumatoid arthritis [78]. The molecule has an unprecedented carbon framework with 11 stereocenters, highly substituted cyclopentane rings, and two monomeric lactone units linked by three quaternary carbons. Potent inhibitory activity shown by this molecule against generation of NO in RAW264.7 stimulated by lipopolysaccharide (LPS) has inspired its total synthesis. The dimeric molecule was accessed by hydrogen assisted [4+2]-hetero-Diels−Alder reaction of dehydrozaluzanin C **228** by Lei and coworkers [79]. The latter was synthesized as shown in Scheme 2.30. α-Santonin (**221**, commercially available) on photoirradiation in acetic acid gave *O*-acetylisophotosantonic lactone **222**. This on hydrogenation, ketone reduction, and hydroxyl elimination and further acyl hydrolysis gave the alcohol **223**. Kinetic dehydration using the optimized elimination conditions (SOCl$_2$/DABCO) provided alkene **224** in more than 12 : 1 ratio with the regioisomer. Compound **224** on selenylation with LDA/PhSeSePh and selenide oxidation/elimination with H$_2$O$_2$/AcOH gave α-alkylidene-γ-butyrolactone **225**. Selective olefin epoxidation to **226** and epoxide ring opening using Al(O-*i*Pr)$_3$ offered alcohol **227**. This on DMP oxidation gave dehydrozaluzanin **228**. The latter was then engaged in self-hetero-[4+2]-Diels−Alder reaction to give dimer **229**. Acid-mediated pyran cleavage provided **230**, which on intramolecular aldol reaction furnished (+)-ainsliadimer (**231**).

2.26   Cannabicyclol, Clusiacyclols A and B, Iso-Eriobrucinols A and B, and Eriobrucinol

Hsung et al. [80] developed a unified strategy toward the total synthesis of cyclobutane-containing family of chromane natural products such as cannabicyclol (**235**), clusiacy-clols A (**238**) and B (**239**), eriobrucinol (**240**), and *iso*-eriobrucinols A (**241**) and B (**242**) as shown in Scheme 2.31 [81, 82]. The strategy was based on oxa-[3+3] annulation to construct the chromane nucleus and a sequential cationic [2+2] cycloaddition to construct the cyclobutane ring or annulation/cycloaddition tandem strategy leading to the chromane nucleus. By following the sequential strategy, the oxa-[3+3] annulation of olivetol (**232**) with citral (**233**) using piperidine and acetic anhydride led to chromene **234** in 50% yield. Further, TFA-promoted cationic [2+2] cycloaddition of **234** delivered chromanyl cyclobutane natural product cannabicyclol (**235**). A similar tandem process of *oxa*-[3+3] annulation and cationic [2+2] cycloaddition of phloroglucinol (**236**) delivered chromanyl cyclobutane **237**. This was regioselectively transformed into

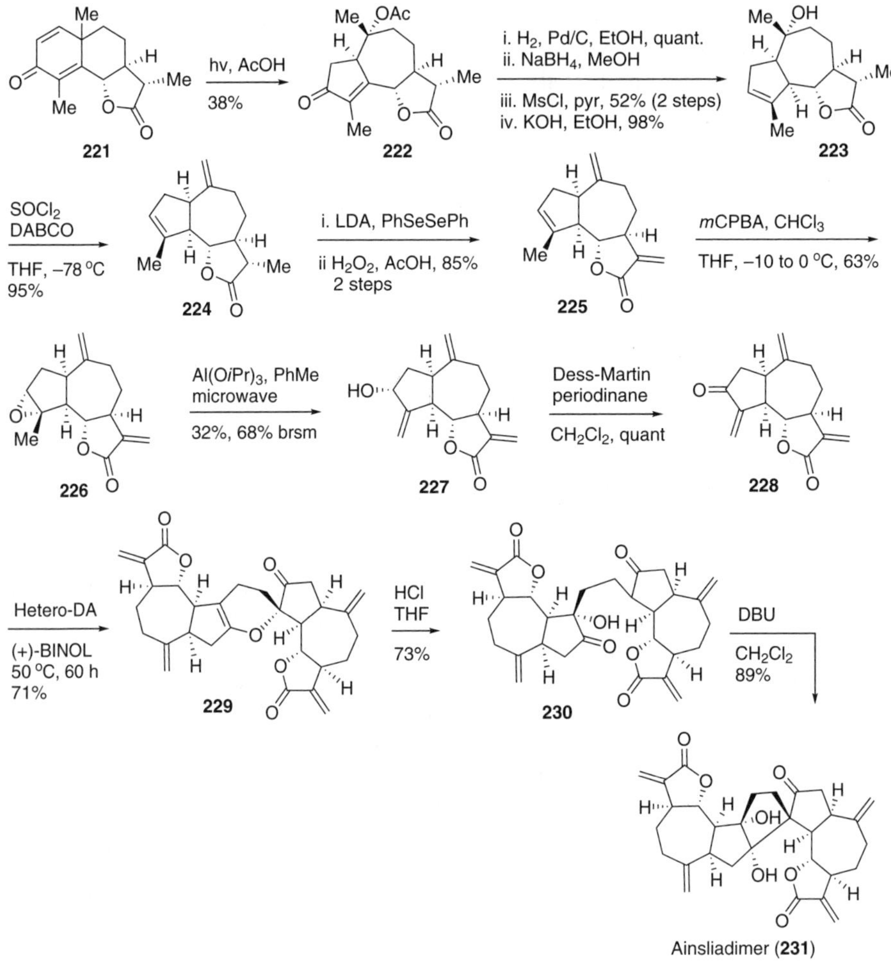

Scheme 2.30 Total synthesis of (+)-ainsliadimer (**231**).

clusiacyclols A (**238**) and B (**239**) via C-benzoylation using AlCl₃ and ZnBr₂, respectively. Using the common intermediate **237**, a Lewis acid, InCl₃, promoted oxa-[3+3] annulation with ethyl propiolate delivered *iso*-eriobrucinols A (**241**) and B (**242**) in 44% yield in 2 : 1 ratio. Changing the Lewis acid InCl₃ to ZnCl₂ delivered eriobrucinol (**240**) regioselectively.

2.27   (−)-Mersicarpine, (−)-Scholarisine G, (+)-Melodinine, (−)-Leuconoxine, and (−)-Leuconolam

Zhu et al. [83] demonstrated a unified strategy toward the enantioselective total synthesis of (−)-mersicarpine, (−)-scholarisine G, (+)-melodinine E, (−)-leuconoxine, and (−)-leuconolam from the common cyclohexenone derivative **243** (Scheme 2.32). Their synthesis started with palladium-catalyzed enantioselective decarboxylative allylation of β-ketoester **243** using the Stoltz conditions [84] followed by a

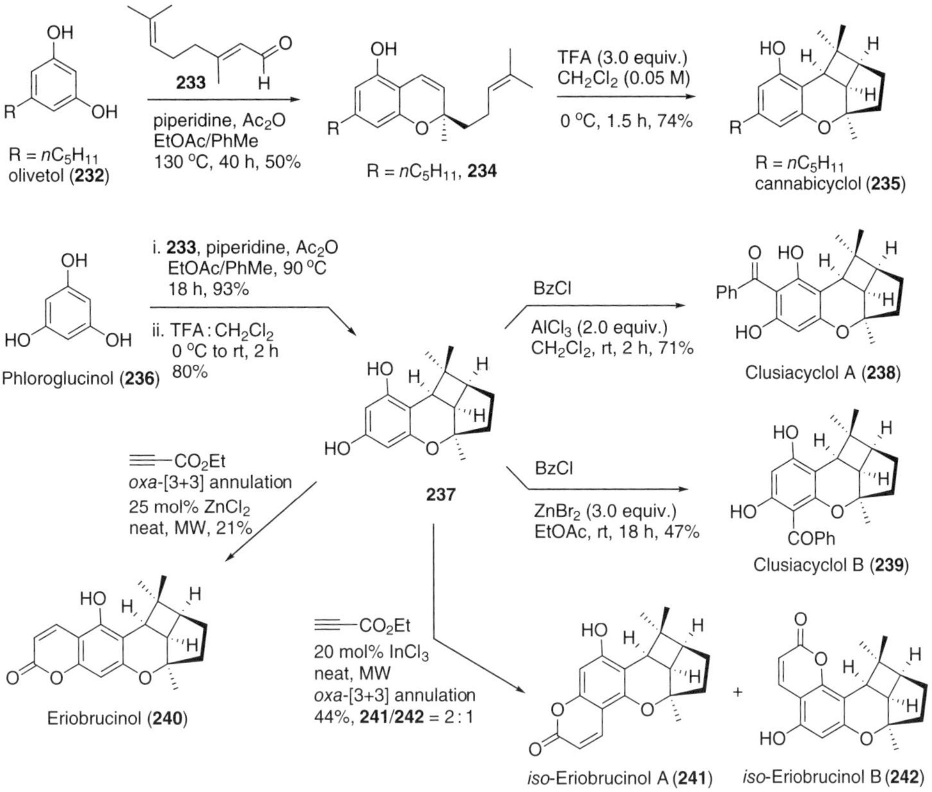

Scheme 2.31 Total synthesis of cannabicyclol (**235**), clusiacyclols A (**238**) and B (**239**), eriobrucinol (**240**), and *iso*-eriobrucinols A (**241**) and B (**242**).

hydroboration–iodination sequence to afford iodide **244**. This was subsequently transformed into azide **245**. Enone **246** was synthesized in three steps from **245**, involving reaction with an excess of IBX in DMSO, followed by vinylic iodination and Suzuki–Miyaura coupling of the resultant iodide with 2-nitrophenylboronic acid. Further oxidation of **246** under ozonolytic conditions afforded the diketone ester **247** in 92% overall yield. Hydrogenation of **247** gave **248**, which cyclized to **249**. This on reaction with KOH underwent lactamization, and bubbling with oxygen gave peroxide **250**, which finally on reduction with dimethyl sulfide furnished (−)-mersicarpine (**251**).

For total synthesis of (−)-scholarisine G, the intermediate **247** was subjected to hydrogenation in the presence of Pd/C and acetic anhydride to give indolin-3-one, which was directly oxidized by purging oxygen into the unstable indol-3-one **252** (Scheme 2.32). This on KOH treatment afforded the 2-ethoxyindoline-3-one lactam as a mixture of two diastereomers (2 : 1), which upon treatment with TFA gave the *N*-acyliminium ion that was attacked by the free amide to give spiro center in **253**. An intramolecular aldol cyclization of **253** using *t*BuOK completed the synthesis of (−)-scholarisine G (**254**). This upon hydroxyl elimination via a mesylate gave (+)-melodinine E (**255**). Heating (+)-melodinine E (**255**) in a solution of THF-3N H_2SO_4 (2 : 1) at 40 °C furnished (−)-leuconolam (**257**) via acyliminium cation **256**. Diastereoselective hydrogenation of (+)-melodinine E (**255**) furnished (−)-leuconoxine (**258**).

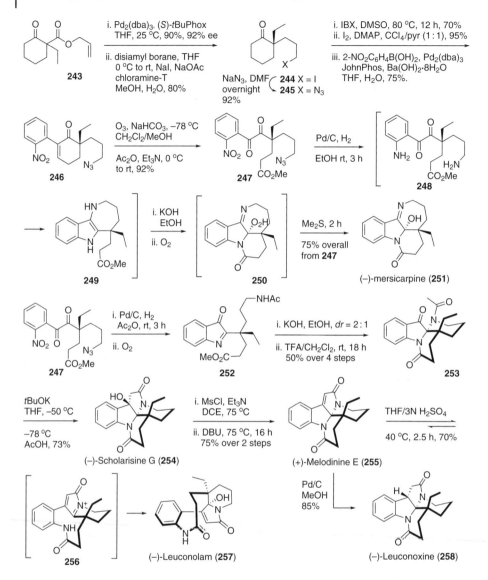

Scheme 2.32 Zhu's total synthesis of (−)-mersicarpine (**251**), (−)-scholarisine G (**254**), (+)-melodinine E (**255**), (−)-leuconolam (**257**), and (−)-leuconoxine (**258**).

2.28 (−)-Lannotinidine B

The first total synthesis of (−)-lannotinidine B has been reported by Yao et al. (Scheme 2.33) [85]. They commenced the total synthesis from chiral enone **259**, which on Morita–Baylis–Hillman reaction with aldehyde **260** afforded β-hydroxyl ketone that on DMP oxidation gave **261**. The Mukaiyama–Michael addition of *tert*-butyl(1-methoxyvinyloxy)dimethylsilane **107** on **261** resulted in an inseparable tautomer

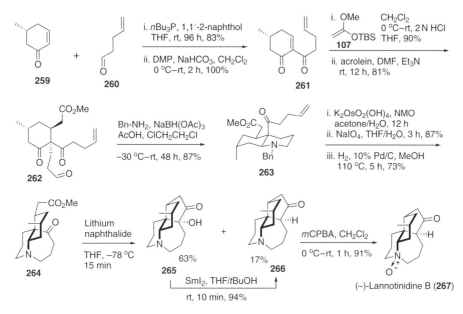

Scheme 2.33 Yao's total synthesis of (−)-lannotinidine B (**267**).

mixture of 1,3-diketone and enone (∼3/7 ratio by ^{1}H NMR). This on Michael addition to acrolein led to **262**. Further the cascade reaction of reductive amination of aldehyde **262** with benzylamine and intramolecular cyclization of resultant secondary amine with carbonyl gave the sole product **263** in good yield and more than 99% ee. Oxidative cleavage of terminal olefin in **263** followed by removal of the benzyl group by hydrogenolysis gave a chance for the third reductive amination sequence, which resulted in tricycle **264** in good yield. For conversion of ester **264** into the corresponding hydroxy ketone **265**, direct reductive ketyl radical anion coupling was explored. After many unsuccessful trials, the desired compound **265** was finally achieved by applying Gössinger conditions [86] using lithium naphthalide, which gave **265** and **266** in 63 and 17% yields, respectively. Treatment of **265** with SmI$_2$ gave **266**, which upon treatment with *m*CPBA gave *N*-oxide natural product (−)-lannotinidine B (**267**).

2.29 (−)-Lycopodine

Bödeker and coworkers isolated lycopodine (**275**) from *Lycopodium complanatum* L. in 1981 [87]. It is a Chinese folk medicine for the treatment of skin disorders and analgesia, and it also possesses antipyretic and anticholinesterase activities. In 2016, She and coworkers reported a PGF total synthesis of (−)-lycopodium (**275**) from chiral pool material (*R*)-(+)-pulegone using alkyne aza-Prins cyclization reaction as a key step (Scheme 2.34) [88]. The starting compound of the synthesis, Wade's enone **268**, was prepared from (*R*)-(+)-pulegone in four steps by literature known method [89]. Michael addition of propargyl indium on enone **268** gave compound **269** in 63% yield

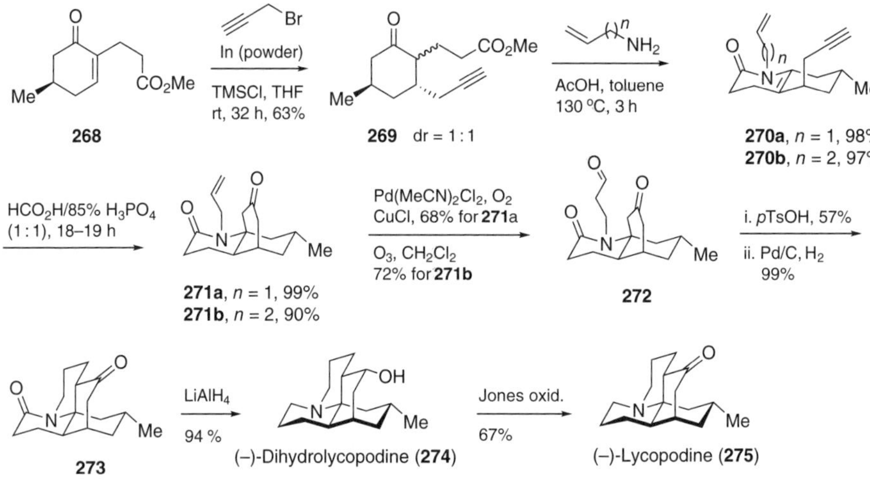

Scheme 2.34 Total synthesis of (−)-lycopodine (**275**).

as a 1 : 1 diastereomeric mixture. Compound **269** was treated with allylamine and acetic acid at 130 °C to furnish enamide **270a** or **270b** in excellent yields. These on aza-Prins cyclization in the presence of HCO$_2$H : 85% H$_3$PO$_4$ (1 : 1) gave **271a** or **271b** in 99 or 90% yields, respectively. The reaction involves the formation of *N*-acyliminium ion that is captured by internal alkyne. Further the anti-Markovnikov oxidation of **271a** gave the aldehyde **272**, which on intramolecular aldol condensation followed by double bond reduction furnished compound **273**. Finally the LiAlH$_4$ reduction of the amide to dihydrolycopodine **274** and Jones' oxidation of the alcohol gave (−)-lycopodine (**275**).

2.30 (−)-Lycospidine A

Lycospidine A (**284**) was isolated by Zhao and coworkers from *L. complanatum* [90]. It has a unique 5-membered A-ring and diosphenol D-ring with four stereocenters. It shows various biological activities such as antipyretic and anticholinesterase activity. She and coworkers [91] reported the first asymmetric total synthesis (−)-lycospidine A (**284**; Scheme 2.35). The known sulfoxide **276** (prepared from pulegone) was added to methyl acrylate, followed by sulfoxide elimination to give unsaturated ketone, which on Michael addition of propargyl indium gave **277** as a 1 : 1 diastereomeric mixture. This on amidation/aza-Prins domino cyclization gave compound **278**. The reaction involves the formation of bicyclic enamide, which gets stereoselectively protonated to give *N*-acyliminium ion that is captured by the pendant alkyne. Further, the tetracyclic skeleton **280** was synthesized by ozonolysis of **278** to **279** followed by intramolecular aldol condensation. While direct methods for α-hydroxylation of ketone failed, the compound **280** was converted into dihydroxy compound **282** by α-chlorination (to **281**), hydrolysis to hydroxyl group, keto, and double bond reduction. Finally reduction of the amide and Swern oxidation of the diol gave the diketone **283**, which spontaneously isomerized to (−)-lycospidine A (**284**).

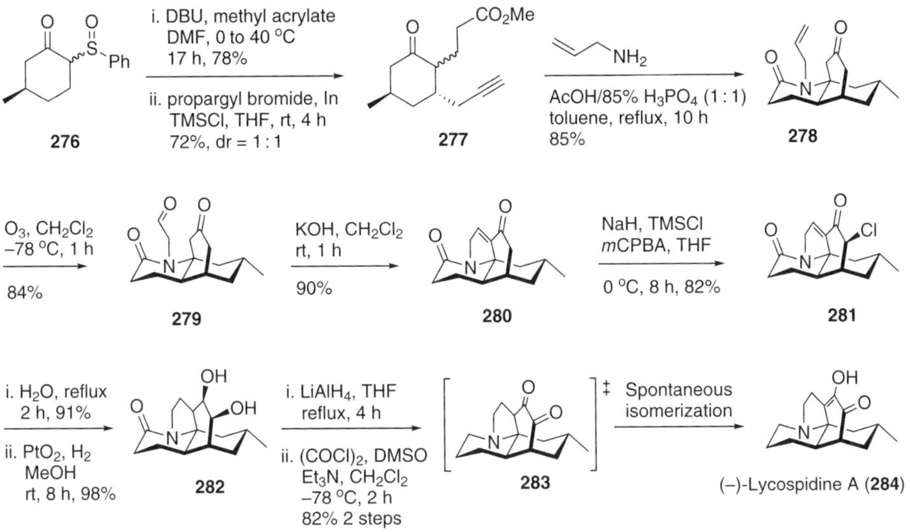

Scheme 2.35 Asymmetric total synthesis of (−)-lycospidine A (**284**).

2.31 Transtaganolides C and D

Transtaganolides are chemical components of *Thapsia* sp. and have sarcoplasmic–endoplasmic reticulum Ca^{2+}-ATPase inhibitory properties [92]. Stoltz et al. [93] achieved the total synthesis of transtaganolides C (**290**) and D (**291**) starting from pyrone ester **286** obtained from (Z)-methyl-3-iodoacrylate **285** in four steps and with 16% overall yield [94] (Scheme 2.36). Negishi cross-coupling of **286** with methoxyethyl zinc chloride gave the desired intermediate **287** in 44% yield. Further the

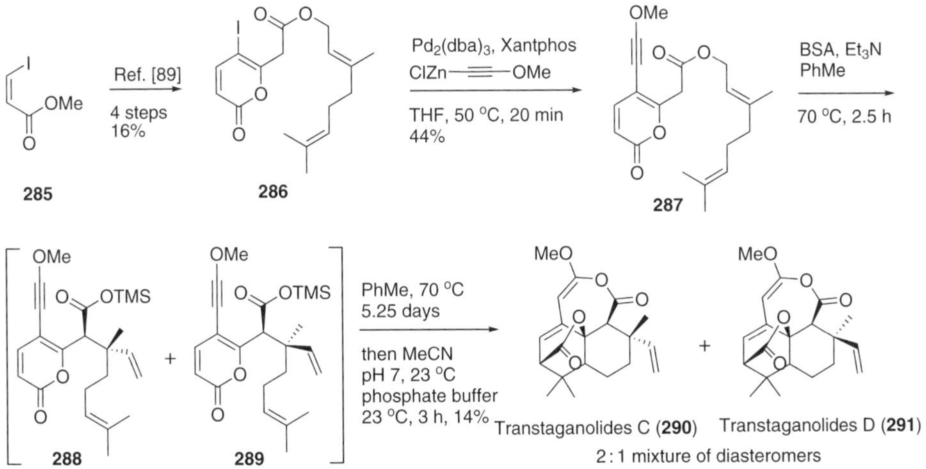

Scheme 2.36 Total synthesis of transtaganolides C (**290**) and D (**291**).

Ireland–Claisen/Diels–Alder cascade reaction gave transtaganolides C (**290**) and D (**291**) as 2 : 1 mixture of diastereomers (via **288** and **289**). These were further purified by normal-phase HPLC, which gave 9 and 5% yields for **290** and **291**, respectively.

2.32 (+)-Chatancin

Sato and coworkers isolated the diterpene chatancin (**299**) from the soft coral *Sarcophyton* sp. off the coast of Okinawa, Japan [95]. It inhibits the platelet-activating factor (PAF)-induced platelet aggregation and PAF receptor binding [95]. Zhao and Maimone [96] reported the enantioselective total synthesis of chatancin (**299**) in seven steps from (*S*)-dihydrofarnesal (**292**) as shown in Scheme 2.37. The Lewis acid-promoted addition of silyl ketene acetal **293** to **292** and *in situ* DMP oxidation delivered ketone **294**. This on heating in toluene released acetone, thereby cyclizing to give hydroxyl pyrone, which on treatment with Tf₂O and Et₃N delivered triflate **295** in 67% yield. Introduction of methyl ester on to the triflate under palladium catalysis followed by intramolecular pyrone/olefin [4+2] cycloaddition produced the diastereomers **296** and **297** (dr = 1 : 1) in 90% yield. The diastereomer **297** was subjected to allylic chlorination (SO₂Cl₂, Na₂CO₃) and subsequently heated with active Zn dust to deliver the single diastereomer **298**. This finally on hydrogenation afforded chatancin (**299**) in 93% yield.

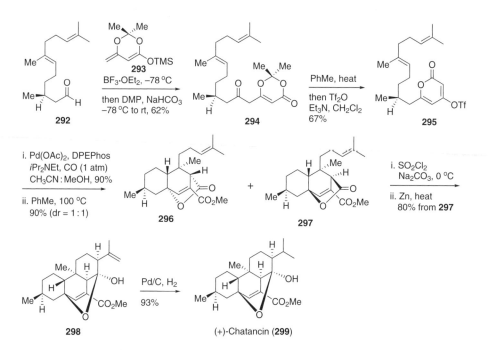

Scheme 2.37 Protecting-group-free enantioselective total synthesis of (+)-chatancin (**299**).

2.33 (−)-Jiadifenolide

In 2009, Fukuyama and coworkers [97] isolated the sesquiterpene (−)-jiadifenolide (**311**) from the dried pericarps of *Illicium jiadifengpi*. Zhang and coworkers [98] reported the PGF total synthesis of (−)-jiadifenolide from cyclic β-ketoester **300** using SmI$_2$/H$_2$O-mediated stereoselective reductive cyclization and unprecedented formal [4+1] annulative THF formation as a key step (Scheme 2.38). Stereoselective allylation of cyclic β-ketoester **300** followed by ozonolytic cleavage of olefin gave the aldehyde **301**. The Baylis–Hillman-type addition of butenolide **302** to **301** provided **303**. This on acetylation/elimination gave **304**. The *in situ* conversion of the ketone and lactone to enolates and reduction of the ester group with DIBAL-H and selective hydrogenation of olefin furnished **305**. Further SmI$_2$/H$_2$O-mediated stereoselective reductive cyclization of compound **305** gave tricyclic compound (80%, dr = 7 : 1) from which **306** was isolated in 70% yield. Primary alcohol oxidation led to aldehyde **307**, which was further treated with TMSC(Li)N$_2$ at −78 °C to afford 3-trimethylsiloxy furanoid **308** in 78% yield. Compound **308** was converted into epoxy compound **309** by oxidative double bond formation followed by epoxidation. Oxidation of 3-trimethylsiloxy furanoid **309** under Sharpless conditions led to α-ketolactone **310**, which on LiOH treatment completed the total synthesis of (−)-jiadifenolide (**311**).

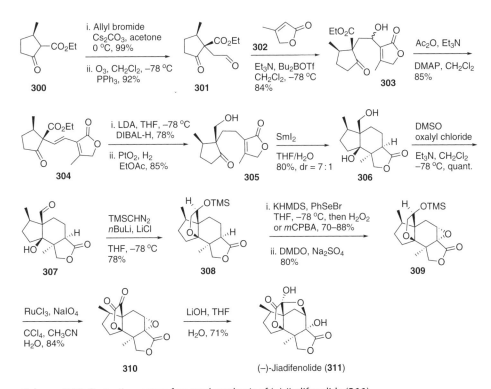

Scheme 2.38 Protecting-group-free total synthesis of (−)-jiadifenolide (**311**).

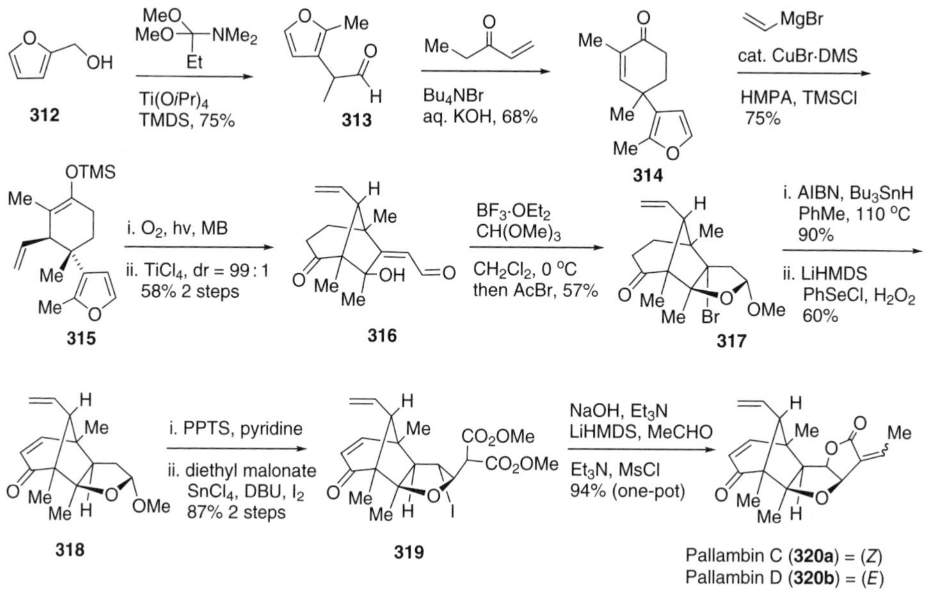

Scheme 2.39 Total synthesis of pallambins C (**320a**) and D (**320b**).

2.34 Pallambins C and D

Lou and coworkers isolated pallambins A–D from the Chinese liverwort *Pallavicinia ambigua* in 2012 [99]. Baran and coworkers [100] synthesized this challenging natural product in 11 steps from readily available furfuryl alcohol by a PGF strategy (Scheme 2.39). Commercially available furfuryl alcohol **312** was converted into aldehyde **313** by tandem Eschenmoser–Claisen rearrangement and reduction followed by acidic work-up. Robinson annulation of aldehyde **313** with ethyl vinyl ketone (EVK) worked efficiently and offered **314** in 68% yield. Michael addition of vinyl Grignard on compound **314** in the presence of CuBr·DMS salt gave compound **315**. Oxidative cleavage of the furan unit in **315** to the ketoaldehyde and Mukaiyama aldol cyclization gave cyclic compound **316** in 58% yield over two steps and dr = 99 : 1. The latter on treatment with CH(OMe)₃/BF₃·OEt₂ and then AcBr efficiently cyclized to form the THF ring compound **317**. Radical removal of bromine and unsaturation by addition–elimination of α-selenide gave **318**. PPTS buffered with pyridine-mediated elimination of the methoxy group afforded the sensitive dihydrofuran, which was further converted into iodo-diester **319**. Finally the sequence of alkaline hydrolysis of the esters followed by addition of TMSCl, Et₃N-induced lactonization, and decarboxylation followed by aldol reaction with acetaldehyde and elimination completed the total synthesis of pallambins C (**320a**) and D (**320b**) in a one-pot reaction in excellent 94% yield.

2.35 (+)-Vellosimine

Vellosimine is a sarpagine alkaloid mainly isolated from the plant family Apocynaceae (genus *Rauvolfia*) [101]. In 2016, Krüger and Gaich [102] reported a concise synthesis of vellosimine (**329a**), *N*-methylvellosimine (**329b**), and 10-methoxyvellosimine (**329c**)

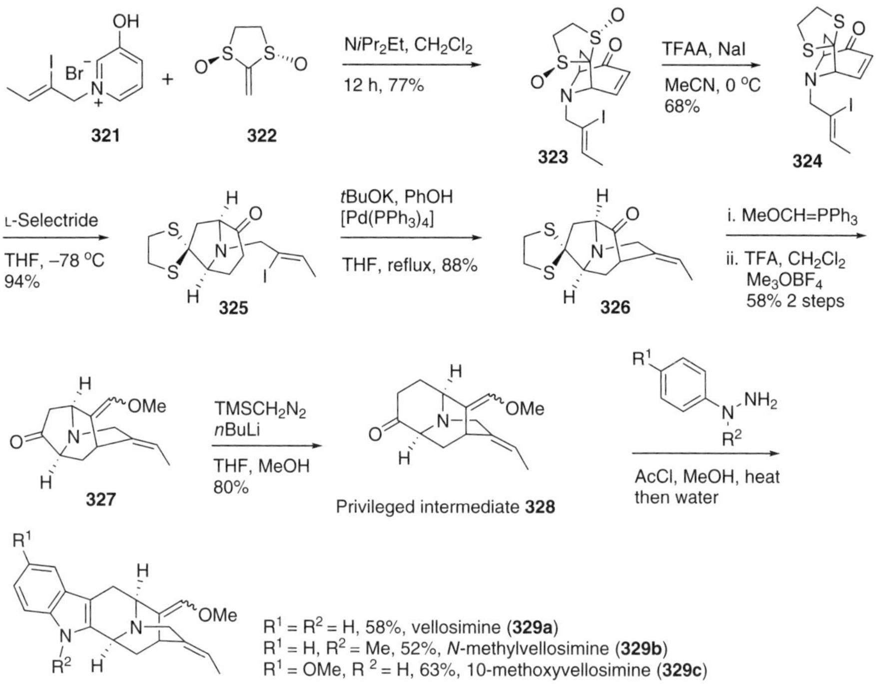

Scheme 2.40 Total syntheses of (+)-vellosimine (**329a**) and analogs.

without using any protecting groups (Scheme 2.40). The synthesis started with [5+2] cycloaddition of oxidopyridinium ion **321** and Aggarwal's chiral ketene equivalent **322** (prepared in four steps) to produce compound **323** in 77% yield with 2 : 1 ratio of regioisomers (93% ee). Reductive removal of the bis(sulfoxide) to **324** followed by α,β-unsaturated double bond reduction offered compound **325**. Palladium (tetrakis) triphenylphosphine-catalyzed intramolecular coupling reaction of compound **325** gave tricyclic ketone **326** in 88% yield. This underwent Wittig reaction and further dithi-olane deprotection gave keto-compound **327**. Ketone **327** was treated with TMSCH$_2$N$_2$ and *n*BuLi for ring enlargement reaction to regioselectively provide the privileged inter-mediate **328** in 80% yield. Finally, the Fischer indole synthesis on compound **328** with various substituted phenylhydrazines completed the total synthesis of vellosimine (**329a**), *N*-methylvellosimine (**329b**), and 10-methoxyvellosimine (**329c**) in 58, 52, and 63% yields, respectively.

2.36 (–)-Pallavicinin and (+)-Neopallavicinin

Secolabdane-type diterpenoids (+)-pallavicinin and (–)-neopallavicinin were first iso-lated by Wu and coworkers from the Taiwanese liverwort *Pallavicinia subciliata* and then by Lou and coworkers from the Chinese liverwort *P. ambigua* [103, 104]. The unique structural features such as ladder-shaped/cage-like [6-5-5-5] tetracyclic skeleton containing seven contiguous stereogenic centers (quaternary carbons) and a bridged ring in the left-part bicyclo[3.2.1] moiety as well as potential bioactivities of these molecules make them intriguing targets for total synthesis. Huang et al. [105] disclosed an elegant PGF asymmetric total synthesis of (−)-pallavicinin (**340a**) and (+)-neopallavicinin (**340b**) involving 15 steps (Scheme 2.41). The synthesis commenced with chiral cyclohexanone **331** prepared by following Stoltz's procedure from **330**, in 85% ee, constructing the first quaternary stereogenic center. The compound **331** was converted into TBS enol ether and subjected to palladium-catalyzed oxidative cyclization to the bicyclic compound **332**. Conjugate addition of the vinyl group to **332** (dr = 10 : 1) and

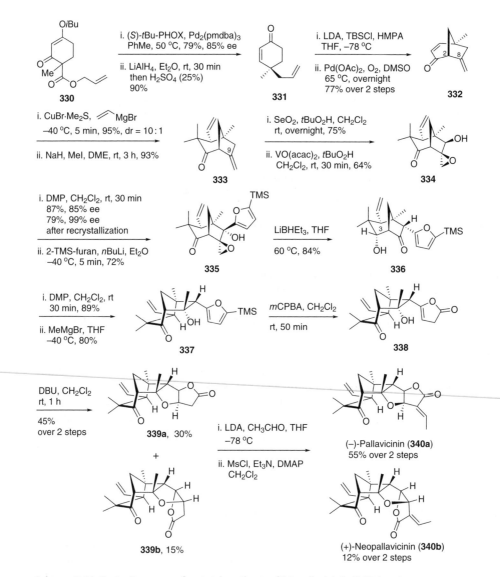

Scheme 2.41 Protecting-group-free total synthesis of (−)-pallavicinin (**340a**) and (+)- neopallavicinin (**340b**).

double α-methylation gave **333**. Allylic oxidation led to alcohol as the sole stereoisomer, and subsequent hydroxyl directed epoxidation furnished single epoxide product **334**. This on DMP oxidation provided the diketone (>99% ee after single recrystallization), which on chemo- and stereoselective addition of trimethylsilyl furan lithium to the C9 carbonyl group produced the desired epoxide **335** in 72% yield. This was then subjected to LiBHEt$_3$-mediated Payne rearrangement [106]. Treatment of **335** with LiBHEt$_3$ at 30 °C afforded the epoxide migration product (not shown) in 35% yield along with the unexpected ketone **336**, the key intermediate of the synthetic strategy in 45% yield. Increasing the reaction temperature to 60 °C led to ketone **336** as the sole product in 84% yield. The ketone **336** on DMP oxidation to diketone and selective monomethylation at the C8 carbonyl group afforded **337** in 80% yield. This was oxidized with *m*CPBA to the unstable β,γ-unsaturated butenolide **338**, which upon DBU treatment underwent double bond migration followed by intramolecular oxa-Michael addition to produce the tetracyclic compounds **339a** and **339b** in 30 and 15% yields, respectively. Finally, the aldol reaction of **339a** with acetaldehyde and elimination of alcohol via the mesylate furnished (−)-pallavicinin (**340a**) in 55% yield over two steps. Similarly, **339b** led to (+)-neopallavicinin (**340b**) in 12% yield.

2.37 Asteriscunolides A–D and Asteriscanolide

Li et al. disclosed a collective synthesis of humulanolides using ring-opening/ring-closing metathesis (ROM/RCM) cascade reaction as the key step (Scheme 2.42) [107]. Several of these molecules have potent anticancer properties, which is higher than cis-platin [108]. Li's synthesis began with the asymmetric preparation of tetraene **345** from readily available material **341**. One-carbon Wittig olefination on **341** followed by oxidation with TEMPO gave the corresponding aldehyde, which on second Wittig olefination gave Weinreb amide **343**. This on Grignard reaction with isopropenylmagnesium bromide afforded triene **344**. Alcohol **344** was esterified with cyclobut-1-enecarboxylic acid under Yamaguchi conditions in moderate yield to afford the tetraene **345**. The latter was then subjected to ROM/RCM cascade reaction using Grubbs–Hoveyda II catalyst, which gave (−)-asteriscunolide D (**346a**) in 36% yield. The synthesis of **346a** was completed in six steps with 5.3% overall yield from **341**. The synthesis of the remaining humulanolides has been achieved from (−)-asteriscunolide D (**346a**). Regio- and stereoselective hydrogenation of **346a** with Wilkinson's catalyst gave 6,7,9,10-tet-rahydroasteriscunolide **347** in 99% yield. Treatment of **347** with DBU gave (+)-asteriscanolide (**348**) via transannular Michael reaction. Conversion of **346a** to **350** was achieved in two steps. The first step involved addition of sodium methoxide, and the presumed tricyclic compound **349** was obtained as a complex mixture of insepara-ble isomers. The reaction goes through nucleophilic addition of methoxide to dienone followed by 1,4-addition of resulting enolate to the butenolide moiety, which forms the eight-membered ring **349**. Subsequent treatment of **349** with BF$_3$·Et$_2$O produced 6,7,9,10-tetradehydroasteriscanolide (**350**) via elimination of methoxy groups in overall 65% yield. Exposure of **346a** to UV light gave asteriscunolides A–C (**346b–d**), which were separated by preparative thin-layer chromatography in 46, 18, and 18% yields, respectively.

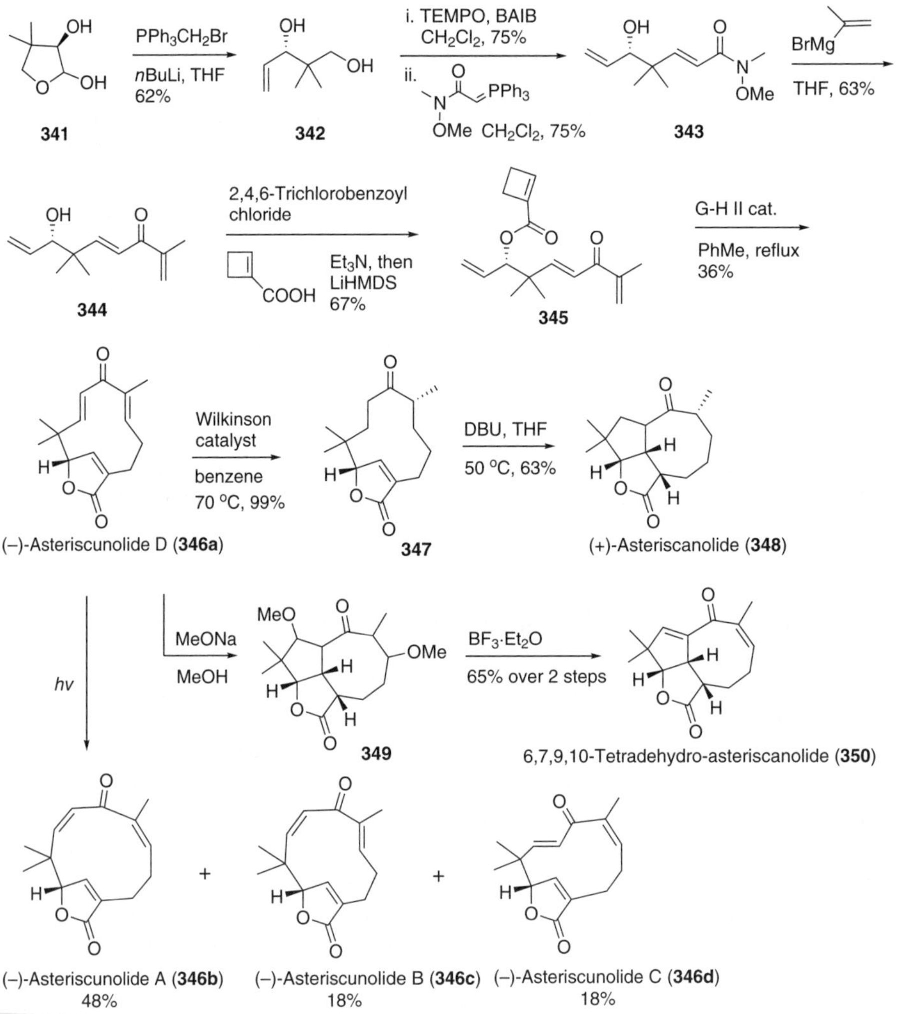

Scheme 2.42 Li's synthesis of various humulanolides.

2.38 (−)- and (+)-Palmyrolide A

(−)-Palmyrolide A, a neuroactive macrolide, was found to be a potent inhibitor of calcium ion oscillations in murine cerebrocortical neurons and possess a sodium ion channel blocking ability in neuroblastoma cells [109]. Since the absolute configurations in (−)-palmyrolide A were not determined, Maio et al. [110] carried out extensive synthetic studies of all possible diastereomers of (−)-palmyrolide A. While this study confirmed the absolute structure, the total synthesis of natural (−)-palmyrolide A (**358**) is shown in Scheme 2.43. L-Proline-catalyzed aldol reaction of pivaldehyde **351** with acetone gave β-hydroxy ketone **352**. This on stereoselective Kiyooka *syn*-reduction using DIBAL-H gave the diol, which on treatment with thionyl chloride in pyridine and oxidation using RuO$_4$ gave *syn*-cyclic sulfate **353**. Nucleophilic ring opening of *syn*-cyclic sulfate by allyl MgBr and CuI gave *anti*-compound **354** as a single diastereomer.

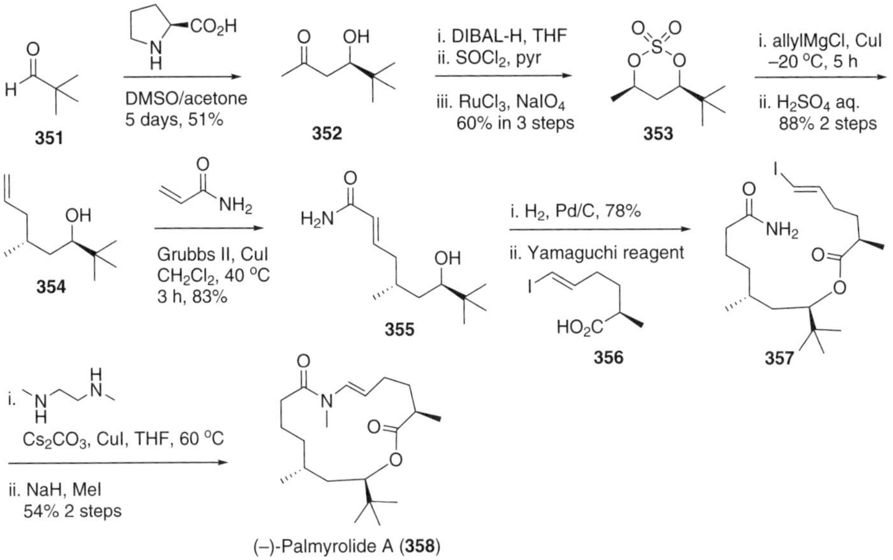

Scheme 2.43 Total synthesis of (–)-palmyrolide A (**358**).

This on cross-metathesis with acrylamide using the modified Lipshutz CuI conditions [111] efficiently provided the enamide **355**. The latter on hydrogenation of olefin and Yamaguchi esterification with **356** gave **357**. The final ring closure using the modified Buchwald conditions furnished the enamide, which was N-methylated to complete the synthesis of (–)-palmyrolide A (**358**).

A year later in 2013, Reddy et al. accomplished the total synthesis of (+)-palmyrolide A (*ent*-**358**) and its *cis*-diastereomer **363** as shown in Scheme 2.44 [112]. The end part was similar to that of Maio's synthesis (Scheme 2.43). Addition of *t*BuMgCl to (*S*)-citronellal (**359**) gave the alcohol, which was subjected to ozonolysis followed by Zhu's oxidative one-carbon homologation to afford the primary amides **360a** and **360b** as a 1 : 1 mixture. This on oxidation–reduction gave the desired alcohol **360a**. Further the Yamaguchi esterification of **360a** with carboxylic acid **361** gave the ester that on Lemieux–Johnson oxidation followed by Takai's reaction furnished Maio's intermediate *ent*-**357**. A modified Buchwald method under high dilution conditions for macrocyclization of *ent*-**357** gave poor yields. Changing the reaction conditions like temperature and prolonged heating surprisingly produced the *cis*-isomer **362** (43%), which upon N-methylation gave (–)-*cis*-palmyrolide A (**363**). The macrocyclization under microwave conditions gave **364** in 70% yield (brsm) in 10 min. This upon N-methylation furnished (+)-palmyrolide A (*ent*-**358**). The conversion of *trans*-enamide **364** to *cis*-enamide **362** under microwave heating for 20 min was also demonstrated.

2.39 (±)-Bipinnatin J

Bipinnatin J (**375**), a diterpene, was isolated from *Pseudopterogorgia bipinnata* [113]. Roethle and Trauner [114] reported a nine-step PGF synthesis using ruthenium-catalyzed Alder-ene reaction, Stille cross-coupling, and an intramolecular Nozaki–Hiyama–Kishi allylation as key steps. The synthesis started from 3-butynol **365**

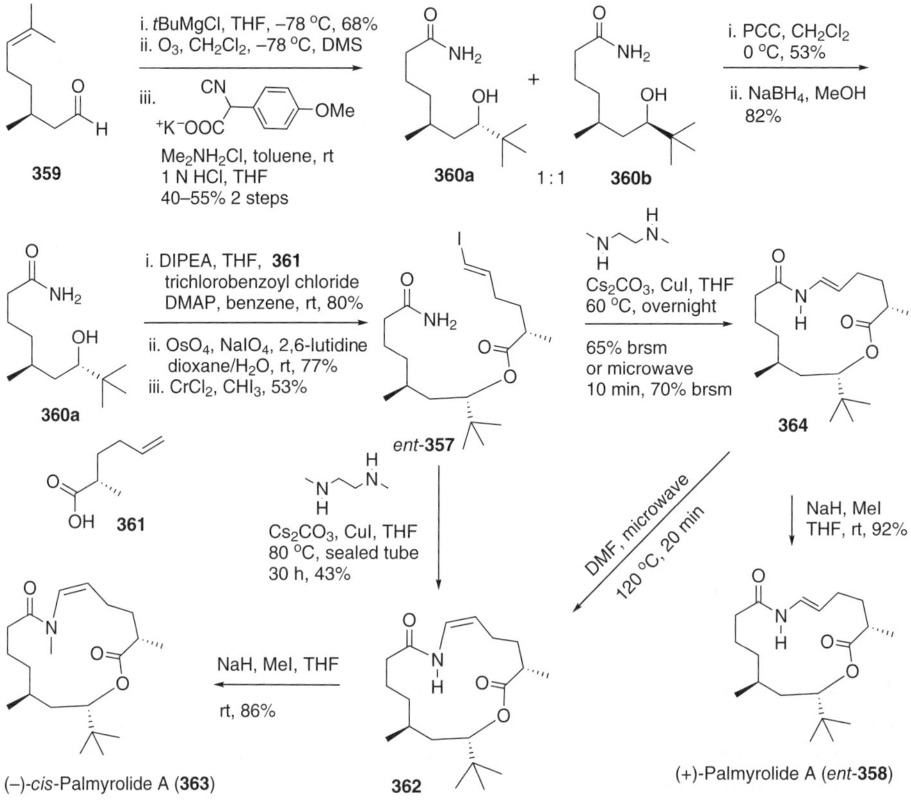

Scheme 2.44 Total synthesis of (+)-palmyrolide A (*ent*-**358**) and (−)-*cis*-palmyrolide A (**363**).

conversion to (*Z*)-vinyl iodide **366** and further oxidation to the sensitive aldehyde **367** (Scheme 2.45). Addition of lithiated ethyl propiolate to **367** afforded the propargylic alcohol **368** as a racemate. The Trost enyne reaction [115] of **368** catalyzed by ruthenium(II) with a slight excess of allyl alcohol gave the desired aldehyde **369**. Reaction of **369** with a stabilized Wittig reagent followed by chemoselective reduction of the resulting aldehyde **370** gave primary allylic alcohol **371**. The furan moiety of the natural product was synthesized from 3-methylfurfural, which gave furylstannane **372** in a single step. Stille coupling of **371** and **372** gave **373** in excellent yield. This allylic alcohol was converted into allylic bromide **374**, which on further Nozaki–Hiyama–Kishi allylation [116] conditions provided bipinnatin J (**375**) almost as a single diastereomer (dr > 9 : 1).

2.40 Cyanolide

Cyanolide A, a 16-membered macrodiolide, exhibits potent molluscicidal activity against the water snail *Biomphalaria glabrata* [117], which is responsible for the human parasitic disease schistosomiasis. Waldeck and Krische [118] achieved the

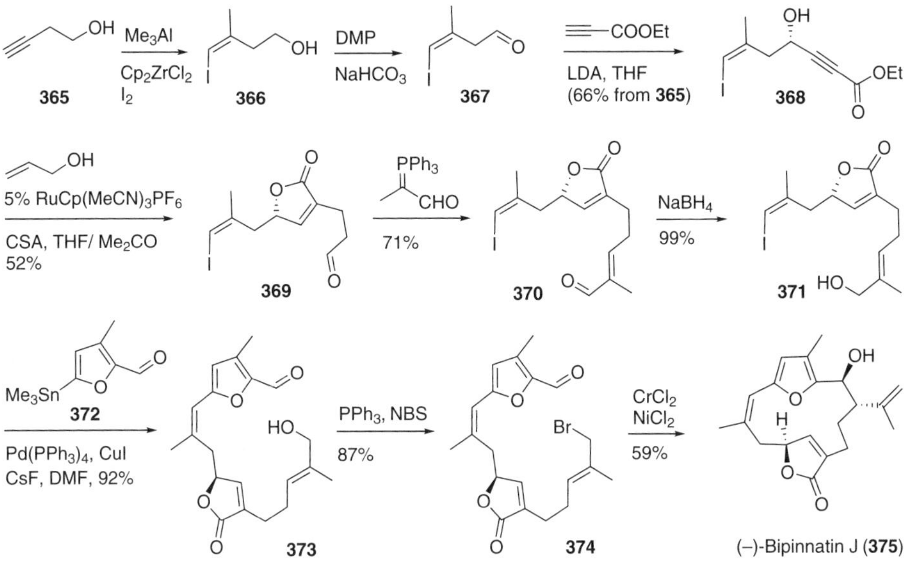

Scheme 2.45 Total synthesis of (±)-bipinnatin J (**375**).

total synthesis of cyanolide A (**383**) without involvement of any protecting groups (Scheme 2.46). The synthesis began with double allylation of neopentane diol (**376**) through iridium catalysis, giving diol **377** as a single enantiomer. This on a cascade metathesis–oxa-Michael reaction with EVK gave the desired *cis*-2,3-disubstituted pyran **379** as a 10 : 1 diastereomeric mixture. This was then subjected to cross-metathesis with ethylene to furnish the terminal olefin, which on glycosylation using phenyl thioglycoside **380** in the presence of methyl triflate gave **381** as a 2 : 1 diastereomeric mixture favoring the β-anomer. Reduction of β-alkoxy ketone by Li(*s*Bu)₃BH furnished the expected 1,3-*syn*-diastereomer as a 4 : 1 mixture. This was further converted into the carboxylic acid **382** by oxidative cleavage of the terminal olefin. Macrodiolide formation from carboxylic acid **382** under Shiina's conditions [119] gave cyanolide A (**383**). A more concise synthesis was also executed from compound **379** by first glycosylation to enone **384**. This upon olefin cleavage followed by Li(*s*Bu)₃BH reduction gave the 5 : 1 diastereomeric alcohol mixture. The latter was subjected to site-selective oxidation to hydroxy acid **382**, and further Yamaguchi reaction conditions led to cyanolide A (**383**) in only six steps from neopentane diol (**376**).

2.41 Conclusions

Total synthesis of natural products has been always an enchanting task. From the planning stage to the bench execution, total synthesis remains full of surprises, sometimes failures and then triumphs. Nevertheless, literature is bountiful with

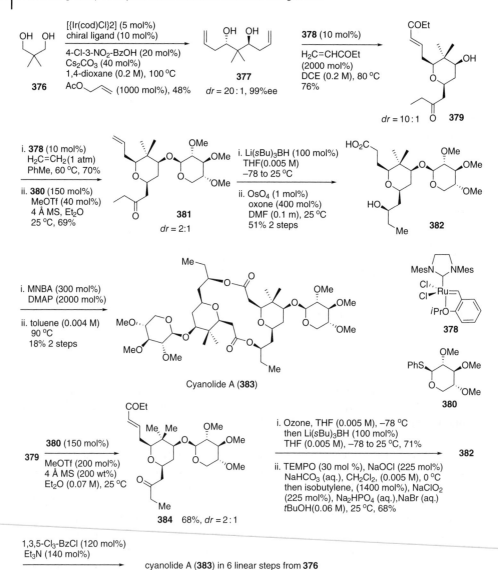

Scheme 2.46 Total synthesis of cyanolide A (**383**).

examples of total syntheses of natural products being achieved with much determination and discloses the art of planning and the hard endeavor of execution. The total synthesis without the use of protecting groups is a remarkable achievement. This chapter has given an overview of many total syntheses of simple to complex natural products being synthesized devoid of protecting groups. This literature will indeed inspire future synthetic efforts.

References

1 (a) Hendrickson, J.B. (1975). *J. Am. Chem. Soc.* 97: 5784. (b) Gaich, T. and Baran, P.S. (2010). *J. Org. Chem.* 75: 4657 and references there in.

2 For reviews on protecting-group-free synthesis see: (a) Hoffmann, R.W. (2006). *Synthesis* 3531. (b) Young, I.A. and Baran, P.S. (2009). *Nat. Chem.* 1: 193. (c) Saicic, R.N. (2014). *Tetrahedron* 70: 8183.

3 (a) Tietze, L.F. and Beifuss, U. (1993). *Angew. Chem. Int. Ed.* 32: 131. (b) Bunce, R.A. (1995). *Tetrahedron* 51: 13103. (c) Nicolaou, K.C., Montagnon, T., and Snyder, S.A. (2003). *Chem. Commun.* 551. (d) Padwa, A. (2004). *Pure Appl. Chem.* 76: 1933. (e) Nicolaou, K.C., Edmonds, D.J., and Bulger, P.G. (2006). *Angew. Chem. Int. Ed.* 45: 7134. (f) Pellissier, H. (2012). *Adv. Synth. Catal.* 354: 237. (g) Clavier, H. and Pellissier, H. (2012). *Adv. Synth. Catal.* 354: 3337.

4 For selected literature see: (a) Lyons, T.W. and Sanford, M.S. (2010). *Chem. Rev.* 110: 1147. (b) Hashiguchi, B.G., Bischof, S.M., Konnick, M.M., and Periana, R.A. (2012). *Acc. Chem. Res.* 45: 885. (c) Crabtree, R.H. and Lei, A. (2017). *Chem. Rev.* 117: 8481.

5 (a) Yamaguchi, J., Yamaguchi, A.D., and Itami, K. (2012). *Angew. Chem. Int. Ed.* 51: 8903. (b) Chen, D.Y.-K. and Youn, S.W. (2012). *Chem. Eur. J.* 18: 9452.

6 (a) Ma, F., Lei, M., and Hu, L. (2016). *Org. Lett.* 18: 2708. (b) Xu, J., Liu, Y., Wang, Y. et al. (2017). *Org. Lett.* 19: 1562. (c) Hong, K., Park, H., and Yu, J.-Q. (2017). *ACS Catal.* 7: 6938. (d) Liu, X.-H., Park, H., Hu, J.-H. et al. (2017). *J. Am. Chem. Soc.* 139: 888. (e) Zhang, X., Zheng, H., Li, J. et al. (2017). *J. Am. Chem. Soc.* 139: 14511. (f) St John-Campbell, S., White, A.J.P., and Bull, J.A. (2017). *Chem. Sci.* 8: 4840.

7 (a) Bedke, D.K. and Vanderwal, C.D. (2011). *Nat. Prod. Rep.* 28: 15. (b) Nilewski, C. and Carreira, E.M. (2012). *Eur. J. Org. Chem.* 1685.

8 Chung, W.-J., Carlson, J.S., Bedke, D.K., and Vanderwal, C.D. (2013). *Angew. Chem. Int. Ed.* 52: 10052.

9 Nilewski, C., Geisser, R.W., and Carreira, E.M. (2009). *Nature* 457: 573.

10 (a) Denmark, S.E., Barsanti, P.A., Wong, K.-T., and Stavenger, R.A. (1998). *J. Org. Chem.* 63: 2428. (b) Denmark, S.E., Barsanti, P.A., Beutner, G.L., and Wilson, T.W. (2007). *Adv. Synth. Catal.* 349: 567.

11 Barrero, A.F., Quílez del Moral, J.F., Herrador, M.M. et al. (2009). *Curr. Org. Chem.* 13: 1164.

12 Morales, C.P., Catalán, J., Domingo, V. et al. (2011). *J. Org. Chem.* 76: 2494.

13 Justicia, J., Jiménez, T., Morcillo, S.P. et al. (2009). *Tetrahedron* 65: 10837.

14 Lee, K., Kim, H., and Hong, J. (2009). *Org. Lett.* 11: 5202.

15 Siririath, S., Tanaka, J., Ohtani, I.I. et al. (2002). *J. Nat. Prod.* 65: 1820.

16 ElMarrouni, A., Joolakanti, S.R., Colon, A. et al. (2010). *Org Lett.* 12: 4074.

17 Chi, Y. and Gellman, S.H. (2005). *Org. Lett.* 19: 4253.

18 Herrmann, A.T., Martinez, S.R., and Zakarian, A. (2011). *Org. Lett.* 13: 3636.

19 Satake, M., Bourdelais, A.J., Van Wagoner, R.M. et al. (2008). *Org. Lett.* 10: 3465.

20 (a) Jin, W., Li, X., Huang, Y. et al. (2010). *Chem. Eur. J.* 16: 8259. (b) Smith, M.B. and March, J. (2007). *March's Advanced Organic Chemistry*, 6e, 1815–1823. Hoboken: Wiley.

21 Achmatowicz, O., Bukowski, P., Szechner, B. et al. (1971). *Tetrahedron* 27: 1973.

22 Hu, Y. and Brown, R.C.D. (2005). *Chem. Commun.* 5636.

23 (a) Cecil, A.R.L., Hu, Y.L., Vicent, M.J. et al. (2004). *J. Org. Chem.* 69: 3368. (b) Trost, B.M., Calkins, T.I., and Bochet, C.G. (1997). *Angew. Chem. Int. Ed.* 36: 2632.

24 Yang, Y., Kinoshita, K., Koyama, K. et al. (1999). *Phytomedicine* 6: 89.

25 Hikino, H., Hikino, Y., Takeshita, Y. et al. (1967). *Chem. Pharm. Bull.* 15: 321.

26 Booker-Milburn, K.I., Jenkins, H., Charmant, J.P.H., and Mohr, P. (2003). *Org. Lett.* 5: 3309.

27 Wang, B., Zhong, Z., and Lin, G.-Q. (2009). *Org. Lett.* 11: 2011.

28 Aoyagi, S., Wang, T.-C., and Kibayashi, C. (1993). *J. Am. Chem. Soc.* 115: 11393.

29 Carr, G., Williams, D.E., Ratnayake, R. et al. (2012). *J. Nat. Prod.* 75: 1189.

30 Stambulyan, H. and Minehan, T.G. (2016). *Org. Biomol. Chem.* 14: 8728.

31 Ito, Y., Hirao, T., and Saegusa, T. (1978). *J. Org. Chem.* 43: 1011.

32 Guilet, D., Guntern, A., Ioset, J.-R. et al. (2003). *J. Nat. Prod.* 66: 17.

33 Mukherjee, P. and Sarkar, T.K. (2012). *Org. Biomol. Chem.* 10: 3060.

34 da Costa, G.M., de Lemos, T.L.G., Pessoa, O.D.L. et al. (1999). *J. Nat. Prod.* 62: 1044.

35 Löbermann, F., Mayer, P., and Trauner, D. (2010). *Angew. Chem. Int. Ed.* 49: 6199.

36 Gavagnin, M., Carbone, M., Nappo, M. et al. (2005). *Tetrahedron* 61: 617.

37 Tang, Y., Liu, J.-T., Chen, P. et al. (2014). *J. Org. Chem.* 79: 11729.

38 Hajos, Z.G. and Parrish, D.R. (1985). *Org. Synth.* 63: 26.

39 (a) Burgess, E.M., Penton, H.R., and Taylor, E.A. (1973). *J. Org. Chem.* 38: 26. (b) Khapli, S., Dey, S., and Mal, D. (2001). *J. Indian Inst. Sci.* 81: 461. (c) Rinner, U., Adams, D.R., dos Santos, M.L. et al. (2003). *Synlett* 1247.

40 Piao, S.-J., Song, Y.-L., Jiao, W.-H. et al. (2013). *Org. Lett.* 15: 3526.

41 McCallum, M.E., Rasik, C.M., Wood, J.L., and Brown, M.K. (2016). *J. Am. Chem. Soc.* 138: 2437.

42 Ruider, S.A., Sandmeier, T., and Carreira, E.M. (2015). *Angew. Chem. Int. Ed.* 54: 2378.

43 (a) Gormisky, P.E. and White, M.C. (2013). *J. Am. Chem. Soc.* 135: 14052. (b) Osberger, T.J. and White, M.C. (2014). *J. Am. Chem. Soc.* 136: 11176. (c) Ammann, S.E., Rice, G.T., and White, M.C. (2014). *J. Am. Chem. Soc.* 136: 10834.

44 (a) Rasik, C.M. and Brown, M.K. (2015). Total synthesis of gracilioether F and (±)-hippolachnin A. Poster presentation at the 44th National Organic Chemistry Symposium, University of Maryland, College Park, MD, USA (28 June 2015). (b) McCallum, M.E. and Wood, J.L. (2015). Total synthesis of hippolachnin A. Poster presentation at the 44th National Organic Chemistry Symposium, University of Maryland, College Park, MD, USA (28 June 2015).

45 Schmidt, T.J., Vossing, S., Klaes, M., and Grimme, S. (2007). *Planta Med.* 73: 1574.

46 Tietze, L.F., Duefert, S.-C., Clerc, J. et al. (2013). *Angew. Chem. Int. Ed.* 52: 3191.

47 Veysoglu, T., Mitscher, L.A., and Swayze, J.K. (1980). *Synthesis* 807.

48 Weinstabl, H., Suhartono, M., Qureshi, Z., and Lautens, M. (2013). *Angew. Chem. Int. Ed.* 52: 5305.

49 Catellani, M., Frignani, F., and Rangoni, A. (1997). *Angew. Chem. Int. Ed.* 36: 119.

50 Adam, J.-M., Foricher, J., Hanlon, S. et al. (2011). *Org. Process Res. Dev.* 15: 515.

51 (a) Gao, W., Chen, A.P.C., Leung, C.H. et al. (2008). *Bioorg. Med. Chem. Lett.* 18: 704. (b) Wang, K., Su, B., Wang, Z. et al. (2010). *J. Agric. Food. Chem.* 58: 2703. (c) Baumgartner, B., Erdelmeier, C.A.J., Wright, A.D. et al. (1990). *Phytochemistry* 29: 3327.

52 Ying, W. and Herndon, J.W. (2013). *Eur. J. Org. Chem.* 3112.

53 (a) Cambeiro, X.C. and Pericas, M.A. (2011). *Adv. Synth. Catal.* 353: 113. (b) Chung, J.Y.L. and Wasicak, J.T. (1990). *Tetrahedron Lett.* 31: 3957. (c) Dickson, H.D., Smith, S.C., and Hinkle, K.W. (2004). *Tetrahedron Lett.* 45: 5597. (d) Barrett, A.G.M., Hopkins, B.T., Love, A.C., and Tedeschi, L. (2004). *Org. Lett.* 6: 835. (e) Miwa, K., Aoyama, T., and Shioiri, T. (1994). *Synlett* 107.

54 Moldvai, I., Dornyei, G., Tamesvari-Major, E., and Szantay, C. (2007). *Org. Prep. Proced.* 39: 503.

55 (a) Wu, C.M., Yang, C.W., Lee, Y.Z. et al. (2009). *Biochem. Biophys. Res. Commun.* 386: 140. (b) Lee, S.K., Nam, K.A., and Heo, Y.H. (2003). *Planta Med.* 69: 21.

56 (a) Staerk, D., Christensen, J., Lemmich, E. et al. (2000). *J. Nat. Prod.* 63: 1584. (b) Gao, W., Lam, W., Zhong, S. et al. (2004). *Cancer Res.* 64: 678.

57 Su, B., Chen, F., and Wang, Q. (2013). *J. Org. Chem.* 78: 2775.

58 Liu, G.-Q., Reimann, M., and Opatz, T. (2016). *J. Org. Chem.* 81: 6142.

59 Vaidya, S.D. and Argade, N.P. (2013). *Org. Lett.* 15: 4006.

60 (a) Jao, C.-W., Lin, W.-C., Wu, Y.-T., and Wu, P.-L. (2008). *J. Nat. Prod.* 71: 1275. (b) Chen, M., Gan, L., Lin, S. et al. (2012). *J. Nat. Prod.* 75: 1167. (c) Danz, H., Stoyanova, S., Wippich, P. et al. (2001). *Planta Med.* 67: 411.

61 (a) Quinghaosu Research Group (1980). *Sci. Sin.* 23: 380. (b) Shen, C.-C. and Zhuang, L.-G. (1984). *Med. Res. Rev.* 4: 47. (c) Klayman, D.L. (1985). *Science* 228: 1049. (d) Klayman, D.L., Lin, A.J., Acton, N. et al. (1984). *J. Nat. Prod.* 47: 715. (e) Butler, A.R. and Wu, Y.-L. (1992). *Chem. Soc. Rev.* 21: 85.

62 Yadav, J.S., Thirupathaiah, B., and Srihari, P. (2010). *Tetrahedron* 66: 2005.

63 Nicolaou, K.C., Wu, T.R., Sarlah, D. et al. (2008). *J. Am. Chem. Soc.* 130: 11114.

64 Vonwiller, S.C., Warner, J.A., Mann, S.T., and Haynes, R.K. (1995). *J. Am. Chem. Soc.* 117: 11098.

65 Wang, Q.Z., Liang, J.Y., and Feng, X. (2010). *Chin. Chem. Lett.* 21: 596.

66 Unsworth, W.P., Kitsiou, C., and Taylor, R.J.K. (2013). *Org. Lett.* 15: 3302.

67 Decker, M. (2005). *Eur. J. Med. Chem.* 40: 305.

68 Jiao, R.-H., Xu, S., Liu, J.-Y. et al. (2006). *Org. Lett.* 8: 5709.

69 (a) Sinder, B.B. and Wu, X.-X. (2007). *Org. Lett.* 9: 4913. (b) Toumi, M., Couty, F., Marrot, J., and Evano, G. (2008). *Org. Lett.* 10: 5027. (c) Malgesini, B., Forte, B., Borghi, D. et al. (2009). *Chem. Eur. J.* 15: 7922. (d) Coste, A., Karthikeyan, G., Couty, F., and Evano, G. (2009). *Synthesis* 2927.

70 Peng, Q.-L., Luo, S.-P., Xia, X.-E. et al. (2014). *Chem. Commun.* 50: 1986.

71 (a) Itokawa, H., Ibraheim, Z.Z., Qiao, Y.F., and Takeya, K. (1993). *Chem. Pharm. Bull.* 41: 1869. (b) Quao, Y.-F. (1991). Compound **180** was also isolated from *Rubia alata*, Phd thesis. Tokyo College of Pharmacy.

72 Lumb, J.-P. and Trauner, D. (2005). *J. Am. Chem. Soc.* 127: 2870.

73 Bloom, B.R. and Murray, C.J.L. (1992). *Science* 257: 1055.

74 Han, J.-C., Liu, L.-Z., Li, C.-C., and Yang, Z. (2013). *Chem. Asian J.* 8: 1972.

75 Wall, M.E., Wani, M.C., Cook, C.E. et al. (1966). *J. Am. Chem. Soc.* 88: 3888.

76 Gallo, R.C., Whang-Peng, J., and Adamson, R.H. (1971). *J. Natl. Cancer Inst.* 46: 789.

77 Xu, P., Chen, D.-S., Xi, J., and Yao, Z.-J. (2015). *Chem. Asian J.* 10: 976.

78 Wu, Z.-J., Xu, X.-K., Shen, Y.-H. et al. (2008). *Org. Lett.* 10: 2397.

79 Li, C., Yu, X., and Lei, X. (2010). *Org. Lett.* 12: 4284.

80 Yeom, H.-S., Li, H., Tang, Y., and Hsung, R.P. (2013). *Org. Lett.* 15: 3130.

81 For isolation of clusiacyclols A and B, see: Gonzalez, J.G., Olivares, E.M., and Monache, F.D. (1995). *Phytochemistry* 38: 485.

82 For isolation of eriobrucinol, see:(a)Jefferies, P.R. and Worth, G.K. (1973). *Tetrahedron* 29: 903. (b) Bukuru, J., Van, T.N., Puyvelde, L.V. et al. (2003). *Tetrahedron* 59: 5905. (c) Rashid, M.A., Armstrong, J.A., Gray, A.I., and Waterman, P.G. (1992). *Phytochemistry* 31: 3583.

83 Xu, Z., Wang, Q., and Zhu, J. (2013). *J. Am. Chem. Soc.* 135: 19127.

84 (a) Behenna, D.C., Mohr, J.T., Sherden, N.H. et al. (2011). *Chem. Eur. J.* 17: 14199. (b) Hong, A.Y. and Stoltz, B.M. (2013). *Eur. J. Org. Chem.* 2745.

85 Ge, H.M., Zhang, L.-D., Tan, R.X., and Yao, Z.-J. (2012). *J. Am. Chem. Soc.* 134: 12323.

86 Auer, E., Gössinger, E., and Graupe, M. (1997). *Tetrahedron Lett.* 38: 6577.

87 Bödeker, K. (1881). *Liebigs Ann. Chem.* 208: 363.

88 Ma, D., Zhong, Z., Liu, Z. et al. (2016). *Org. Lett.* 18: 4328.

89 (a) Reusch, W. and Johnson, C.K. (1963). *J. Org. Chem.* 28: 2557. (b) Katsuhara, J. (1967). *J. Org. Chem.* 32: 797. (c) Caine, D., Procter, K., and Cassell, R.A. (1984). *J. Org. Chem.* 49: 2647. (d) Kozak, J.A. and Dake, G.R. (2008). *Angew. Chem. Int. Ed.* 47: 4221.

90 Cheng, J.-T., Liu, F., Li, X.-N. et al. (2013). *Org. Lett.* 15: 2438.

91 Xu, S., Zhang, J., Ma, D. et al. (2016). *Org. Lett.* 18: 4682.

92 (a) Saouf, A., Guerra, F.M., Rubal, J.J. et al. (2005). *Org. Lett.* 7: 881. (b) Appendino, G., Prosperini, S., Valdivia, C. et al. (2005). *J. Nat. Prod.* 68: 1213.

93 Gordon, J.R., Nelson, H.M., Virgil, S.C., and Stoltz, B.M. (2014). *J. Org. Chem.* 79: 9740.

94 Larsson, R., Sterner, O., and Johansson, M. (2009). *Org. Lett.* 11: 657.

95 Sugano, M., Shindo, T., Sato, A. et al. (1990). *J. Org. Chem.* 55: 5803.

96 Zhao, Y.-M. and Maimone, T.J. (2015). *Angew. Chem. Int. Ed.* 54: 1223.

97 Kubo, M., Okada, C., Huang, J.-M. et al. (2009). *Org. Lett.* 11: 5190.

98 Shen, Y., Li, L., Pan, Z. et al. (2015). *Org. Lett.* 17: 5480.

99 Wang, L.-N., Zhang, J.-Z., Li, X. et al. (2012). *Org. Lett.* 14: 1102.

100 Martinez, L.P., Umemiya, S., Wengryniuk, S.E., and Baran, P.S. (2016). *J. Am. Chem. Soc.* 138: 7536.

101 (a) Lim, S.-H., Low, Y.-Y., Sinniah, S.K. et al. (2014). *Phytochemistry* 98: 204. (b) Lounasmaa, M., Hanhinen, P., and Westersund, M. (1999). *Alkaloids* 52: 103. (c) Manske, R.H.F. ed. (1950–1968). *The Alkaloids*, vol. I–IX. New York: Academic Press. (d) Glasby, J.S. (1975). *Encyclopedia of the Alkaloids*, vol. I–III. New York: Plenum. (e) Taylor, W.I., Sklar, R., and Bartlett, M.F. (1960). *J. Am. Chem. Soc.* 82: 3790.

102 Krüger, S. and Gaich, T. (2015). *Angew. Chem. Int. Ed.* 54: 315.

103 (a) Wu, C.-L., Liu, H.-J., and Uang, H.-L. (1994). *Phytochemistry* 35: 822. (b) Liu, H.-J. and Wu, C.-L. (1999). *J. Asian Nat. Prod. Res.* 1: 177.

104 Li, Z.-J., Lou, H.-X., Yu, W.-T. et al. (2005). *Helv. Chim. Acta* 88: 2637.

105 Huang, B., Guo, L., and Jia, Y. (2015). *Angew. Chem. Int. Ed.* 54: 13599.

106 (a) Dobler, M., Anderson, J.C., Juch, M., and Borschberg, H.-J. (1995). *Helv. Chim. Acta* 78: 292. (b) Hanson, R.M. (2002). *Organic Reactions*, vol. 60 (ed. L.E. Overman, P.G. Humphreys and G.S. Welmaker), 1–156. New York: Wiley.

107 Han, J.-C., Li, F., and Li, C.-C. (2014). *J. Am. Chem. Soc.* 136: 13610.

108 Rauter, A.P., Branco, I., Bermejo, J. et al. (2001). *Phytochemistry* 56: 167.

109 Perreira, A.R., Cao, Z., Engene, N. et al. (2010). *Org. Lett.* 12: 4490.

110 Tello-Aburto, R., Newar, T.D., and Maio, W.A. (2012). *J. Org. Chem.* 77: 6271.

111 (a) Voigtritter, K., Ghorai, S., and Lipshutz, B.H. (2011). *J. Org. Chem.* 76: 4697. (b) Venukadasula, P.K.M., Chegondi, R., Suryn, G.M., and Hanson, P.R. (2012). *Org. Lett.* 14: 2634.

112 Philkhana, S.C., Seetharamsingh, B., Dangat, Y.B. et al. (2013). *Chem. Commun.* 49: 3342.

113 Rodriguez, A.D. and Shi, J.-G. (1998). *J. Org. Chem.* 63: 420.

114 Roethle, P.A. and Trauner, D. (2006). *Org. Lett.* 8: 345.

115 Trost, B.M. and Toste, F.D. (1999). *Tetrahedron Lett.* 40: 7739.

116 (a) Fürstner, A. (1999). *Chem. Rev.* 99: 991. (b) Still, W.C. and Mobilio, D. (1983). *J. Org. Chem.* 48: 4785. (c) Rayner, C.M., Astles, P.C., and Paquette, L.A. (1992). *J. Am. Chem. Soc.* 114: 3926.

117 Pereira, A.R., McCue, C.F., and Gerwick, W.H. (2010). *J. Nat. Prod.* 73: 217.

118 Waldeck, A.R. and Krische, M.J. (2013). *Angew. Chem. Int. Ed.* 52: 4470.

119 Shiina, I., Kubota, M., Oshiumi, H., and Hashizume, M. (2004). *J. Org. Chem.* 69: 1822.

3

Protecting-Group-Free Synthesis of Natural Products and Analogs, Part II

Hiroyoshi Takamura and Isao Kadota

Department of Chemistry, Graduate School of Natural Science and Technology, Okayama University, Japan

3.1 Introduction

When synthetic chemists carry out selectively the molecular transformation at one reactive moiety of the multifunctional compound, the other reactive portions are usually masked with protecting groups. The protecting group must meet several demands. It must selectively react with the target functional group, be stable during the transformation of other functional groups in the molecule, and be selectively removed under the reaction conditions wherein the other portions in the molecule are not affected. Therefore, it is an important issue in multistep synthesis to select the appropriate protecting group depending on the functional group to be temporarily or almost permanently blocked in the molecule.

Wuts and Greene stated that "All things considered, no protective group is the best protective group" in the first chapter of *Greene's Protective Groups in Organic Synthesis* [1]. There is no doubt that efficiency is a central issue in modern organic synthesis, including multistep synthesis. There are several terminologies to describe this issue: atom economy [2], step economy [3], redox economy [4], and pot economy [5]. Atom economy was proposed by Trost and suggests designing the synthetic method so as to maximize the incorporation of all materials used in the process into the final product. The term of "step economy" was coined by Wender and is a fundamental economy to reduce the number of steps toward the target molecule, solvent, other wastes, separation method, time, and costs. It is clear that the operation of protection and deprotection lengthens the total sequence by two steps and results in the loss of materials. Redox economy, which was proposed by Baran and Hoffmann, recommends us to minimize the number of redox reactions that do not form carbon–carbon or carbon–heteroatom bonds and escalate linearly the oxidation state of intermediates during the synthesis. Pot economy, which was recently proposed by Hayashi, is a concept to reduce the number of pots required for each reaction method. One-pot synthesis includes cascade,

Protecting-Group-Free Organic Synthesis: Improving Economy and Efficiency, First Edition.
Edited by Rodney A. Fernandes.
© 2018 John Wiley & Sons Ltd. Published 2018 by John Wiley & Sons Ltd.

domino, and tandem reactions. Protecting-group-free synthesis is conceptionally related to redox and pot economy, as well as step economy.

Considering the retrosynthetic analysis of the target molecule without protecting group may generate a certain amount of risk due to the unexpected reactivity of the functional groups and instability of the intermediates. In other words, the innate characteristics of the functional groups and the molecules should be utilized to the maximum and new synthetic methodologies and chemical reagents must be invented in several cases [6]. Protecting-group-free synthesis is not the purpose but the expedient for performing efficient synthesis. The purpose of this chapter is not to belittle the use of protecting groups in the synthetic chemistry. Actually, total synthesis of polyketides, polypeptides, polysaccharides, and polynucleotides necessitates the use of protective groups due to the lack of chemoselectivity and purification issues. In this chapter, the total synthesis of natural products avoiding the use of protecting groups, which were reported during the period from 2005 to 2015 and selected by the authors of this chapter, will be presented to introduce the depthful concept and ingenious attempts.

3.2 Hapalindole U and Ambiguine H

Hapalindole U (**5**) and ambiguine H (**10**) are naturally occurring indole alkaloids isolated from the *Stigonemataceae* family of cyanobacteria [7]. Baran and coworkers at the Scripps Research Institute reported an efficient and short route for the total synthesis of **5** and **10** without using protecting groups [8, 9]. First, the ketone **1** (which was synthesized in optically pure form in four steps from commercially available material) was connected to indole **2** by the oxidative coupling developed by the same group [10] to yield the product **3** as a single diastereomer (Scheme 3.1). Next, the 6-*exo*-trig cyclization of **3** to **4** by the radical- and palladium-mediated reactions was investigated. Although the former gave the undesired 7-*endo*-trig and debrominated products, the reductive Heck reaction [11] with Herrmann's catalyst provided the desired cyclized product **4**. With the tetracyclic framework of hapalindole U (**5**) and ambiguine H (**10**) in hand, the ketone **4** was transformed to hapalindole U (**5**) by reductive amination, formylation, and dehydration. The remaining task in the conversion of hapalindole U (**5**) to ambiguine H (**10**) was to install the *t*-prenyl group in the C2 position of hapalindole U (**5**). Many attempts to introduce the *t*-prenyl unit in **5**, and its synthetic intermediates failed due to the incompatibility of the isonitrile moiety and the unusual reactivity of the unprotected indole part. To overcome the encountered problems, Baran's group tried to make use of the natural reactivity of these two functional groups. Thus, treatment of hapalindole U (**5**) with *t*-BuOCl followed by reaction with prenyl-9-BBN produced cyclic chloroimidate **7**. This is a cascade transformation consisting of (i) electrophilic chlorination of the isonitrile moiety, (ii) nucleophilic addition of the C3 position of the indole unit to the isonitrile group, (iii) nucleophilic reaction of the resulting imine to prenyl-9-BBN to give intermediate **6**, and (iv) delivery of the *t*-prenyl moiety to the C2 position. The final photolytic tandem conversion of **7**, involving (i) homolytic cleavage to afford intermediate **8**, (ii) hydrogen abstraction for reconstruction of the indole moiety to yield intermediate **9**, and (iii) reinstallation of the isonitrile group and removal of the BBN unit, was successfully carried out to furnish ambiguine H (**10**) in 63% yield.

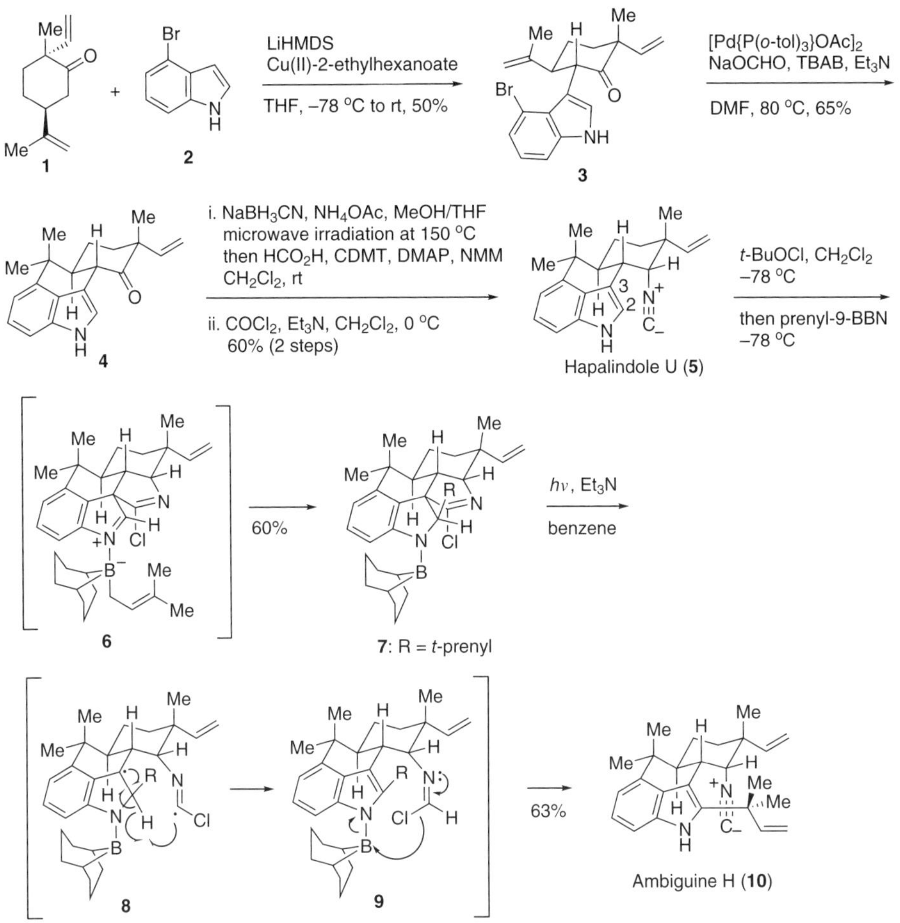

Scheme 3.1 Total synthesis of hapalindole U (**5**) and ambiguine H (**10**).

3.3 Stenine

Extracts of *Stemona* and *Croomia* species have been used in Chinese and Japanese folk medicine. The *Stemona* alkaloids are attractive targets for synthetic chemists owing to their structural diversity and wide range of biological activities [12]. The structural features of stenine (**20**), which is one of the *Stemona* alkaloids, are tetracyclic skeleton and contiguous seven chiral centers. Zeng and Aubé at the University of Kansas investigated the total synthesis of stenine (**20**) via domino reactions, which could be applied to the total synthesis of other *Stemona* alkaloids [13]. The Horner–Wadsworth–Emmons reagent **11** was coupled to aldehyde **12** to give the corresponding enone, which was converted into the corresponding trimethylsilyloxy diene **13** (Scheme 3.2a). Next, the domino transformation that combines Diels–Alder reaction with Schmidt reaction was examined [14]. After the detailed investigation of the reaction conditions, it was found that the use of SnCl$_4$ as a Lewis acid was effective and the desired product **16a**, which bears the stereochemistry corresponding to that of stenine (**20**), was produced in 70%

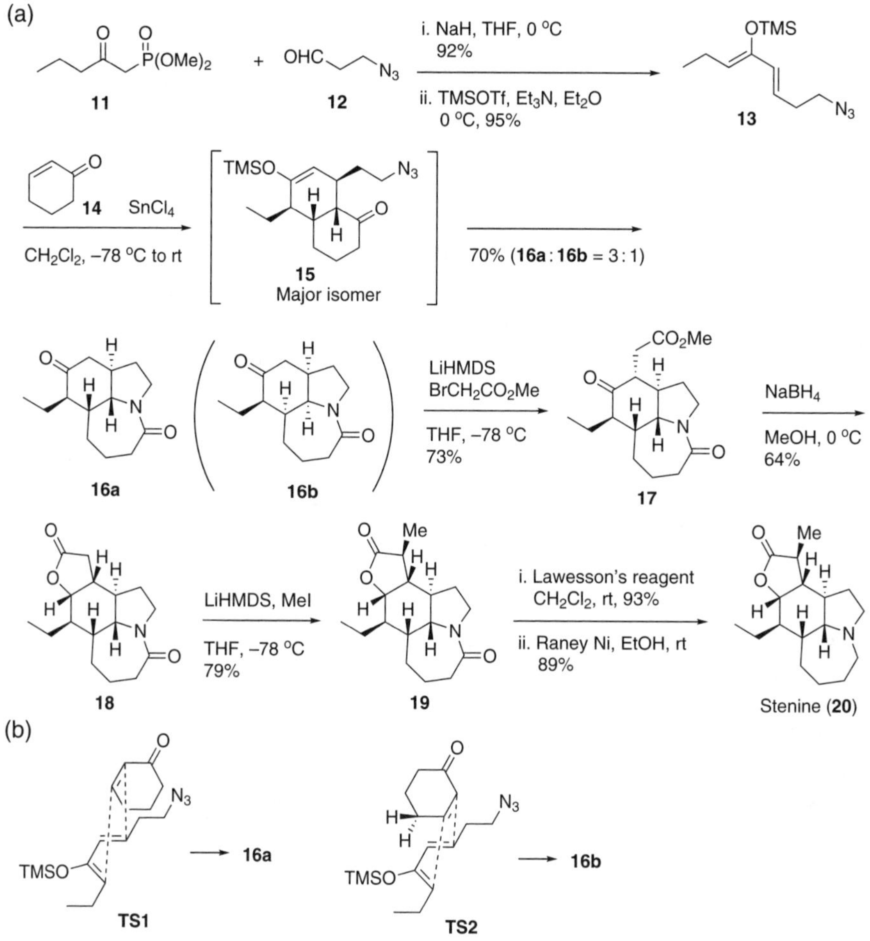

Scheme 3.2 (a) Total synthesis of stenine (**20**). (b) Transition states in the Diels–Alder reaction of **13** and cyclohexenone (**14**).

yield with a 3 : 1 diastereomeric ratio. The tricyclic compound **16a**, which is the ring expansion product, would be formed by an intramolecular Schmidt reaction of the Diels–Alder product **15**. In the Diels–Alder reaction, presumably, an *exo*-selectivity (**TS1** leading to **16a**; Scheme 3.2b) is favorable because there would be steric repulsion between one of the γ-protons with the silyl enol ether in an *endo* reaction mode (**TS2** leading to **16b**). Subsequently, additional stereocenters required for the total synthesis of stenine (**20**) were stereoselectively introduced by the substrate-controlled reactions. Thus, treatment of **16a** with LiHMDS/BrCH$_2$CO$_2$Me followed by chemoselective reduction of ketone **17** with NaBH$_4$ afforded γ-butyrolactone **18**. Diastereoselective methylation of **18** proceeded from the convex face to yield the desired product **19**. Finally, chemoselective reduction of the lactam carbonyl moiety of **19** in two steps was carried out to provide stenine (**20**). In this synthesis, the seven-membered ring along with the fused pyrrolidine was introduced by Schmidt reaction. The stereoselective construction of the β-methyl butyrolactone was completely substrate controlled and required no external stereoinduction. Since the chiral centers were introduced as required, the synthesis was efficiently achieved in just eight steps and 14% overall yield from the phosphonate **11**.

3.4 Neostenine

Neostenine (**33**) is a *Stemona* alkaloid that was isolated from *Stemona tuberosa* in 2003 [15]. Booker-Milburn and coworkers at the University of Bristol examined the total synthesis of neostenine (**33**) via [5 + 2] photocycloaddition [16]. Diels–Alder cycloaddition between furan (**21**) and fumaryl chloride (**22**) followed by reduction with LiAlH$_4$ gave diol **23** (Scheme 3.3). Carbon elongation of **23** with KCN and hydrolysis of the resulting dinitrile afforded dicarboxylic acid **24**. The dicarboxylic acid **24** was treated with *p*-TSA to give dilactone **25**, which was exposed to the S$_N$2′ conditions with EtMgCl/CuBr·SMe$_2$ to produce carboxylic acid **26**, stereoselectively.

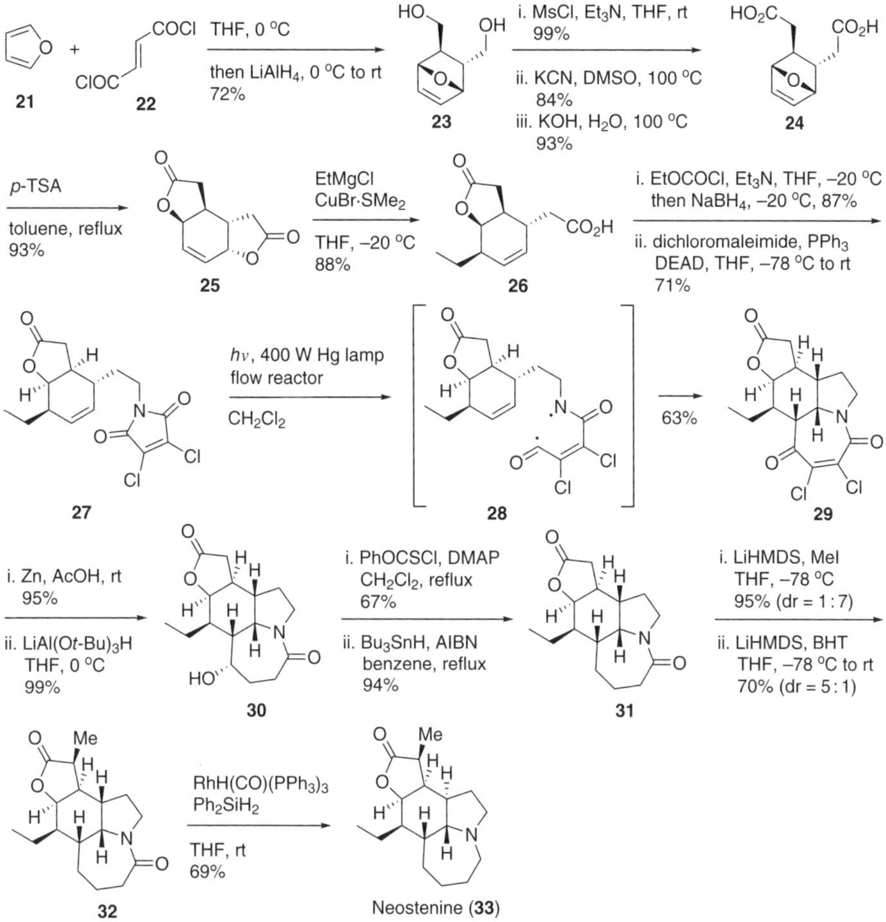

Scheme 3.3 Total synthesis of neostenine (**33**).

Chemoselective reduction of the carboxylic acid moiety of **26** in the presence of the lactone part and subsequent installation of dichloromaleimide provided photocyclization precursor **27**. The detailed investigation of [5 + 2] photocyclization by using several maleimides (unsubstituted, methylated, and chlorinated) in batch and flow reactions revealed that irradiation of the dichloromaleimide **27** with a 400 W Hg lamp in flow reactor [17] furnished the desired product **29** in 63% yield. The mechanistic study of this transformation using tunable UV lasers and time-dependent DFT calculations has indicated the presence of diradical **28** as a reaction intermediate [18]. This flow reaction enabled the supply of 1.3 g of the photoadduct **29** in a single 9 h run. Reduction/dechlorination of **29** with Zn/AcOH followed by chemoselective reduction of the resulting ketone gave alcohol **30**, which was subjected to the deoxygenation process [19] to afford **31**. Reaction of the lactone **31** with LiHMDS/MeI provided the methylated products as a 1 : 7 epimeric mixture in favor of the undesired epimer. The epimeric mixture was subsequently exposed to the epimerization via enolization/protonation to yield the desired product **32** as a major isomer with a 5 : 1 diastereomeric ratio. Finally, chemoselective reduction of the amide **32** with RhH(CO)(PPh$_3$)$_3$/Ph$_2$SiH$_2$ furnished neostenine (**33**) [20].

3.5 Englerin A

The genus *Phyllanthus* is one of the largest genera of the Euphorbiaceae, which contains approximately 700 species, and recognized as a rich source of biologically active compounds, which are widely used in various traditional medicine systems. A guaiane sesquiterpene englerin A (**48**) was isolated from *Phyllanthus engleri* growing in East Africa by the bioassay-guided fractionation [21]. Englerin A (**48**) has potent ability to inhibit the growth of various renal cancer cells with GI$_{50}$ values of 1–87 nM. Interestingly, its congener englerin B possessing the C9-hydroxy group displays the low activity against renal cancer cells, indicating that the growth inhibitory activity of englerin A (**48**) against renal cancer cells is dependent on its substituent at the C9 position. Ma and coworkers at Shanghai Institute of Organic Chemistry took an interest in biological activity of englerin A and launched its total synthesis [22].

The synthesis started by the bromination of (*R*)-citronellal **34** to give *gem*-dibromide **35** (Scheme 3.4). Elimination of dibromide in **35** with *t*-BuOK/18-crown-6 followed by selective allylic oxidation provided alcohol **36**, which was oxidized with IBX to afford α,β-unsaturated aldehyde **37**. Boron-mediated asymmetric aldol reaction between the aldehyde **37** and 3-methyl-2-butanone proceeded at −78 °C to provide the expected and desired alcohol **38**. After the detailed examination of the gold-catalyzed cyclization of **38**, it was proven that the cyclization of the alcohol **38** in the presence of AuCl proceeded to produce the desired oxotricyclic product **41** in 48% yield [23]. The formation of **41** is understandable by the AuCl-catalyzed cascade reaction. Thus, the gold-catalyzed cyclization of **38** would afford cyclopropyl intermediate **39**, in which the carbonyl group would attack the cyclopropyl carbon to afford intermediate **40**, leading to the formation of the oxotricyclic product **41**. Next, stereoselective construction of the *trans*-fused ring of englerin A (**48**) and completion of its total synthesis were explored. Thus, the alkene **41** reacted with *m*CPBA to give β-epoxide **42** stereoselectively, which was treated with CSA to afford alkene **43** possessing the opposite stereochemistry at the C6 position to that of englerin A (**48**). The absolute configuration at the C6 position of

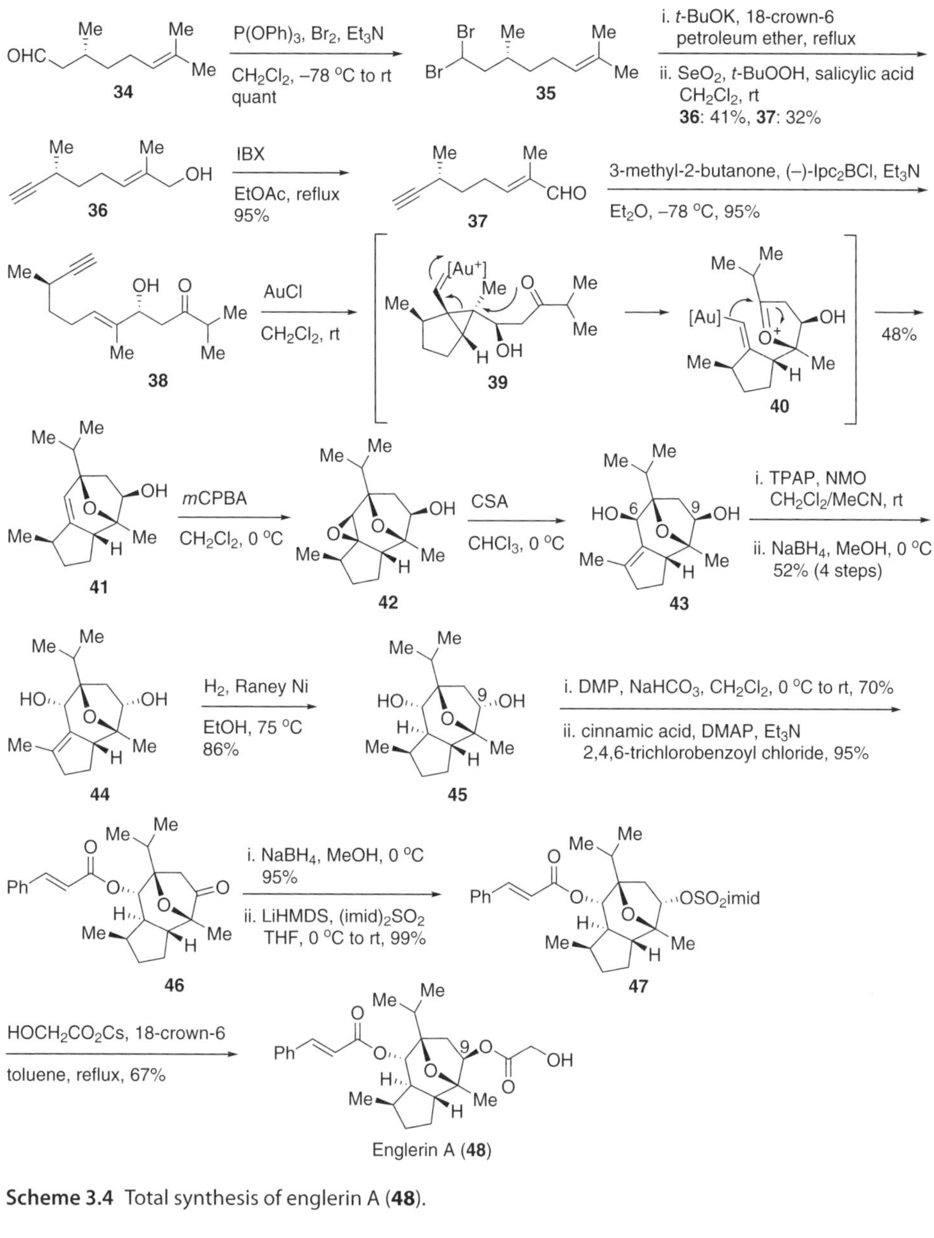

Scheme 3.4 Total synthesis of englerin A (**48**).

43 was inverted via oxidation–reduction sequence with the C9 stereoinversion to provide diol **44**. The stereochemistry at the *trans*-fused cyclic system was introduced by hydrogenation with Raney Ni to furnish **45**. After the hydroxy group at the C9 position was selectively oxidized, the resulting alcohol was esterified under Yamaguchi conditions to afford ester **46**, which was transformed into sulfonate **47** in two steps. The stereoinverted reaction of **47** with HOCH$_2$CO$_2$Cs/18-crown-6 completed the total synthesis of englerin A (**48**). The total synthesis of **48** was achieved in 15 steps and 8.1% overall yield starting from (*R*)-citronellal (**34**). The tricyclic system was beautifully assembled by gold-catalyzed cascade reaction.

3.6 Shimalactones A and B

Natural products featuring a bicyclo[4.2.0]octane or octadiene framework have been reported since endiandric acids D and E were identified as the first examples of this class in 1982. It has been proposed that these natural products are biosynthetically supplied by $8\pi/6\pi$ cascade electrocyclization of the tetraene bearing the (Z,Z)-geometries. Shimalactones A (66) and B (67) are neuritogenic polyketides isolated from the cultured marine fungus *Emericella variecolor* GF10 [24]. Trauner et al. at the University of California, Berkeley, embarked on the total synthesis of shimalactones A (66) and B (67) by utilizing biomimetic transformations [25].

The synthesis of shimalactones started with *anti*-aldol addition of the boron enolate derived from propionyl oxazolidinone 50 and iodoenal 49 to give alcohol 51 (Scheme 3.5) [26]. Acylation of 51 followed by construction of the diene moiety by Stille coupling afforded 53. After Dieckmann cyclization of 53 with KHMDS, treatment of resulting β-ketolactone 54 with CSA produced oxabicyclo[2.2.1]heptanes 55a and 55b at a 2 : 1 diastereomeric ratio. Since the direct allylic oxidation of 55a failed, radical bromination of 55a was carried out to give allylic bromide 56a along with its isomer 56b. Oxidation of 56a with IBX and subsequent reaction of the resultant α,β-unsaturated aldehyde with phosphonate 57 provided 58. The aldehyde 59, derived from 58 in two steps, was treated with 60 to afford vinyl iodide 61. The latter reacted with vinyl stannane 62 under Stille conditions at room temperature to furnish shimalactones A (66) and B (67) in 55 and 11% yields, respectively. The reaction conditions presumably gave (E,E,Z,Z,E)-conjugated polyene 63, which could not be isolated and subsequently underwent 8π electrocyclization to yield cyclooctatrienes 64 and 65, which were transformed into shimalactones A (66) and B (67), respectively, via 6π electrocyclization.

3.7 Cyanthiwigin F

Cyanthiwigin natural products belong to a larger class of bioactive molecules known as the cyathins, which exhibit a wide range of biological activities [27]. Cyanthiwigin F (78) has cytotoxicity against human primary tumor cells with IC_{50} value of 3.1 μM. The structural feature of diterpene cyanthiwigins is a 5,6,7-tricarbocyclic core with the contiguous four stereocenters. Stoltz's research group at California Institute of Technology demonstrated the double catalytic enantioselective alkylation in the total synthesis of cyanthiwigin F (78) [28]. They envisioned the initial construction of the central six-membered ring introducing two quaternary stereocenters (at the C6 and C9 positions) and to construct the seven-membered and five-membered rings at a later stage, sequentially. Stoltz and coworkers developed the asymmetric decarbonylative allylation of β-ketoesters catalyzed by a chiral palladium complex in 2005 [29]. They tried to apply this methodology to the total synthesis of cyanthiwigin F (78). Thus, the self-condensation of diallyl succinate 68 by Claisen–Dieckmann reaction followed by methylation with K_2CO_3/MeI gave bis(β-ketoester) 69 as a 1 : 1 mixture of racemic [(S,S)-69 and (R,R)-69] and meso diastereomers (Scheme 3.6a). Treatment of the 1 : 1 diastereomeric mixture of 69 with Pd(dmdba)$_2$/chiral *t*-butyl phosphinooxazoline 70 produced double allylated products (R,R)-71 (99% ee) and meso-71 in 78% yield as a 4.4 : 1 diastereomeric mixture. Since the six-membered diketone (R,R)-71 possessing the two quaternary

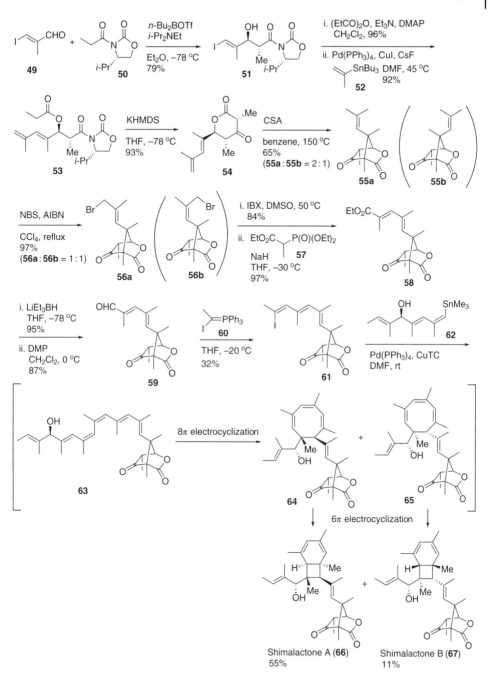

Scheme 3.5 Total synthesis of shimalactones A (**66**) and B (**67**).

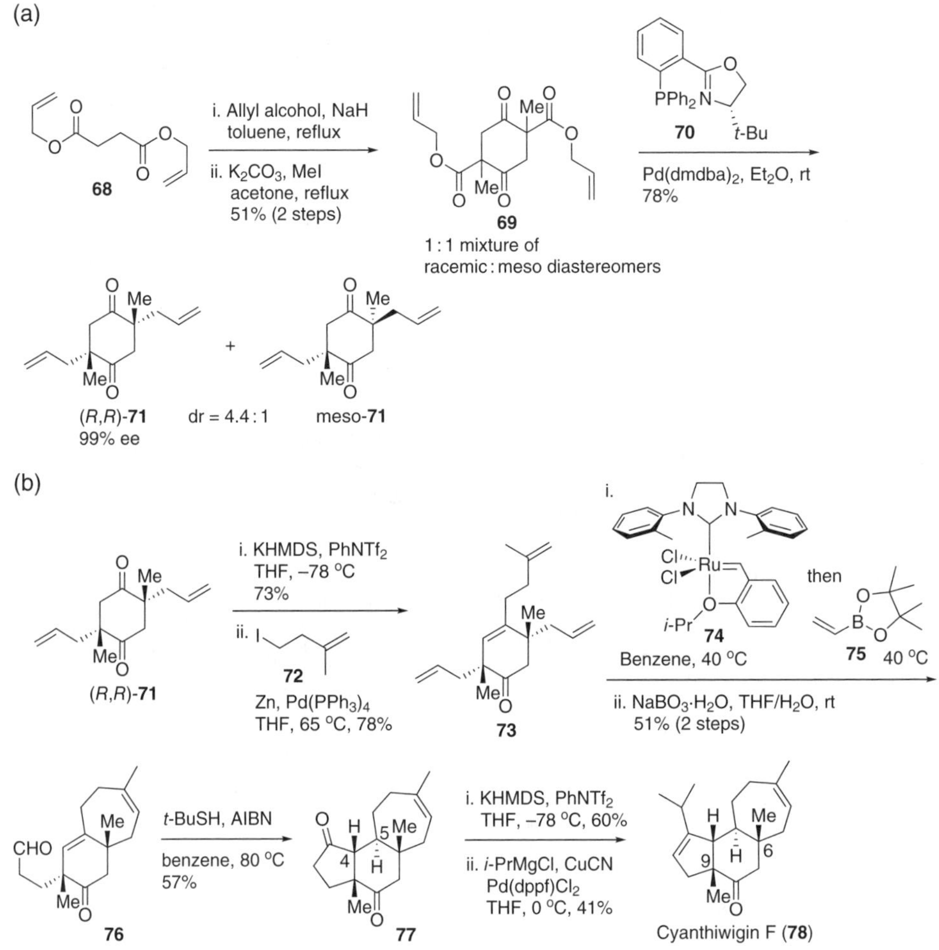

(a)

(b)

Scheme 3.6 (a) Double asymmetric alkylation. (b) Total synthesis of cyanthiwigin F (**78**).

stereocenters was synthesized as an enantioenriched form, they next focused on the stereoselective construction of the seven- and five-membered rings. Monodeprotonation of the diketone (*R,R*)-**71** with KHMDS followed by trapping the resulting enolate with PhNTf$_2$ gave the enol triflate, which was subjected to palladium-catalyzed Negishi coupling conditions with alkyl iodide **72** to yield alkylated product **73** (Scheme 3.6b). The next transformation was the construction of the seven-membered ring and introduction of the aldehyde moiety. After the detailed optimization, a single use of second-generation Hoveyda–Grubbs catalyst (**74**) was found to be effective. Thus, ring-closing metathesis of **73** and subsequent cross-metathesis of the terminal alkene with vinyl boronate **75** followed by oxidative work-up with NaBO$_3$·H$_2$O afforded aldehyde **76**. Ring closure and stereochemical installation at the C4 and C5 positions of the alkenyl aldehyde **76** were achieved by radical reaction with *t*-BuSH/AIBN to provide tricyclic compound **77** as a single diastereomer [30]. The diastereoselectivity in this conversion was presumably generated by the kinetic cyclization and kinetic hydrogen abstraction, which led to the formation of *cis-* and *trans*-fused 5,6,7-tricyclic system. Finally, the alkenyl isopropyl moiety was introduced by the selective formation of enol triflate from diketone **77** and subsequent coupling with *i*-PrMgCl/CuCN/Pd(dppf)Cl$_2$ to give cyanthiwigin F (**78**).

3.8 Sintokamides A, B, and E

The chlorinated peptides sintokamides were isolated from the marine sponge *Dysidea* sp. collected in Indonesia [31]. Sintokamide A (**91**) inhibits the N-terminus transactivation of the androgen receptor in prostate cancer cells. Zakarian's research group at the University of California, Santa Barbara, reported the unified total synthesis of sintokamides A (**91**), B (**92**), and E (**93**), wherein the stereoselective radical chloroalkylation was efficiently utilized as the key step [32]. They planned to apply the stereoselective chloroalkylation [33] to the construction of di- and trichloroleucine subunit and introduce the tetramic acid moiety at the late stage of the synthesis. Thus, treatment of titanium enolate derived from oxazolidinone **79** with BrCCl$_3$/Ru(PPh$_3$)$_3$Cl$_2$ afforded trichloromethylated product **80**, stereoselectively (Scheme 3.7a). The chiral auxiliary in **80** was reductively removed with LiAlH$_4$ to provide alcohol **81**, which was elongated by reacting triflate prepared from **81** with Et$_4$NCN. After nitrile **82** was transformed to sulfinimine **84**, a three-step sequence – (i) diastereoselective Strecker reaction, (ii) hydrolysis of the nitrile and sulfinimine portions, and (iii) *N*-propionylation – produced carboxylic acid **86**. Amine **88a**, which is a coupling partner of **86**, was synthesized by the radical dichloromethylation of the oxazolidinone **79** with BrCHCl$_2$/Cp*Ru(PPh$_3$)$_2$Cl followed by the transformations similar to those used for the synthesis of **86** (Scheme 3.7b). Amidation between the carboxylic acid **86** and the amine **88a** was performed with EDCI/HOAt to yield amide **89**, wherein the epimerization at the C10 position was not observed (Scheme 3.7c). The final task for completion of the total synthesis of sintokamide A (**91**) was to construct the tetramic acid moiety. Alkaline hydrolysis of the methyl ester **89** was followed by reacting the resulting carboxylic acid with Meldrum's acid in the presence of isopropenyl chloroformate and DMAP to furnish **90** [34]. Thermolysis of **90** in MeCN at reflux caused the release of acetone and carbon dioxide to afford the corresponding tetramic acid, which was methylated under Mitsunobu conditions to provide sintokamide A (**91**). Sintokamides B (**92**) and E (**93**) were also synthesized by a reaction sequence analogous to that used for the synthesis of sintokamide A (**91**).

3.9 Ecklonialactones A and B

Ecklonialactones, which feature the *trans*-fused cyclopentane and 14- or 16-membered macrocycle in the structure, were isolated from various brown algae [35]. Although they are regarded as chemical defensive substances in the producing organisms, their physiological properties have not been clarified. Fürstner et al. at Max Planck Institute investigated the total synthesis of ecklonialactones based on the metal-catalyzed transformations [36, 37]. Butenolide **94** underwent a rhodium-catalyzed 1,4-addition [38] of alkenylboronate **95** in the presence of the (−)-carvone-derived chiral ligand **96**

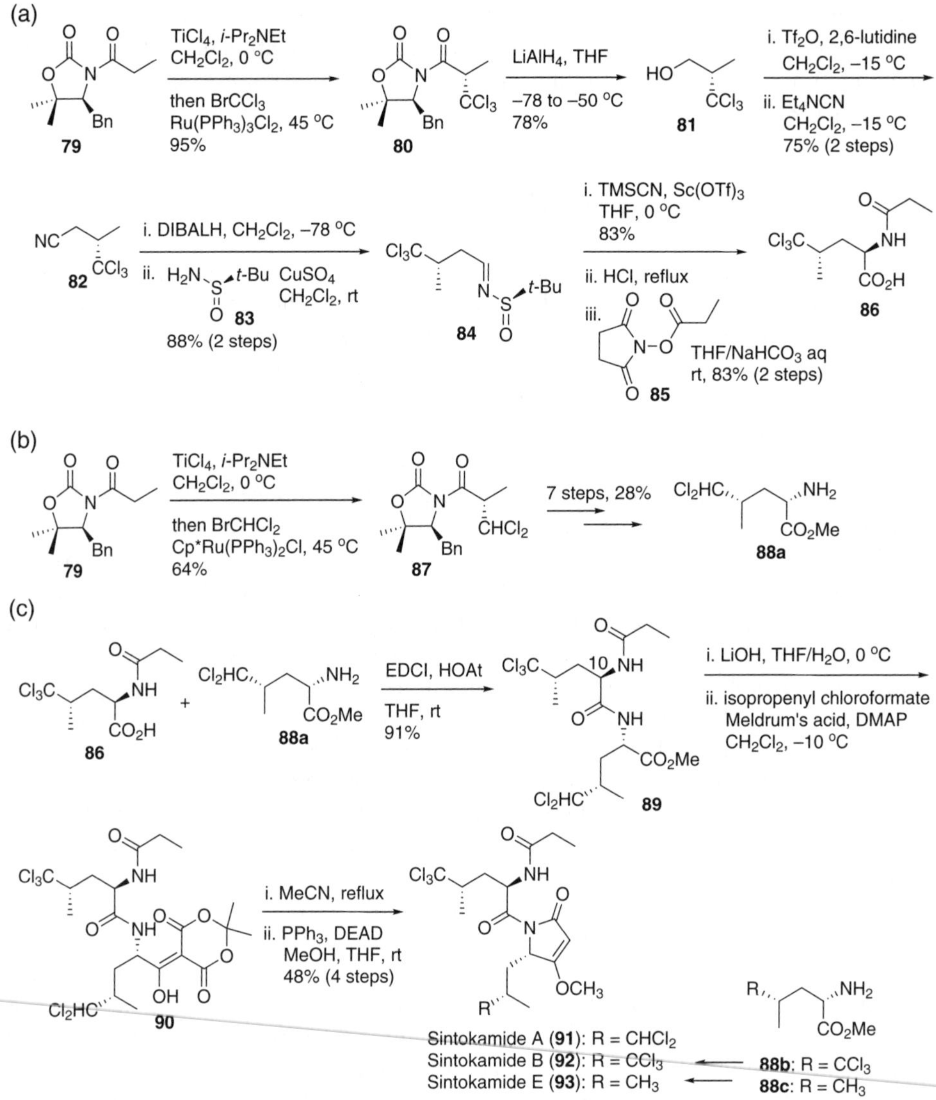

Scheme 3.7 (a) Synthesis of carboxylic acid **86**. (b) Synthesis of amine **88a**. (c) Total synthesis of sintokamides A (**91**), B (**92**), and E (**93**).

to give **97**, and the optical purity of **97** was increased to 93% ee after recrystallization (Scheme 3.8). Diastereoselective allylation of the lactone **97** and subsequent lactone opening with HN(OMe)Me·HCl/AlMe$_3$ afforded Weinreb amide **98**. Ring-closing metathesis of the diene **98** with ruthenium complex **99**, which was developed as a cost-effective catalyst by the same group [39], was performed to provide cyclopentene **100**. After the alkyne moiety in **102** was introduced with Ohira–Bestmann reagent **101**, methylation of the terminal alkyne **102**, addition of the ethyl group to the Weinreb amide moiety, and diastereoselective reduction of the resulting ketone were conducted

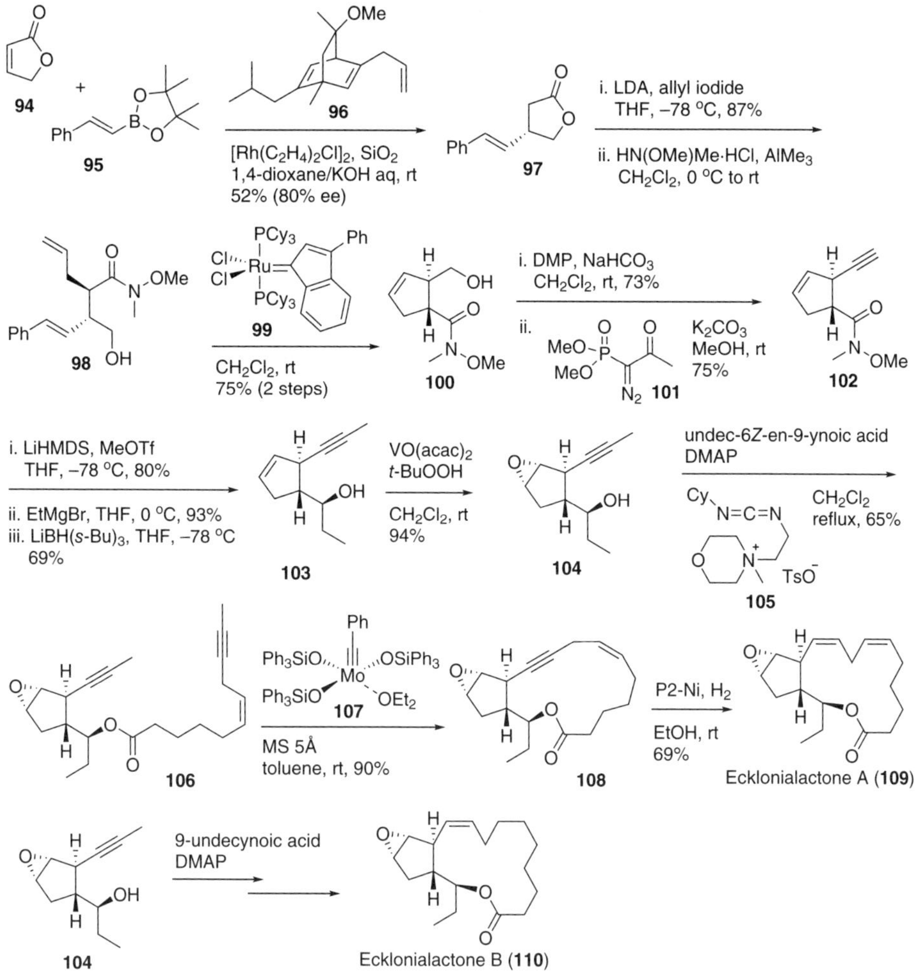

Scheme 3.8 Total synthesis of ecklonialactones A (**109**) and B (**110**).

to provide alcohol **103**. Stereoselective epoxidation was successfully carried out by the hydroxy group direction with VO(acac)$_2$/t-BuOOH to yield epoxy alcohol **104**. The next task for completion of the total synthesis of ecklonialactone A (**109**), esterification and macrocyclization, was found to be challenging and needed the detailed examination owing to the high reactivity of the epoxide portion. Esterification between the alcohol **104** and undec-6Z-en-9-ynoic acid proceeded with the aid of carbodiimide **105** to produce ester **106**. Investigation of ring-closing alkyne metathesis of **106** with various catalysts revealed that the use of molybdenum complex **107** bearing Ph$_3$SiO ligands was effective [40]. The final semireduction of alkyne **108** with P2-Ni produced ecklonialactone A (**109**). The total synthesis of ecklonialactone B (**110**) was also achieved by esterification of the alcohol **104** with 9-undecynoic acid in place of undec-6Z-en-9-ynoic acid and further transformation similar to that for the total synthesis of ecklonialactone A (**109**).

3.10   (E)- and (Z)-Alstoscholarines

The genus *Alstonia* found in Africa and Asia comprises about 60 species. *Alstonia scholaris* is widely distributed in South and Southeast Asia and used traditionally as a medicinal plant by local people. (*E*)- and (*Z*)-Alstoscholarines (**121** and **122**) are monoterpenoid indole alkaloids isolated from *Alstonia scholaris* in 2007 [41]. These natural products have the pentacyclic structure including bicyclo[3.3.1]nonane part, indole ring, and pyrrole moiety. Neuville and Zhu research group has achieved the protecting-group-free total synthesis of (*E*)- and (*Z*)-alstoscholarines (**121** and **122**) in 2011 [42].

Desymmetrization of meso-anhydride **111** was successfully carried out by treating with MeOH in the presence of chiral catalyst **112** to give monoester **113** in 93% ee (Scheme 3.9) [43]. The pyrrole ring was introduced by reacting the carboxylic acid **113** with 2,2′-dipyridyldisulfide/PPh$_3$ followed by pyrrylmagnesium bromide to afford keto-pyrrole **114**. Dihydroxylation of the alkene **114** under Upjohn conditions and oxidative cleavage of the resulting diol provided the corresponding dialdehyde, which cyclized selectively to form six-membered hemiaminal **115**. Next, the construction of the indole unit was examined. Thus, condensation of the aldehyde **115** and *o*-iodoaniline (**116**)

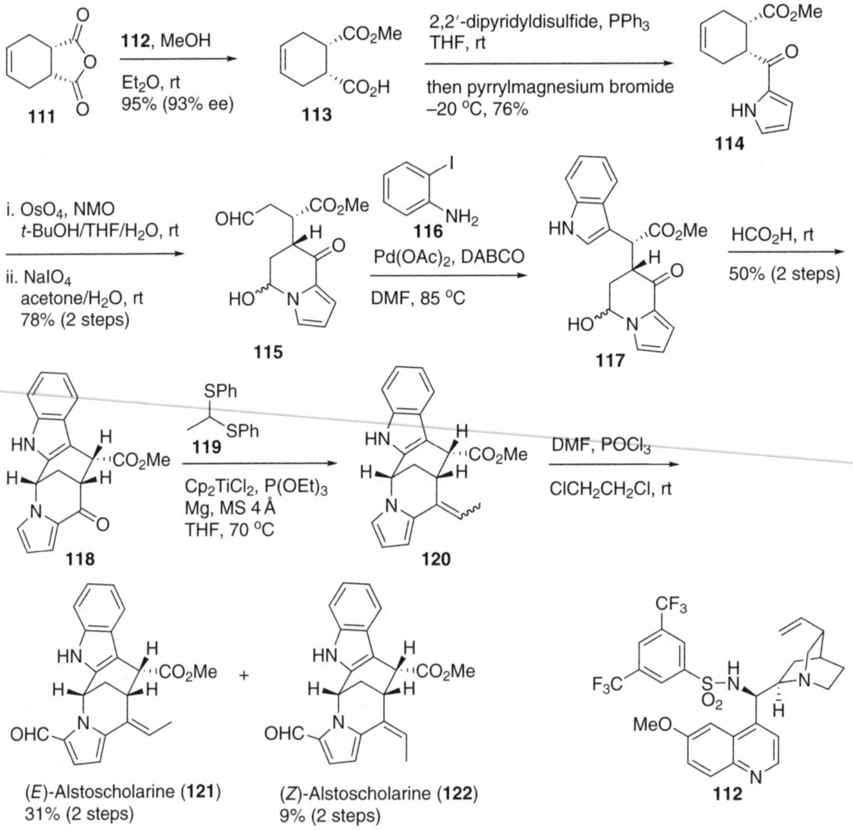

Scheme 3.9 Total synthesis of (*E*)- and (*Z*)-alstoscholarines (**121** and **122**).

and subsequent annulation took place in the presence of Pd(OAc)$_2$ and DABCO to produce indole **117** [44]. The Pictet–Spengler reaction of **117** in HCO$_2$H proceeded smoothly to furnish pentacyclic compound **118**. The final task for completion of the total synthesis was the introduction of ethylidene and formyl moieties. After the investigation for constructing the ethylidene moiety, it was found that the reaction of **118** with **119** under Takeda conditions [45] gave the desired product **120** as the *E/Z* mixture. Finally, the pyrrole **120** was exposed to Vilsmeier–Haack formylation conditions to provide (*E*)- and (*Z*)-alstoscholarines (**121** and **122**) in 31 and 9% yields in two steps, respectively.

3.11 Berkelic Acid

The "Berkeley Pit" is a former open-pit copper mine located in Butte, Montana (USA). There are over 30 billion gallons of water in the pit, which is acidic (pH 2.5) and contaminated with high concentrations of metals including copper, iron, arsenic, cadmium, cobalt, manganese, and zinc. The microorganisms surviving in these extreme situations produce unique and precious secondary metabolites. The bioassay-guided fractionation of compounds from one of the *Penicillium* species collected from the surface waters of the pit led to the isolation of berkelic acid (**134**) [46]. This natural product inhibits the matrix metalloproteinase MMP-3 (GI$_{50}$ = 1.87 μM), cysteine protease caspase-1 (GI$_{50}$ = 98 μM), and human ovarian cancer OVCAR-3 (GI$_{50}$ = 91 nM). Fañanás and Rodríguez research group achieved the total synthesis of berkelic acid (**134**) based on the metal-catalyzed cascade reaction forming the spiroacetal framework [47].

The synthesis of key chiral building block **125** started from (*S*)-3-butyn-2-ol (**123**) as shown in Scheme 3.10. Thus, mesylation of the alcohol **123** followed by substitution by diethyl malonate in the presence of CsF with stereoinversion provided diester **124**. The ester group of **124** was reduced with LiAlH$_4$ to give diol **125**. The authors group envisioned that the palladium-catalyzed one-pot three-component coupling reaction developed by the same group previously [48] could be applied to the construction of the framework of berkelic acid (**134**). After the survey of reaction conditions, they found that the reaction between the alkyndiol **125** and alkynal **126** proceeded in the presence of AgOTf (5 mol%) to afford the desired spiroacetal **129**. To suppress the possible side reactions due to the resulting reactive dihydropyran ring, hydrogenation of **129** was carried out to produce the 2 : 1 mixture of two diastereomers, wherein the desired tetrahydropyran **130** was the major product. In this transformation, presumably, two cyclizations of the alkyndiol **125** and alkynal **126** occurred in the presence of AgOTf to form enol ether **127** and pyran **128**, respectively, which underwent the formal cycloaddition to furnish the spiroacetal **129**. It is noteworthy that just one chiral center in **125** controlled the newly formed three stereochemistries in **129** including the desymmetrization of **125** and only two diastereomers were produced, wherein the major diastereomer has the stereostructure corresponding to that of the natural product [49]. Having succeeded in the stereoselective construction of the tetracyclic skeleton of berkelic acid (**134**), the authors next tried to install the side-chain moiety. To avoid any redox step, alkyl iodide **131**, which was derived from the alcohol **130**, was coupled to cyanohydrin **132** to form diester **133**. Finally, the ester group on aryl ring in **133** was selectively cleavaged to produce berkelic acid (**134**).

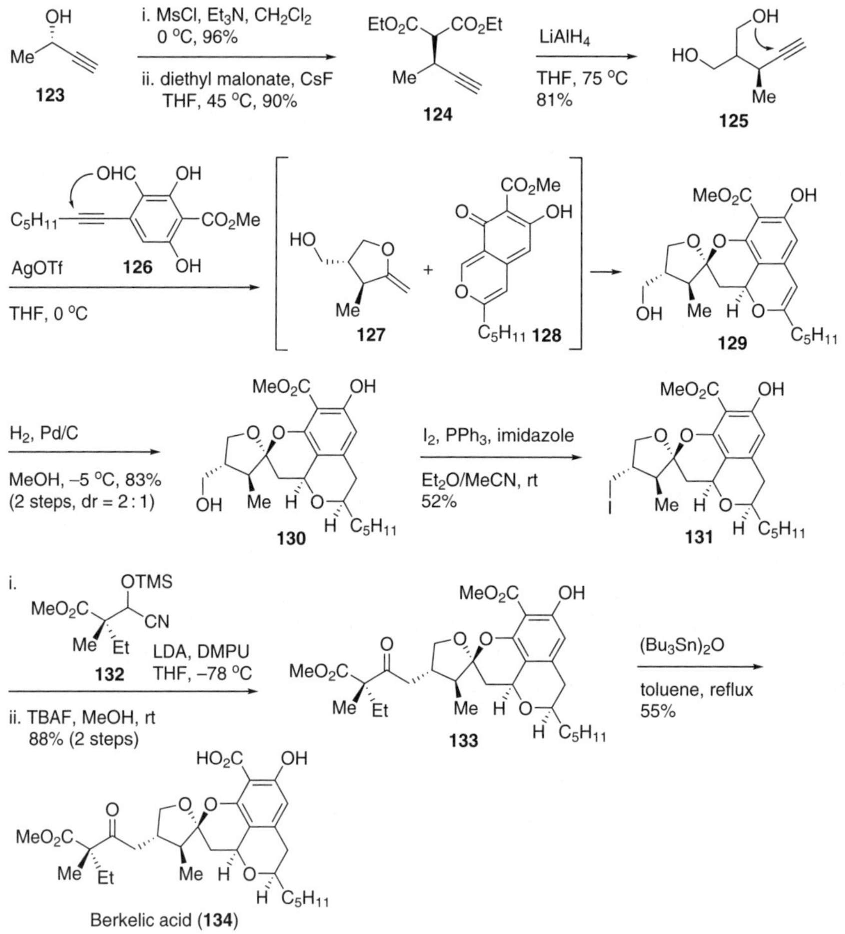

Scheme 3.10 Total synthesis of berkelic acid (**134**).

3.12 Myxalamide A

Myxalamides are polyene antibiotics isolated from the gliding bacterium *Myxococcus xanthus* in 1983 [50]. They have the labile (*E,E,Z,E,E*)-pentaene moiety as the structural feature, which has attracted the attention of the synthetic community. Kobayashi and coworkers at Tokyo University of Science investigated the total synthesis of myxalamide A (**143**) by using vinylogous Mukaiyama aldol reaction and cascade Stille/Suzuki–Miyaura cross-coupling [51].

Kobayashi et al. planned to construct the pentaene portion of myxalamide A (**143**) by performing the one-pot Stille/Suzuki–Miyaura cross-coupling at the late stage of the synthesis to minimize the handling of the labile (*Z*)-alkene. First, as described in Scheme 3.11a, the vinylogous Mukaiyama aldol reaction between aldehyde **135** and *N,O*-acetal **136** was conducted in the presence of TiCl$_4$ to give *anti*-aldol adduct **137** in 86% yield with a high diastereoselectivity (dr = >20 : 1) [52]. The chiral auxiliary in **137** was

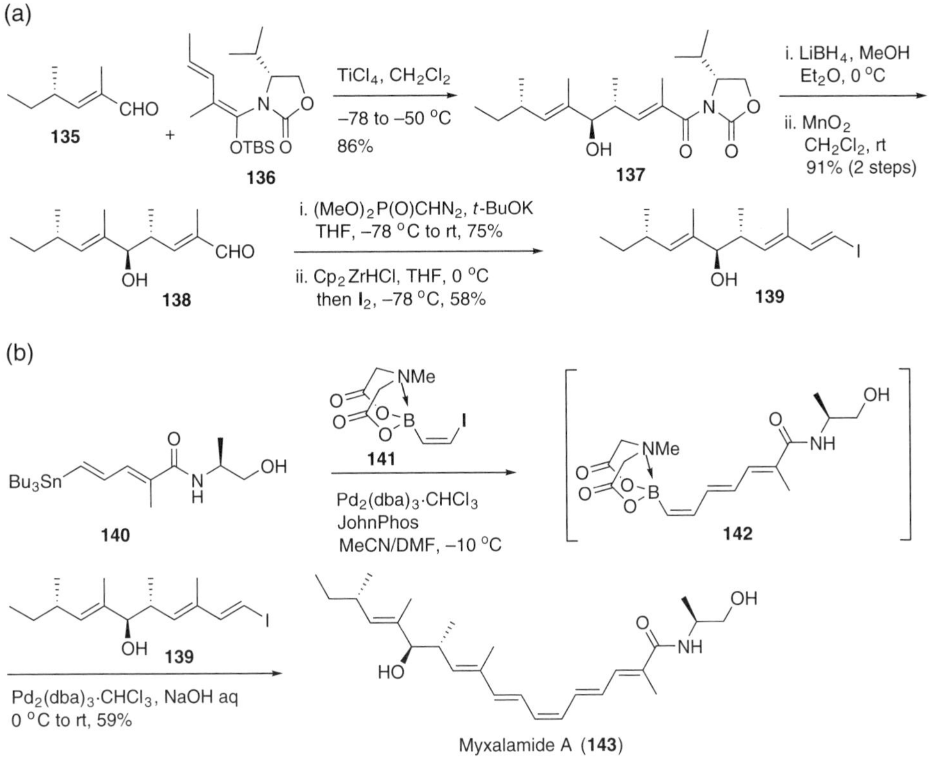

(a)

(b)

Scheme 3.11 (a) Synthesis of vinyl iodide **139**. (b) Total synthesis of myxalamide A (**143**).

removed with LiBH$_4$, and oxidation of the resulting alcohol afforded α,β-unsaturated aldehyde **138**. The Takai olefination of the aldehyde **138** with CHI$_3$/CrCl$_2$ leading to vinyl iodide **139** gave low chemical yield and poor E/Z selectivity. Therefore, the Seyferth–Gilbert homologation of **138** was carried out, and the resulting alkyne was treated with Schwartz reagent to provide the vinyl iodide **139**. The stepwise Stille and Suzuki–Miyaura coupling of vinyl stannane **140**, (Z)-MIDA boronate **141** [53], and vinyl iodide **139** was investigated before tackling the one-pot conversion (Scheme 3.11b). The Stille cross-coupling between **140** and **141** under the standard conditions – Pd(PPh$_3$)$_4$/CuI, Pd(PPh$_3$)$_4$/CuTC, and Pd(MeCN)$_2$Cl$_2$/AsPh$_3$ – resulted in the decomposition of the (Z)-MIDA boronate **141**. When the reaction was carried out in the presence of Pd$_2$(dba)$_3$·CHCl$_3$ and P(2-furyl)$_3$ at −20 °C, the desired product **142** was obtained in 51% yield. Further optimization revealed that the use of Pd$_2$(dba)$_3$·CHCl$_3$ and JohnPhos at −10 °C improved the chemical yield of **142** to 99% yield. The authors next tried the Suzuki–Miyaura coupling between **142** and **139** and found that the reaction conditions similar to those used in the Stille coupling, the use of Pd$_2$(dba)$_3$·CHCl$_3$ and JohnPhos, produced myxalamide A (**143**) in 56% yield. To minimize the isomerization of the (Z)-alkene part of **143**, manipulations were conducted in the dark, and all solvents for the reaction and for chromatography were degassed prior to use. Having succeeded in the stepwise Stille and Suzuki–Miyaura coupling, Kobayashi research group applied the reaction conditions optimized in the stepwise reaction to one-pot reaction. Thus, treatment of **140** and **141** with $Pd_2(dba)_3 \cdot CHCl_3$ (5 mol%)/JohnPhos (20 mol%) followed by addition of **139**, $Pd_2(dba)_3 \cdot CHCl_3$ (5 mol%, to accelerate Suzuki–Miyaura coupling), and aqueous NaOH furnished myxalamide A (**143**) in 59% yield.

3.13 Pipercyclobutanamide A

Cyclobutane-containing natural products are the diverse family with a variety of biological activities [54]. It is generally regarded that these kinds of natural products would be produced by the direct coupling of two alkenes in nature. From the synthetic point of view, while the [2 + 2] cycloaddition of the alkenes is efficient and elegant methodology for the construction of cyclobutane motifs, this strategy would include several problems to be solved, homodimerization and orientation (head-to-head or head-to-tail). Pipercyclobutanamides were first isolated from the fruits of Indian black pepper *Piper nigrum* in 2001 [55]. Baran and coworkers at the Scripps Research Institute demonstrated the sequential C–H arylation and olefination in total synthesis of the proposed structure of pipercyclobutanamide A (**158**) [56, 57].

The synthesis started from methyl coumalate (**144**) as described in Scheme 3.12. Photochemical 4π electrocyclization of **144** led to the formation of cyclized product **145**, which was hydrogenated with Pt/C as a catalyst to yield *cis*-disubstituted cyclobutane **146** as a single diastereomer. Subsequently, 8-aminoquinoline motif was introduced to the carboxylic acid **146** as the directing group working in the C–H arylation step to provide amide **147**. These three transformations were conducted in one-pot, and the cyclobutane **147** was obtained in 54% yield from **144**. Baran research group next examined the C–H mono-olefination of **147** by using (2-iodovinyl)benzene **148** as a model compound. Thus, treatment of **147** with **148** in the presence of $Pd(OAc)_2$ and AgOAc afforded bis-olefinated product **149** as the major product in 50% yield. Although the reason for this bis-olefination is unclear, presumably, the small molecular size of **148** caused a facile second reaction. Therefore, the authors tried to demonstrate the C–H arylation prior to the olefination. When the standard mono-arylation conditions, which was found in the previous work [57], was used in the reaction of the cyclobutane **147** and 1-iodo-3,4-methylenedioxybenzene (**150**), poor conversion was observed. After optimizing the reaction conditions, the use of pivalic acid as an additive and *t*-BuOH as a solvent turned out to be effective, and mono-arylated product **151** was obtained in 54% yield. Subsequent C–H olefination of **151** with vinyl iodide **152** was performed in the presence of $Pd(OAc)_2$ and AgOAc in toluene to afford olefinated product **153** possessing the all-*cis* oriented substituents in 59% yield. Having introduced contiguous four substituents on the cyclobutane ring, the authors turned their attention to alter the all-*cis* stereochemistry to all-*trans* one found in the natural product. The epimerization at the C1 and C3 positions of **153** with NaOMe proceeded at room temperature and 45 °C, respectively, to produce all-*trans* cyclobutane **154**, which was hydrolyzed by aqueous NaOH to furnish **155**. The amide **155** was reduced with DIBALH to the corresponding aldehyde, wherein the chelating of the aminoquinoline moiety with the reagent is likely to be formed. The tolerated carboxylic acid portion in the DIBALH reduction reacted with piperidine to give amide **156**. The final transformation, installation of the (*Z*)-α,β-unsaturated

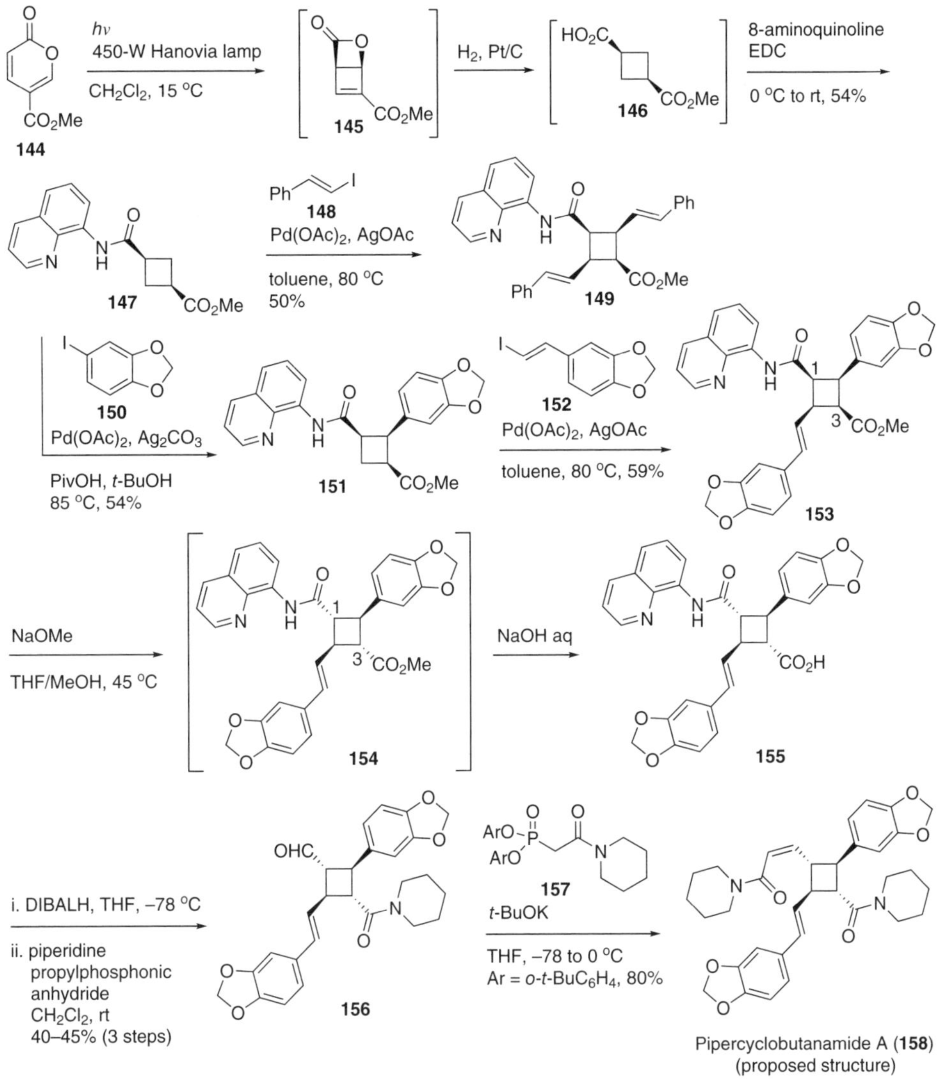

Scheme 3.12 Total synthesis of the proposed structure of pipercyclobutanamide A (**158**).

amide motif, was achieved with Ando phosphonate **157** to produce **158**. The ^{1}H and ^{13}C NMR data of the synthetic product **158** did not agree with those of the natural product, indicating that the structural assignment of the natural product should be reinvestigated. Tang and coworkers also accomplished the total synthesis of the proposed structure of pipercyclobutanamide A, whose NMR data did not match those of the natural product [58]. They analyzed the difference of NMR data between the synthetic and natural products, surveyed known related natural products isolated from *P. nigrum* and *Piper chaba*, and elucidated that the real pipercyclobutanamide A is actually cyclohexene natural product chabamide.

3.14 Fusarisetin A

Fusarisetin A (**173**) was isolated from the soil fungus *Fusarium* sp. collected in Daejeon, Korea [59]. The natural product **173** inhibits acinar morphogenesis, cell migration, and invasion in MDA-MB-231 cells. The molecular structure of this natural product was determined by NMR spectroscopy, X-ray crystallographic analysis (relative configuration), and exciton chirality circular dichroism method (absolute configuration). After that, the natural product was revised to be the antipode of the originally assigned absolute stereostructure by the total synthesis [60]. From the structural perspective, fusarisetin A (**173**) possesses a 6,6,5,5,5-pentacyclic system with ten chiral centers. Theodrakis research group at the University of California, San Diego, examined the total synthesis of fusarisetin A (**173**) inspired by its proposed biosynthesis [61].

The authors planned to construct the *trans*-decalin motif of **173** by intramolecular Diels–Alder reaction. First, cross-metathesis of (*S*)-citronellal (**159**) with methacrolein (**160**) in the presence of second-generation Grubbs catalyst (**161**) afforded alkene **162** (Scheme 3.13). Differentiation of the two aldehyde groups of **162** was succeeded by Wittig reaction with phosphonium bromide **163** to yield tetraene **164** as an *E/Z* mixture. Treatment of this diastereomixture with iodine under a sunlamp led to the isomerization to give exclusively the (*E*)-alkene, which was subjected to Et_2AlCl-mediated intramolecular Diels–Alder reaction to provide the desired *trans*-decalin **165**. Two-carbon elongation of the aldehyde **165** by Reformatsky reaction followed by IBX oxidation produced β-ketoester **166**. Deprotonation of **166** by LiHMDS and subsequent addition of TEMPO/Cp_2FePF_6 resulted in the formation of **167** as a diastereomeric mixture. Next, construction of the five-membered ring inspired by biosynthesis was examined. Thus, heating the alkene **167** at 90 °C in toluene in the presence of amino alcohol **168** gave cyclized product **171** and its diastereomer **172** in about 1 : 1 ratio. Plausible reaction mechanism in this transformation is the generation of radical **169**, 5-*exo*-trig radical cyclization of **169** followed by trapping with TEMPO as a reactive oxygen species, and aminolysis of the resulting ester **170** by the amino alcohol **168**. After the reductive cleavage of the N–O bond of **171** under Zn/AcOH conditions, Dieckmann condensation and subsequent hemiacetalization were performed in one-pot by NaOMe in MeOH to furnish fusarisetin A (**173**). The tricyclic compound **172**, which is the C5 epimer of **171**, was also converted to fusarisetin A (**173**) by the three-step sequence: (i) oxidative cleavage of the N–O bond of **172** by *m*CPBA to form ketone **174**, (ii) diastereoselective reduction of the C5-ketone **174** by $NaBH_4$, and (iii) one-pot Dieckmann condensation/hemiacetalization.

3.15 Rhazinilam

Rhazinilam (**191**) is an *Aspidosperma* alkaloid that was first isolated from the leaves of *Rhazya stricta* Decaisne in 1970 [62]. The molecular structure of this natural product is characterized by a nine-membered lactam ring, tetrahydroindolizine skeleton, and a quaternary chiral center. Tokuyama and coworkers at Tohoku University worked on the total synthesis of rhazinilam (**191**) using a gold-catalyzed cascade cyclization [63].

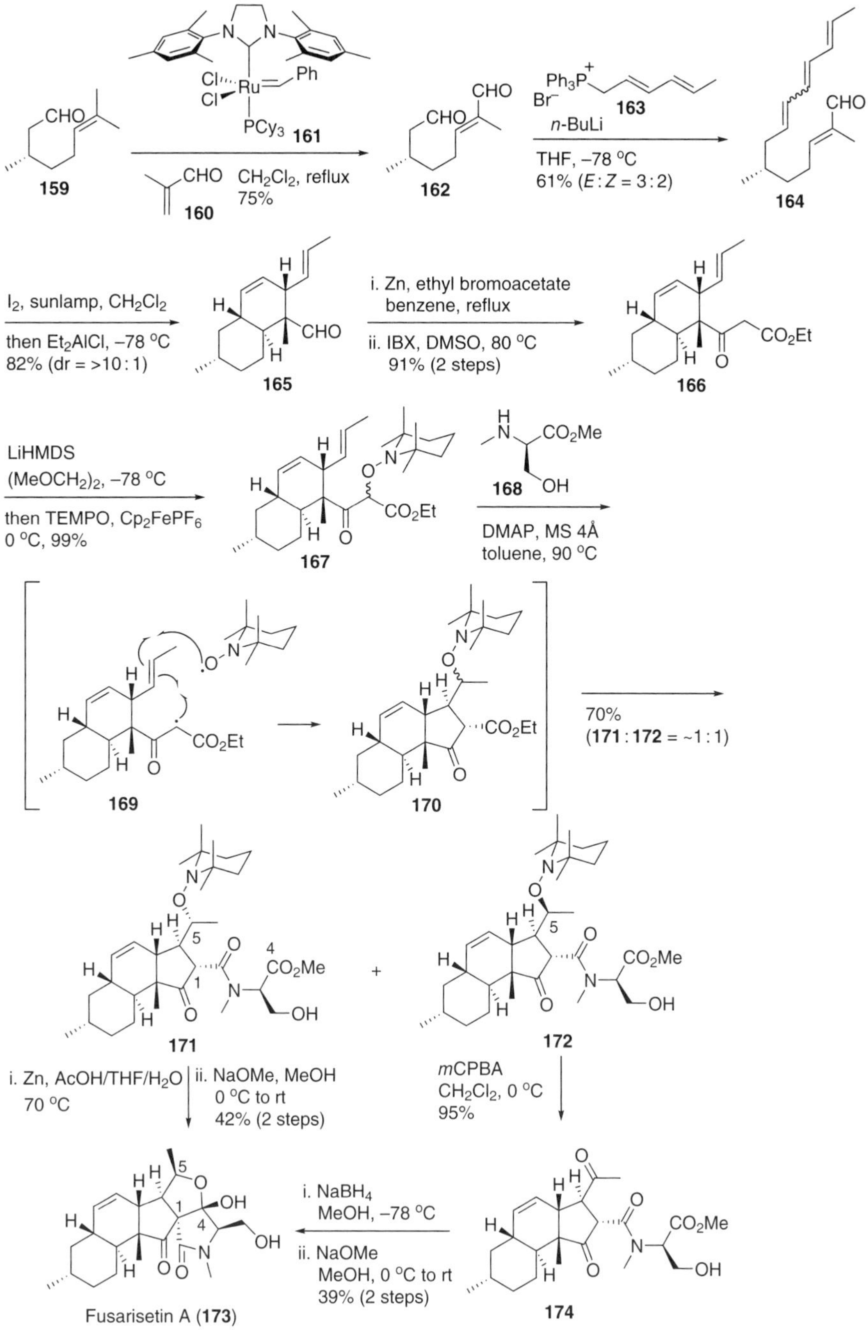

Scheme 3.13 Total synthesis of fusarisetin A (**173**).

The quaternary chiral center of rhazinilam (**191**) was introduced by the reaction of chiral enamine, which was derived from 2-ethylcyclohexanone (**175**) and (*S*)-1-phenethylamine, and methyl acrylate (**176**) to form ketoester **177** in over 99% ee (Scheme 3.14) [64]. After the conversion from **177** to α,β-unsaturated ketone **178** in two steps, the resulting alkene **178** reacted with H_2O_2/aq. NaOH to give epoxy ketone **179** as a diastereomeric mixture. Treatment of **179** with semicarbazide hydrochloride (**180**) followed by oxidative cleavage with $Pb(OAc)_4$ afforded acetylenic aldehyde **181** [65], which was oxidized under Pinnick conditions to provide carboxylic acid **182**. Condensation of the carboxylic acid **182** and amine **183** was performed with EDCI/HOBt/DMAP to yield amide **184**. The Sonogashira reaction with 2-bromoiodobenzene was utilized for construction of the alkynyl aryl moiety to produce cyclization precursor **186**. Next, the authors explored the cascade cyclization of **186** for one-pot construction of the indolizinone skeleton. After extensive

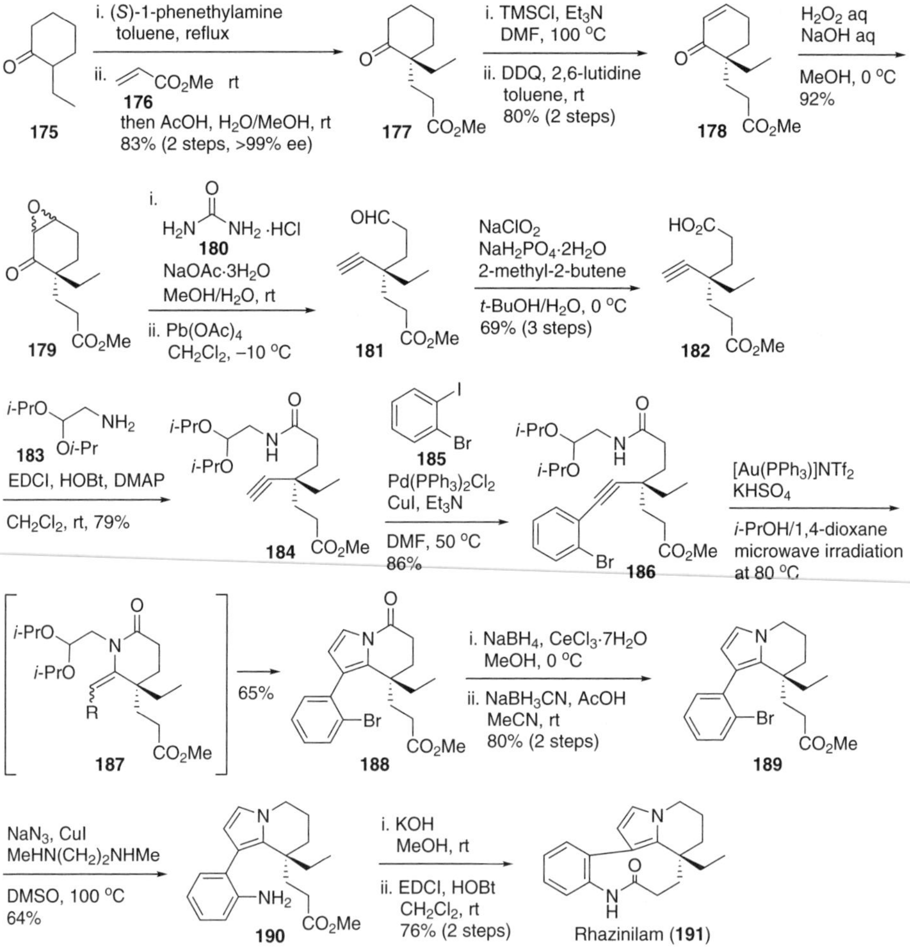

Scheme 3.14 Total synthesis of rhazinilam (**191**).

experimentations, it was found that the alkyne **186** reacted with [Au(PPh₃)]NTf₂ (30 mol%)/KHSO₄ (30 mol%) under microwave irradiation conditions at 80 °C to furnish the desired product **188** in 65% yield. In this reaction, the alkyne portion of **186** was presumably activated by π-philic metal catalyst [Au(PPh₃)]NTf₂ to give 6-*exo*-dig cyclized product **187**, which underwent cyclization and subsequent aromatization to produce the indolizinone **188**. Having constructed the indolizinone core by the cascade reaction, the authors next tried to form the nine-membered lactam ring and complete the total synthesis. Thus, *N*-acyl portion of the pyrrole **188** was reduced in two steps to give tetrahydroindolizine **189**. The aryl bromide **189** was transformed to aniline **190** by one-pot copper-mediated reaction by way of an aryl azide [66]. Finally, alkaline hydrolysis of the ester **190** followed by lactamization produced rhazinilam (**191**).

3.16 Yezo'otogirin C

The genus *Hypericum* is used as a traditional medicine for the treatment of anxiety, bruises, burns, inflammation, and swelling. Yezo'otogirins are tricyclic terpenoids isolated from *Hypericum yeziense* in 2009 [67]. Lee et al. at Peking University Shenzhen Graduate School explored the total synthesis of yezo'otogirin C (**201**) via oxidative cascade cyclization strategy inspired by the plausible biogenetic proposal [68].

Treatment of cyclohexanone **192** with LDA followed by trapping the resulting enolate with MeI gave the desired product **193** as a single diastereomer (Scheme 3.15). The conjugate addition of the cuprate prepared from Grignard reagent **194** and CuBr·SMe₂ and subsequent reaction with methyl cyanoformate provided β-ketoester **195** as a diastereomixture. With the key synthetic intermediate **195** in hand, Lee and coworkers investigated the oxidative radical cyclization of **195** inspired by the biological pathway. The detailed survey of the reaction conditions elucidated that the use of Mn(OAc)₂·4H₂O (2 equiv, oxidant) and Mn(OAc)₃·2H₂O (0.2 equiv, additive) was effective to result in the formation of the desired tricyclic peroxide **198** in 55% yield [69]. The radical species **196**, which was generated from the β-ketoester **195**, seems to undergo 5-*exo*-trig cyclization to provide radical intermediate **197** that is likely to react with oxygen to furnish the tricyclic peroxide **198**. To complete the total synthesis, reduction and dehydration of the peroxide **198** was performed by thiourea in MeOH at reflux to provide enol ether **199**. The reaction of **199** with *i*-PrLi toward yezo'otogirin C (**201**) resulted in failure, and the decarboxylation product was obtained in this case. Therefore, the ester **199** was transformed to aldehyde **200** by DIBALH reduction and subsequent TPAP oxidation, and treatment of **200** with *i*-PrLi followed by Dess–Martin oxidation of the resulting alcohol produced yezo'otogirin C (**201**).

3.17 Clavosolide A

Clavosolides were isolated from the marine sponge *Myriastra clavosa* in 2002 [70]. They are dimeric 16-membered macrolides that possess the cyclopropyl, glycosidic, and tetrahydropyranyl moieties as the structural feature. Breit and coworkers at the

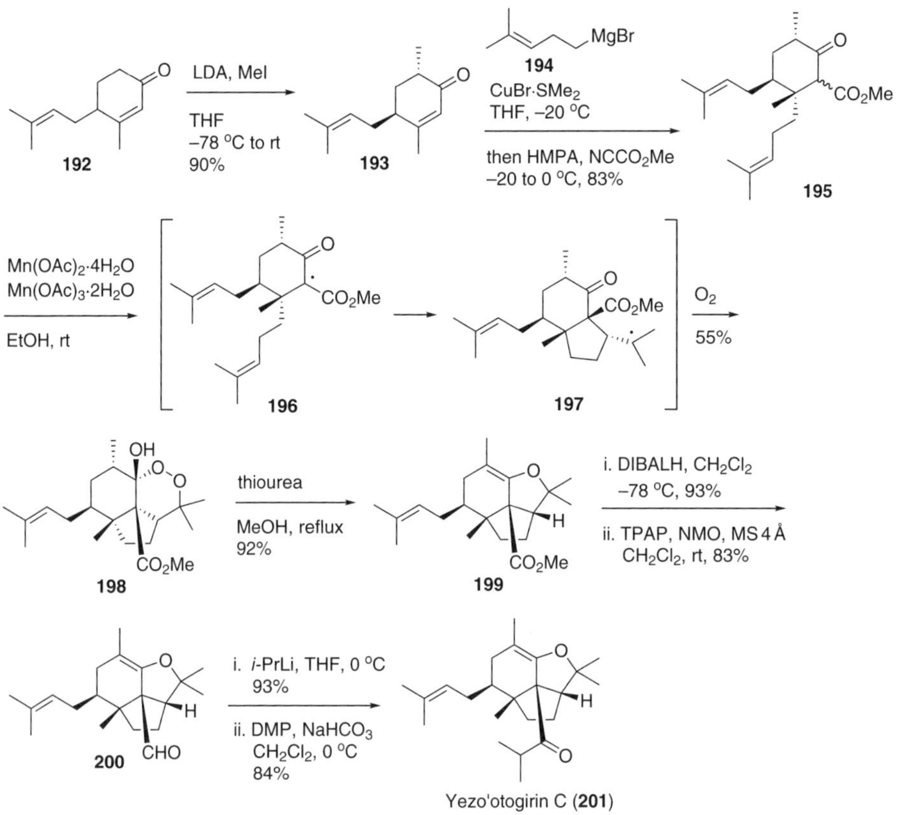

Scheme 3.15 Total synthesis of yezo'otogirin C (**201**).

University of Freiburg demonstrated the rhodium-catalyzed addition of carboxylic acids to allenes, which was originally developed by the same group [71], to result in the total synthesis of clavosolide A (**215**) [72].

Crotyl-transfer reaction of homoallenyl aldehyde **202** with the menthone-based chiral reagent **203** afforded homoallylic alcohol **204** over 99% ee (Scheme 3.16) [73]. Cascade oxa-Michael/Prins cyclization of **204** was carried out by reacting with methyl propiolate (**205**) and quinuclidine followed by addition of TFA to form the corresponding trifluoroacetate, which underwent methanolysis to give alcohol **206**. Glycosylation of **206** with phenyl thioglycoside **207** was performed in the presence of MeOTf to produce glycosylated ester **208** with a 77 : 23 diastereomeric ratio at the anomeric position. The methyl ester **208** was saponified to give carboxylic acid **209**. Before trying the dimerization of the allenyl carboxylic acid **209**, the authors initially investigated the rhodium-catalyzed reaction of the methyl ester **206** with benzoic acid. After the detailed optimization of reaction conditions, the authors found that the treatment of **206** with benzoic acid in the presence of [Rh(cod)Cl]$_2$ (5 mol%), (*R,R*)-DIOP (**210**, 10 mol%), and Cs$_2$CO$_3$ (10 mol%) in 1,2-dichloroethane (0.1 M) at 0 °C produced the desired allylic benzoate **211** in 95% yield at a 95 : 5 diastereomeric ratio. Next, head-to-tail dimerization of the allenyl carboxylic acid **209** was carried out by using the reaction conditions

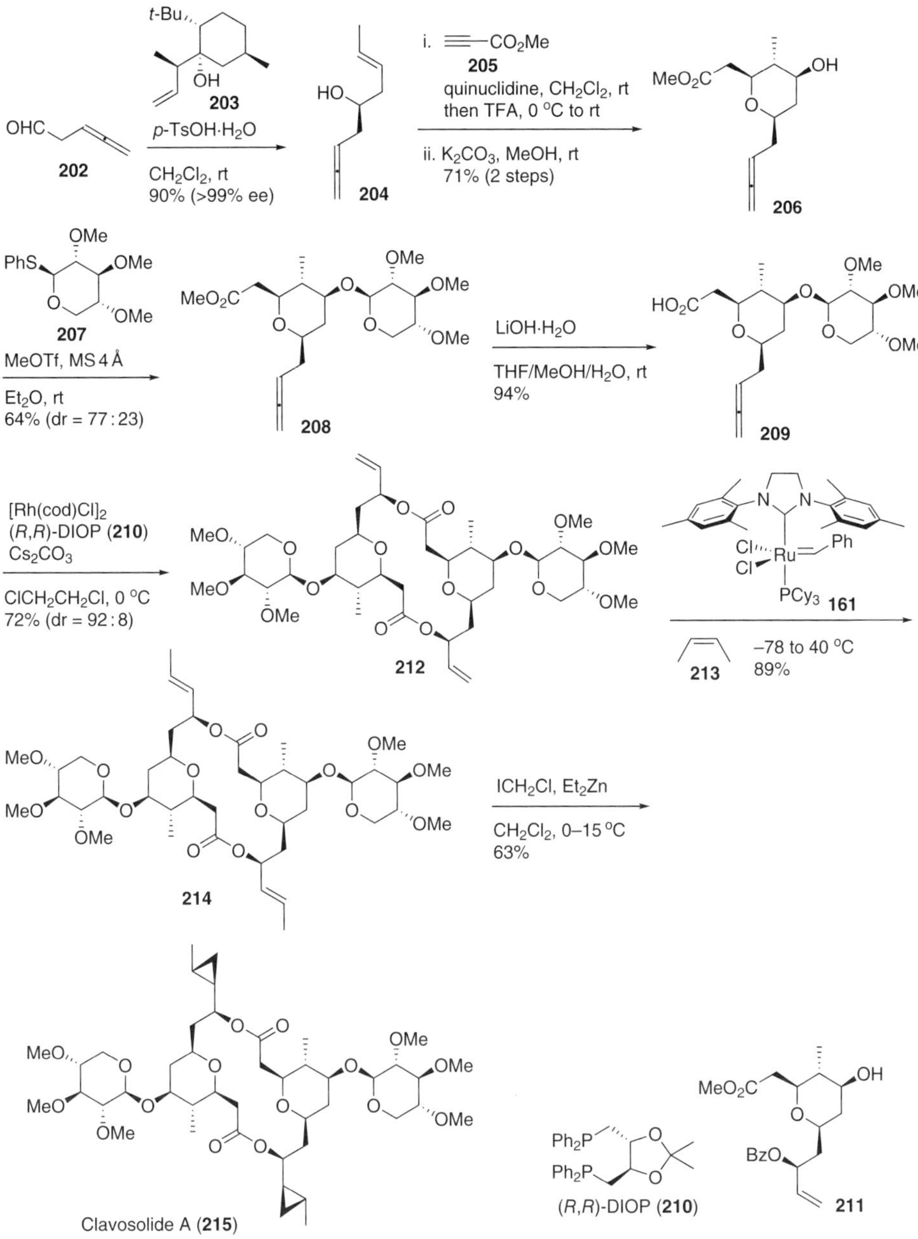

Scheme 3.16 Total synthesis of clavosolide A (**215**).

optimized in the reaction of **206** to result in the formation of dimeric 16-membered macrolide **212** in 72% yield with a 92 : 8 diastereomeric ratio. Subsequently, the obtained diene **212** was subjected to the cross-metathesis with (Z)-2-butene (**213**) by the second-generation Grubbs catalyst (**161**) to furnish (E,E)-diene **214**. Finally, Simmons–Smith cyclopropanation of **214** furnished clavosolide A (**215**).

3.18 Conclusion

The use of protecting groups is needed in multistep synthesis, especially in the polyketide, polypeptide, polysaccharide, and polynucleotide synthesis. On the other hand, protecting-group-free synthesis is an effective avenue to achieve the step-, redox-, and pot-economical synthesis, which is strongly demanded in modern organic synthesis. In addition, novel synthetic methodologies are invented, and unexpected reactions are discovered in pursuing the synthesis without protective groups. Metal-catalyzed or metal-mediated, radical, and cascade reactions and biomimetic transformations are helpful and utilized in the total synthesis of target molecules avoiding the use of protecting groups. In many cases, as a common feature of protecting-group-free synthesis, the inherent character and reactivity of the molecules are brought out and chemoselective transformations are realized. Protecting-group-free synthesis will continue to be a promising scene for the efficient organic synthesis and for invention and discovery.

References

1 Wuts, P.G.M. and Greene, T.W. (2006). *Greene's Protective Groups in Organic Synthesis*, 4e. Hokoben, NJ: Wiley.
2 Trost, B.M. (1995). *Angew. Chem. Int. Ed.* 34: 259.
3 Wender, P.A., Verma, V.A., Paxton, T.J., and Pillow, T.H. (2008). *Acc. Chem. Res.* 41: 40.
4 Burns, N.Z., Baran, P.S., and Hoffmann, R.W. (2009). *Angew. Chem. Int. Ed.* 48: 2854.
5 Hayashi, Y. (2016). *Chem. Sci.* 7: 866.
6 Young, I.S. and Baran, P.S. (2009). *Nat. Chem.* 1: 193.
7 (a) Moore, R.E., Cheuk, C., Yang, X.-Q.G. et al. (1987). *J. Org. Chem.* 52: 1036. (b) Raveh, A. and Carmeli, S. (2007). *J. Nat. Prod.* 70: 196.
8 Baran, P.S., Maimone, T.J., and Richter, J.M. (2007). *Nature* 446: 404.
9 For related publications by the same group, see: Richter, J.M., Ishihara, Y., Masuda, T. et al. (2008). *J. Am. Chem. Soc.* 130: 17938.
10 Baran, P.S. and Richter, J.M. (2004). *J. Am. Chem. Soc.* 126: 7450.
11 Nicolaou, K.C., Bulger, P.G., and Sarlah, D. (2005). *Angew. Chem. Int. Ed.* 44: 4442.
12 Pilli, R.A. and Ferreira de Oliveira, M.C. (2000). *Nat. Prod. Rep.* 17: 117.
13 Zeng, Y. and Aubé, J. (2005). *J. Am. Chem. Soc.* 127: 15712.
14 Zeng, Y., Reddy, D.S., Hirt, E., and Aubé, J. (2004). *Org. Lett.* 6: 4993.
15 Chung, H.-S., Hon, P.-M., Lin, G. et al. (2003). *Planta Med.* 69: 914.
16 Lainchbury, M.D., Medley, M.I., Taylor, P.M. et al. (2008). *J. Org. Chem.* 73: 6497.
17 Hook, B.D.A., Dohle, W., Hirst, P.R. et al. (2005). *J. Org. Chem.* 70: 7558.
18 Davies, D.M.E., Murray, C., Berry, M. et al. (2007). *J. Org. Chem.* 72: 1449.
19 Barton, D.H.R. and Jaszberenyi, J.C. (1989). *Tetrahedron. Lett.* 30: 2619.
20 Kuwano, R., Takahashi, M., and Ito, Y. (1998). *Tetrahedron. Lett.* 39: 1017.
21 Ratnayake, R., Covell, D., Ransom, T.T. et al. (2008). *Org. Lett.* 11: 57.
22 Zhou, Q., Chen, X., and Ma, D. (2010). *Angew. Chem. Int. Ed.* 49: 3513.
23 For the total synthesis of englerin A by using the gold-catalyzed cyclization, see: Molawi, K., Delpont, N., and Echavarren, A.M. (2010). *Angew. Chem. Int. Ed.* 49: 3517.
24 (a) Wei, H., Itoh, T., Konishita, M. et al. (2005). *Tetrahedron* 61: 8054. (b) Wei, H., Itoh, T., Kotoku, N., and Kobayashi, M. (2006). *Heterocycles* 68: 111.

25 Sofiyev, V., Navarro, G., and Trauner, D. (2008). *Org. Lett.* 10: 149.

26 (a) Walker, M.A. and Heathcock, C.H. (1991). *J. Org. Chem.* 56: 5747. (b) Raimundo, B.C. and Heathcock, C.H. (1995). *Synlett* 1213.

27 Peng, J., Walsh, K., Weedman, V. et al. (2002). *Tetrahedron* 58: 7809.

28 Enquist, J.A. Jr and Stoltz, B.M. (2008). *Nature* 453: 1228.

29 Mohr, J.T., Behenna, D.C., Harned, A.M., and Stoltz, B.M. (2005). *Angew. Chem. Int. Ed.* 44: 6924.

30 Yoshikai, K., Hayama, T., Nishimura, K. et al. (2005). *J. Org. Chem.* 70: 681.

31 Sadar, M.D., Williams, D.E., Mawji, N.R. et al. (2008). *Org. Lett.* 10: 4947.

32 Gu, Z. and Zakarian, A. (2010). *Angew. Chem. Int. Ed.* 49: 9702.

33 Beaumont, S., Ilardi, E.A., Monroe, L.R., and Zakarian, A. (2010). *J. Am. Chem. Soc.* 132: 1482.

34 Jouin, P. and Castro, B. (1987). *J. Chem. Soc. Perkin Trans. 1* 1177.

35 (a) Kurata, K., Taniguchi, K., Shiraishi, K. et al. (1989). *Chem. Lett.* 18: 267. (b) Kurata, K., Taniguchi, K., Shiraishi, K., and Suzuki, M. (1993). *Phytochemistry* 33: 155. (c) Todd, J.S., Proteau, P.J., and Gerwick, W.H. (1994). *J. Nat. Prod.* 57: 171.

36 Hickmann, V., Alcarazo, M., and Fürstner, A. (2010). *J. Am. Chem. Soc.* 132: 11042.

37 For a full paper, see:Hickmann, V., Kondoh, A., Gabor, B. et al. (2011). *J. Am. Chem. Soc.* 133: 13471.

38 Hayashi, T. and Yamasaki, K. (2003). *Chem. Rev.* 103: 2829.

39 Fürstner, A., Guth, O., Düffels, A. et al. (2001). *Chem. Eur. J.* 7: 4811.

40 Heppekausen, J., Stade, R., Goddard, R., and Fürstner, A. (2010). *J. Am. Chem. Soc.* 132: 11045.

41 Cai, X.-H., Du, Z.-Z., and Luo, X.-D. (2007). *Org. Lett.* 9: 1817.

42 Gerfaud, T., Xie, C., Neuville, L., and Zhu, J. (2011). *Angew. Chem. Int. Ed.* 50: 3954.

43 Oh, S.H., Rho, H.S., Lee, J.W. et al. (2008). *Angew. Chem. Int. Ed.* 47: 7872.

44 Jia, Y. and Zhu, J. (2006). *J. Org. Chem.* 71: 7826.

45 Horikawa, Y., Watanabe, M., Fujiwara, T., and Takeda, T. (1997). *J. Am. Chem. Soc.* 119: 1127.

46 Stierle, A.A., Stierle, D.B., and Kelly, K. (2006). *J. Org. Chem.* 71: 5357.

47 Fañanás, F.J., Mendoza, A., Arto, T. et al. (2012). *Angew. Chem. Int. Ed.* 51: 4930.

48 Barluenga, J., Mendoza, A., Rodríguez, F., and Fañanás, F.J. (2009). *Angew. Chem. Int. Ed.* 48: 1644.

49 For the total synthesis of berkelic acid by using the similar strategy, see:Bender, C.F., Yoshimoto, F.K., Paradise, C.L., and De Brabander, J.K. (2009). *J. Am. Chem. Soc.* 131: 11350.

50 (a) Jansen, R., Reifenstahl, G., Gerth, K. et al. (1983). *Liebigs Ann. Chem.* 1081. (b) Jansen, R., Sheldrick, W.S., and Höfle, G. (1984). *Liebigs Ann. Chem.* 78.

51 Fujita, K., Matsui, R., Suzuki, T., and Kobayashi, S. (2012). *Angew. Chem. Int. Ed.* 51: 7271.

52 Shimokawa, S., Kamiyama, M., Nakamura, T. et al. (2004). *J. Am. Chem. Soc.* 126: 13604.

53 Lee, S.J., Gray, K.C., Paek, J.S., and Burke, M.D. (2008). *J. Am. Chem. Soc.* 130: 466.

54 Dembitsky, V.M. (2008). *J. Nat. Med.* 62: 1.

55 Fujiwara, Y., Naithou, K., Miyazaki, T. et al. (2001). *Tetrahedron Lett.* 42: 2497.

56 Gutekunst, W.R., Gianatassio, R., and Baran, P.S. (2012). *Angew. Chem. Int. Ed.* 51: 7507.

57 For the total synthesis of cyclobutane natural products by using C–H arylation by the same group, see: Gutekunst, W.R. and Baran, P.S. (2011). *J. Am. Chem. Soc.* 133: 19076.

58 Liu, R., Zhang, M., Wyche, T.P. et al. (2012). *Angew. Chem. Int. Ed.* 51: 7503.

59 Jang, J.-H., Asami, Y., Jang, J.-P. et al. (2011). *J. Am. Chem. Soc.* 133: 6865.

60 Deng, J., Zhu, B., Lu, Z. et al. (2012). *J. Am. Chem. Soc.* 134: 920.

61 Xu, J., Caro-Diaz, E.J.E., Trzoss, L., and Theodorakis, E.A. (2012). *J. Am. Chem. Soc.* 134: 5072.

62 Banerji, A., Majumder, P.L., and Chatterjee, A. (1970). *Phytochemistry* 9: 1491.

63 Sugimoto, K., Toyoshima, K., Nonaka, S. et al. (2013). *Angew. Chem. Int. Ed.* 52: 7168.

64 d'Angelo, J., Desmaële, D., Dumas, F., and Guingant, A. (1992). *Tetrahedron: Asymmetry* 3: 459.

65 Macalpine, G.A. and Warkentin, J. (1978). *Can. J. Chem.* 56: 308.

66 Markiewicz, J.T., Wiest, O., and Helquist, P. (2010). *J. Org. Chem.* 75: 4887.

67 Tanaka, N., Kakuguchi, Y., Ishiyama, H. et al. (2009). *Tetrahedron Lett.* 50: 4747.

68 He, S., Yang, W., Zhu, L. et al. (2014). *Org. Lett.* 16: 496.

69 Qian, C.-Y., Yamada, T., Nishino, H., and Kurosawa, K. (1992). *Bull. Chem. Soc. Jpn.* 65: 1371.

70 (a) Rao, M.R. and Faulkner, D.J. (2002). *J. Nat. Prod.* 65: 386. (b) Erickson, K.L., Gustafson, K.R., Pannell, L.K. et al. (2002). *J. Nat. Prod.* 65: 1303.

71 Koschker, P., Lumbroso, A., and Breit, B. (2011). *J. Am. Chem. Soc.* 133: 20746.

72 Haydl, A.M. and Breit, B. (2015). *Angew. Chem. Int. Ed.* 54: 15530.

73 Nokami, J., Nomiyama, K., Matsuda, S. et al. (2003). *Angew. Chem. Int. Ed.* 42: 1273.

4

Protecting-Group-Free Synthesis of Natural Products and Analogs, Part III

Alakesh Bisai[1] and Vishnumaya Bisai[2]

[1]*Department of Chemistry, Indian Institute of Science Education and Research Bhopal Bhauri, Bhopal, India*
[2]*Department of Chemistry, Indian Institute of Science Education and Research Tirupati, India*

4.1 Introduction

Development of ideal synthetic strategy to access complex natural products of great biological relevance poses utmost synthetic challenge to organic chemists [1]. Especially, a strategy avoiding protecting groups directly enhances the step economy [2] of the overall process [3], by avoiding at least two steps in the synthetic sequence, viz. protection and deprotection. Although protecting groups provide a very smooth synthetic handle in total syntheses [4], strategies avoiding protecting groups are always preferred to make the synthesis more ideal as nature does in biological systems. In this regard, for instance, domino processes represent the best way to affect step efficient synthesis comprising several events happening in a single flask. However, it is also important to have better understanding of the reaction sequence to make sure that cascade propagation does not hamper the synthesis. In this book chapter, protecting-group-free (PGF) synthetic approaches to some of the architecturally intriguing and biologically active natural products based on terpenes and alkaloids are discussed.

Among the natural products studied in the nineteenth and early twentieth centuries, the tropane alkaloids (**1a–d**, Figure 4.1) gathered immense synthetic interest owing to their wide range of potent biological activities, including the use of plants containing tropane alkaloids for their hallucinogenic effects [5]. Cocaine (**1b**), first isolated in 1860 from the leaves of the *Peruvian erythroxylum* coca plant by Niemann, is a potent neurostimulant with notoriously deleterious addictive properties. Hyoscyamine (**1c**, atropine) has a powerful dilatory effect and is a widely used ophthalmic drug, while scopolamine (**1d**) is used for a variety of medical applications, such as control of emesis and resuscitation. Notably, modern synthesis witnesses the tropane alkaloids, most commonly scopolamine (**1d**), as starting materials for the synthesis of valuable commercial drugs [6]. Structurally, the tropane alkaloids share the common [3.2.1]-bicyclic amine substructure

Protecting-Group-Free Organic Synthesis: Improving Economy and Efficiency, First Edition.
Edited by Rodney A. Fernandes.
© 2018 John Wiley & Sons Ltd. Published 2018 by John Wiley & Sons Ltd.

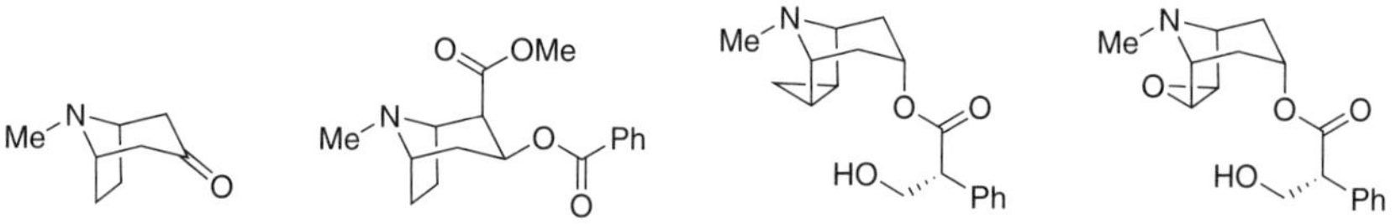

Figure 4.1 Tropinone and related alkaloids.

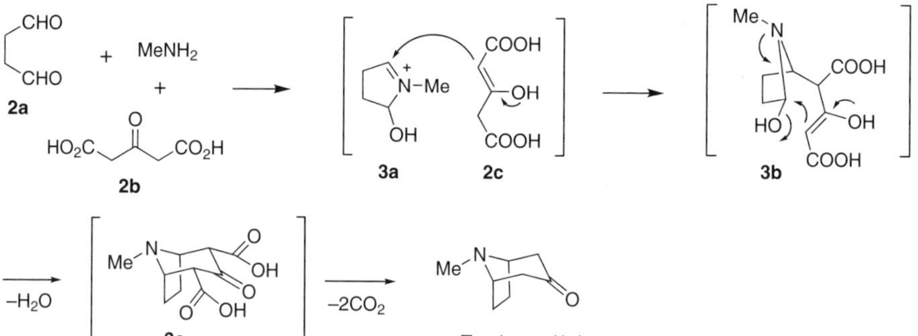

Tropinone (**1a**) Cocaine (**1b**) Hyoscyamine (**1c**) Scopolamine (**1d**)

Scheme 4.1 Robinson's synthesis of tropinone (**1a**).

exemplified by tropinone (**1a**). The structural elucidation of such valuable natural alkaloids has been achieved with great reliance upon laboratory synthesis [7].

Robinson's biomimetic approach to the synthesis of tropinone (**1a**) [8a] is one of the greatest accomplishments in total synthesis. The strategy comprising a cascade of tandem events [9] greatly enhances the overall efficiency. In a remarkable single-step transformation, succinaldehyde (**2a**), methylamine hydrochloride, and acetone dicarboxylic acid **2b** were combined at physiological pH to afford tropinone (**1**) in an impressive yield of 42%. The overall transformation represents a judicious and strategic composition of several fundamental reactions: condensation of methylamine with dialdehyde **2a** to afford cyclic iminium **3a**, Mannich reaction of **3a** with the enol tautomer of calcium acetone dicarboxylate **2c** to afford **3b**, subsequent Mannich cyclization to afford bicyclic intermediate **3c**, and decarboxylations to afford tropinone (**1**). Notably, replacing acetone for acetone dicarboxylic acid under basic conditions afforded tropinone in very poor yield. In 2003, Das and coworkers [8b] have reported detailed theoretical studies on the mechanism of Robinson's tropinone synthesis by ab initio quantum mechanical calculation. The study revealed that a six-membered piperidine moiety is generated during the formation of the second C—C bond (transition state of formation of **3c** from **3b**) and the overall conformation of this ring formation is the chair form (Scheme 4.1).

4.2 Syntheses of Naturally Occurring Alkaloids

In efforts toward the synthesis of *Aristotelia* alkaloid, Heathcock realized the synthesis of a number of related alkaloids en route to the proposed aristone precursor, (−)-alloaristoteline (**5e**) [10]. The strategy followed the proposed biosynthetic hypotheses [11]

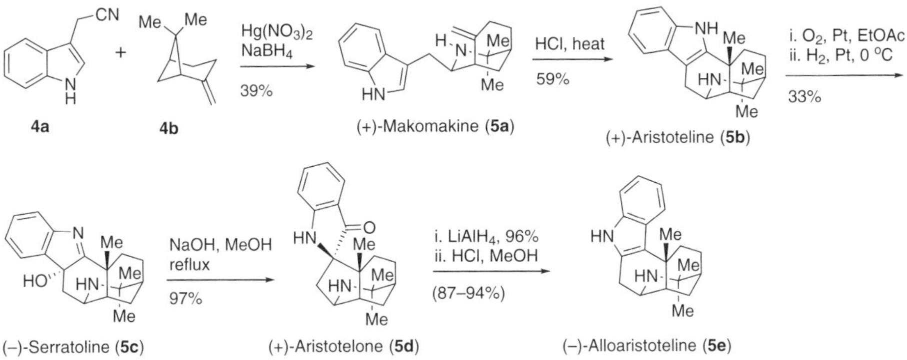

Scheme 4.2 Protecting-group-free total syntheses of *Aristotelia* alkaloids.

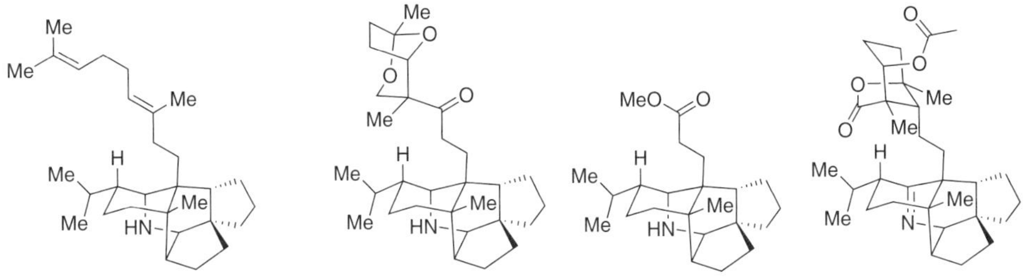

Dihydro-protodaphniphylline (**6a**) Secodaphniphylline (**6b**) Methyl homoseco-daphniphyllate (**6c**) Daphnioldhanine I (**6d**)

Figure 4.2 Selected daphniphyllum alkaloids.

and used acid- and base-induced rearrangements of intermediate natural products (Scheme 4.2). An initial Hg(NO$_3$)$_2$-mediated Ritter reaction of (lS)-(−)-β-pinene (**4b**) with 3-indolylacetonitrile **4a** to synthesize (+)-makomakine (**5a**) followed an acid-catalyzed Friedel–Crafts reaction to yield (+)-aristoteline (**5b**) through a known sequence [12]. Oxidation of the indole C3 position afforded (−)-serratoline (**5c**), which upon subsequent base-induced ring contraction yielded the natural product (+)-aristotelone (**5d**). Carbonyl reduction, followed by acid-induced ring expansion, produced (−)-alloaristoteline (**5e**).

The daphniphyllum alkaloids comprising more than 200 natural products [13] have gained enormous attention both due to their diverse biological activities and to structural complexities (**6a–d**, Figure 4.2) [14]. In the late 1980s, Heathcock's synthesis of the *daphniphylline* alkaloids is a fantastic example of nature's approach to architecturally intriguing structures [15]. It was envisioned that the carbon skeleton of the daphniphyllum alkaloids can be directly mapped onto squalene. The synthesis demonstrates a remarkable organocatalytic stereoselective cascade transforming dialdehyde **7a** to dihydroprotodaphniphylline (**6a**) in a single operation (Scheme 4.3) [16].

As per Heathcock's hypothesis, the natural sequence is believed to involve condensation of an alkylamine (such as methylamine) with dihydrosqualene dialdehyde **7a** to give an intermediate enamine, which would be highly competent for conjugate addition into the unsaturated aldehyde (or the corresponding iminium **7b**). Subsequent

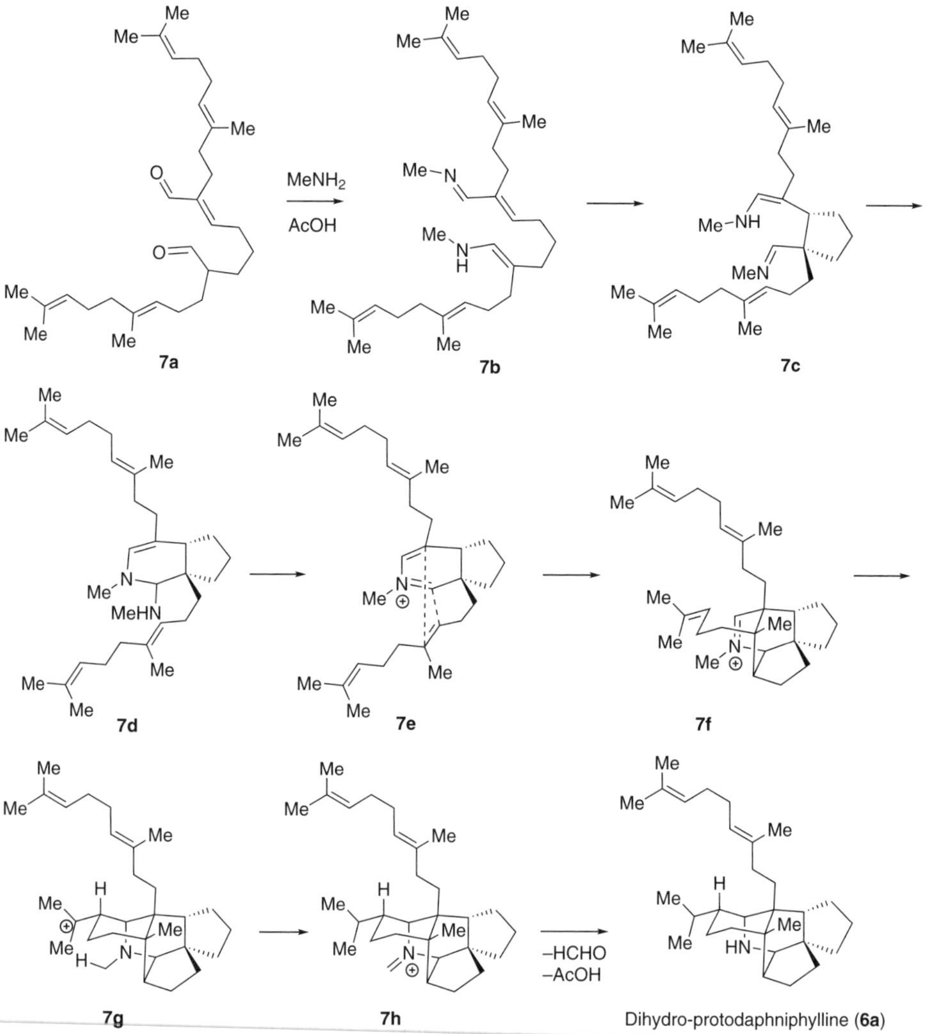

Scheme 4.3 Total synthesis of dihydroprotodaphniphylline (**6a**).

intramolecular aminal formation followed by acid-promoted iminium ion formation in the form of azadiene **7e** sets the stage for an intramolecular cycloaddition with the proximal angular prenyl moiety to provide iminium ion **7f**. Trapping of the iminium ion with the terminal trisubstituted alkene then gives rise to the final C—C bond in the sequence while simultaneously positioning the tertiary carbocation in proximity to the methylamine. At this stage an usual intramolecular hydride transfer from *N*-methyl group generates an iminium intermediate **7h**, which hydrolyzes upon work-up to yield dihydroprotodaphniphylline (**6a**) in 65% overall yield [16a].

Kobayashi and coworkers reported the isolation of a series of structurally fascinating indole monoterpenoid alkaloids, subincanadines A–C and F–G (**9a–e**, Figure 4.3), from the barks of the Brazilian medicinal plant *Aspidosperma subincanum* Mart [17]. Later, they carried out structure determination and preliminary biological properties of

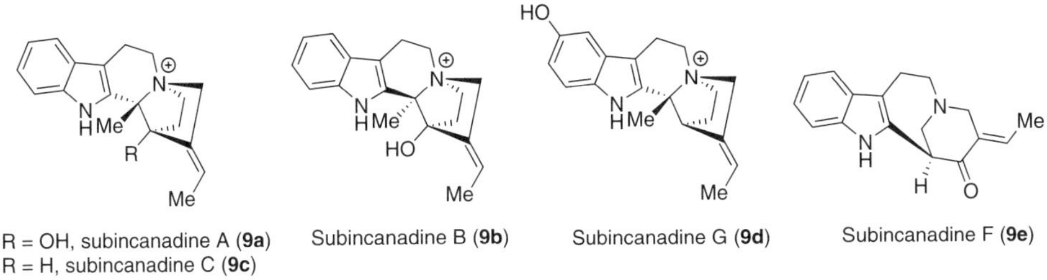

R = OH, subincanadine A (**9a**)   Subincanadine B (**9b**)   Subincanadine G (**9d**)   Subincanadine F (**9e**)
R = H, subincanadine C (**9c**)

Figure 4.3 Indole monoterpene alkaloids, subincanadines A–C and F–G (**9a–e**).

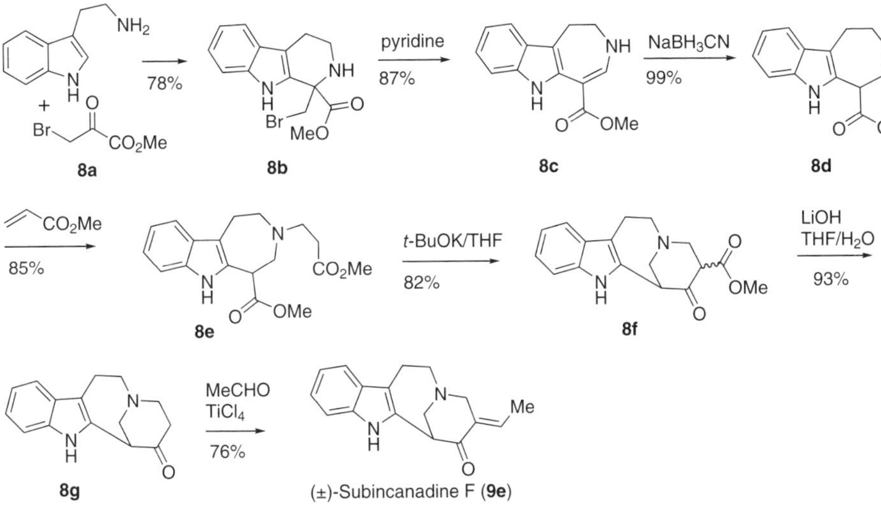

Scheme 4.4 Protecting-group-free total synthesis of (±)-subincanadine F (**9e**).

subincanadine F (**9**) as well concluding its unique 1-azabicyclo[4.3.1]decane skeleton [17]. *In vitro* pharmacological evaluations of subincanadine F (**9**) revealed cytotoxic activities against murine lymphocytic leukemia (L1210) and human epidermoid carcinoma (KB) cell lines with IC$_{50}$ values of 2.4 and 4.8 μg ml^{-1}, respectively [17c].

Owing to their unique structural characteristics in addition to their impressive pharmacological activities, subincanadines gained a huge synthetic interest by several research groups. In this regard, the PGF total synthesis of (±)-subincanadine F (**9e**) by Li and coworkers [18] is worth mentioning (Scheme 4.4). The synthesis includes a Pictet–Spengler condensation of tryptamine with bromopyruvate **8a**, giving the bromomethylated compound **8b** in 78% yield, which was then subjected to treatment with pyridine at refluxing temperature to get ring expansion product **8c** (87% yield). Reduction of **8c** with NaBH$_3$CN furnished **8d** quantitatively. Next, aza-Michael reaction of **8d** with methyl acrylate followed by a Dieckmann condensation of diester in the presence of 2.0 equiv of *t*BuOK afforded condensation product **8f** in 82% yield [19]. The treatment of **8f** with LiOH in aqueous THF at reflux afforded the de-esterification product **8g** in high yield [20a, b]. Eventually, aldol condensation of **8g** with acetaldehyde in presence of TiCl$_4$ and N-ethyldiisopropylamine [20c–e] afforded (±)-subincanadine F (**9**) in excellent (*E*)-selectivity in 76% yield [18].

In 2011, Zhai and coworkers reported the PGF total synthesis of (±)-subincanadine C (**9c**) from tryptamine in only seven steps and with 39% overall yield [21]. The synthesis features a key Ni(COD)$_2$-mediated intramolecular Michael addition. The Pictet–Spengler reaction of tryptamine hydrochloride with ethyl pyruvate afforded compound **10a** in 85% yield. This was alkylated with known bromide **10b** in the presence of K$_2$CO$_3$ to give vinyl iodide **10c** in 93% yield. Upon partial reduction of the ester group and subsequent Horner–Wadsworth–Emmons (HWE) olefination, unsaturated ester **10e** was produced from **10d** in 85% overall yield. The key intramolecular Michael addition of **10e** with 5 equiv of Ni(COD)$_2$ and 10 equiv of Et$_3$N in MeCN at room temperature provided the desired cyclization product **10f** in 63% yield. The *cis*-relationship between the hydrogen atom at C15 and the methyl (C17) as well as the (*E*)-configuration of the double bond in **10f** was unambiguously confirmed by X-ray crystallographic analysis. Finally, reduction of **10f** with DIBAL-H afforded the primary alcohol **10g**, which was treated with PPh$_3$ and N-chlorosuccinimide (NCS) to furnish (±)-subincanadine C (**9c**) via sequential chlorination and intramolecular nucleophilic substitution (Scheme 4.5) [21].

Till date, over 80 types of benzo[*c*]phenanthridine alkaloids produced by nature have been isolated and characterized from different sources [22]. For instance, oxyavicine (**11a**) isolated from *Zanthoxylum avicennae* root bark [23a], *Broussonetia papyrifera* fruits [23b], and *Zanthoxylum nitidum* root [23c] finds usage in the treatment of ophthalmic disorders [23b] (Figure 4.4). On the other hand, oxynitidine (**11b**) isolated from *Zanthoxylum rhoifolium* [24a], *Z. nitidum* root [24b], and *Fagara macrophylla* bark

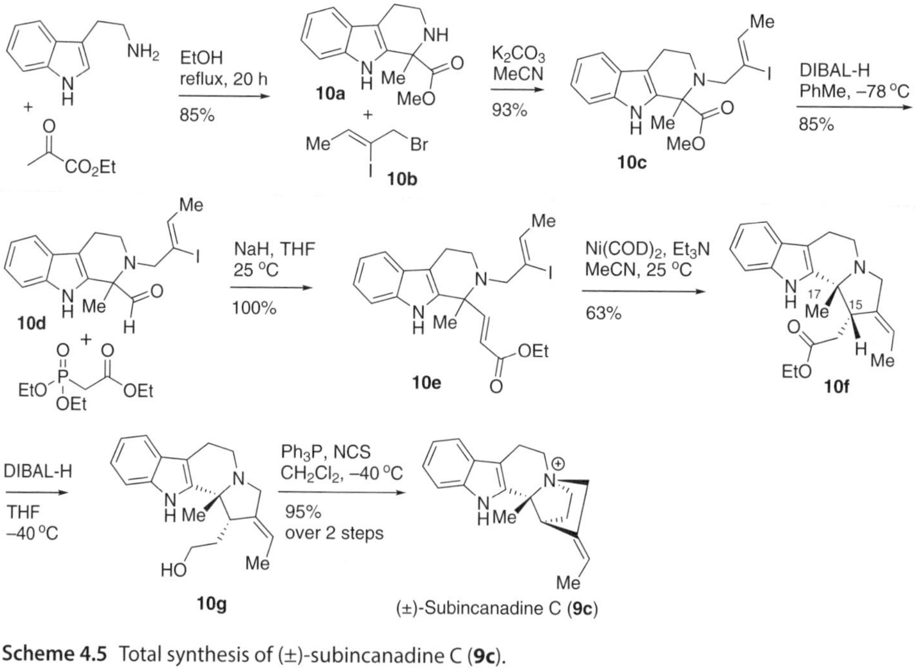

Scheme 4.5 Total synthesis of (±)-subincanadine C (**9c**).

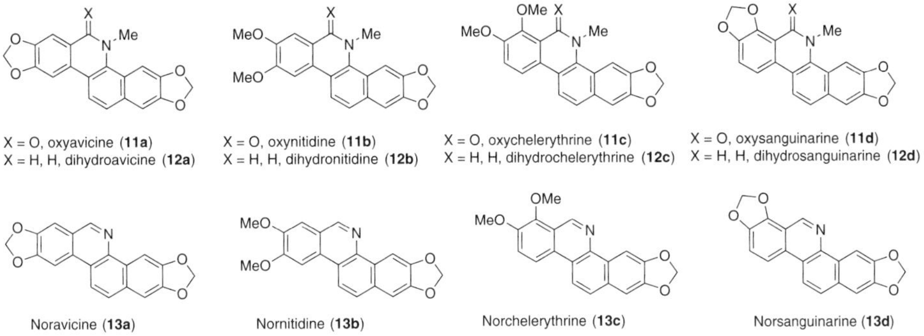

X = O, oxyavicine (**11a**)
X = H, H, dihydroavicine (**12a**)

X = O, oxynitidine (**11b**)
X = H, H, dihydronitidine (**12b**)

X = O, oxychelerythrine (**11c**)
X = H, H, dihydrochelerythrine (**12c**)

X = O, oxysanguinarine (**11d**)
X = H, H, dihydrosanguinarine (**12d**)

Noravicine (**13a**)

Nornitidine (**13b**)

Norchelerythrine (**13c**)

Norsanguinarine (**13d**)

Figure 4.4 Selected benzo[*c*]phenanthridine alkaloids.

R = —CH$_2$—, **14a**
R = Me, **14b**

R = —CH$_2$—, **16a**
R = Me, **16b**

R = —CH$_2$—, **17a**
R = Me, **17b**

R = —CH$_2$—, dihydroavicine (**12a**)
R = Me, dihydronitidine (**12b**)

R = —CH$_2$—, oxyavicine (**11a**)
R = Me, oxynitidine (**11b**)

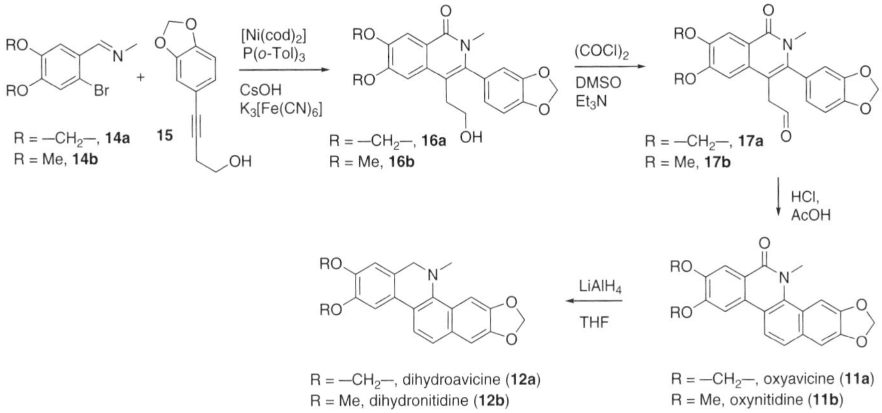

Scheme 4.6 Total syntheses of benzo[*c*]phenanthridine alkaloids.

[24c] possesses potent inhibitory activity against hepatitis B virus [24d]. Recently, oxysanguarine (**11d**) isolated from the Ranunculaceae (seeds of Coptis) makes an important drug constituent in China and Japan.

A PGF total synthesis of a number of benzo[*c*]phenanthridine alkaloids reported by Cheng and coworkers [25] involves a Ni-catalyzed cyclization of imines **14a–b** with alkyne **15** in the formation of corresponding isoquinolinium salts, which upon subsequent oxidation provided amides **16a–b** in 69–78% yields (Scheme 4.6). Later, oxidation of the alcohol functionality in **16a–b** yielded the corresponding aldehydes **17a–b**, followed by cyclization using HCl (aq), and acetic acid completed the total syntheses of oxyavicine (**11a**) and oxynitidine (**11b**) [25] in 60–65% yields [26]. Alkaloids oxyavicine (**11a**) and oxynitidine (**11b**) can be conveniently transformed into the corresponding naturally occurring dihydro derivatives dihydroavicine (**12a**) and dihydronitidine (**12b**) [25] upon reduction with LiAlH$_4$.

Bisai and coworkers have demonstrated highly regioselective organocatalytic biaryl coupling via a homolytic aromatic substitution (HAS). This process [27, 28] utilizes *t*BuOK as the sole coupling promoter (in the absence or presence of a catalytic amount of organic molecules) for the total syntheses of benzo[*c*]phenanthridines [29]. The substrate of the type 2-bromo-*N*-(α-naphthyl)benzylamine **19** (Scheme 4.7) poses

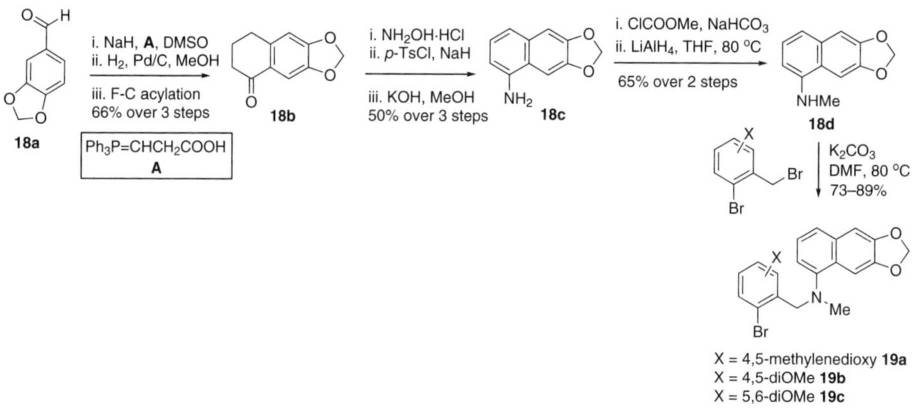

Scheme 4.7 Synthesis of 2-bromo-*N*-(α-naphthyl)amines **19a–c**.

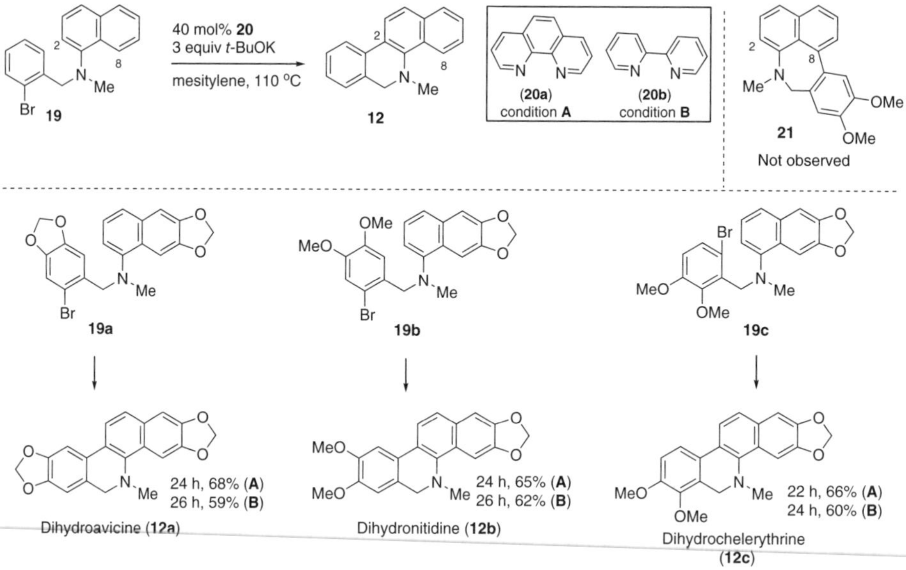

Scheme 4.8 Syntheses of benzo[c]phenanthridine alkaloids **12** via an HAS.

challenge in this transformation due to two possible regioisomeric products depending upon two different C–H activation pathways. In one pathway, it could react at C2 position to afford more stable dihydrophenanthridine **12**, or alternatively it could also react at C8 position to afford naphthobenzazepine structure **21** (Scheme 4.8) [30]. In the synthetic sequence, they prepared α-tetralone **18b** starting from piperonal aldehyde **18a** via Wittig olefination and hydrogenation followed by Friedel–Crafts acylations (Scheme 4.7). α-Tetralone **18b** was then converted into α-naphthylamine **18c** in three-step procedure (viz. formation of oxime, tosylation followed by detosylation/aromatization sequence). The latter was then converted into **18d** in two-step sequence involving reaction with chloromethyl formate followed by reduction using LiAlH$_4$. Few 2-bromo-*N*-(α-naphthyl)benzylamines **19a–c** were prepared from N-benzylation of **18d**.

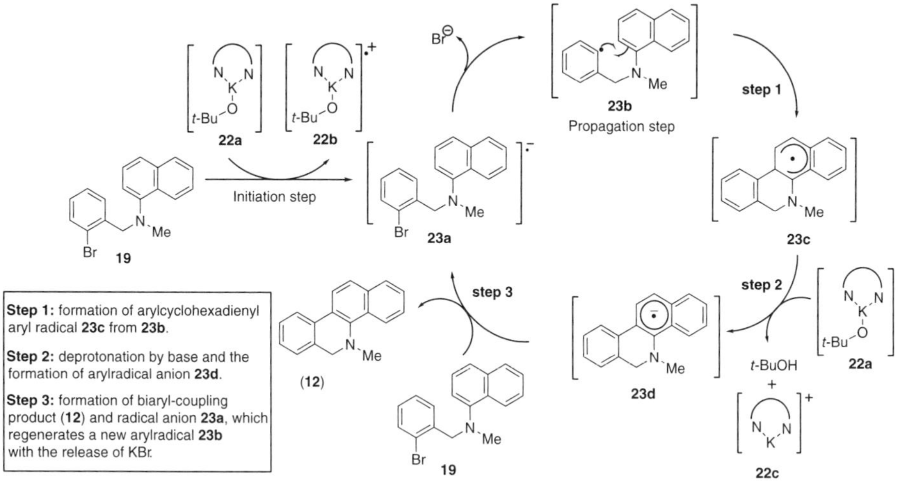

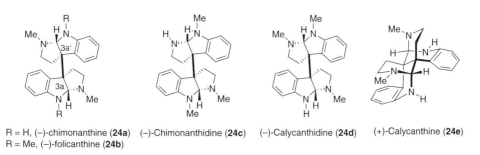

Scheme 4.9 Proposed mechanism of benzo[c]phenanthridine synthesis.

R = H, (−)-chimonanthine (**24a**) (−)-Chimonanthidine (**24c**) (−)-Calycanthidine (**24d**) (+)-Calycanthine (**24e**)
R = Me, (−)-folicanthine (**24b**)

Figure 4.5 Bis-cyclotryptamine alkaloids in active form.

Mechanistically, it was believed that a single electron transfer (SET) from **22** (or *t*BuOK) onto **19a–i** (Scheme 4.9) provides a radical **23b** through the intermediacy of radical anion **23a**, which could then follow a propagation step providing arene-annulated cyclohexadienyl radical **23c** (step 1). A similar proton transfer (PT) from **23c** in the presence of **22a** (or *t*BuOK) leads to the formation of radical anion **23d**, *t*BuOH, and K⁺ (step 2). Finally, a radical anion transfer from **23d** to the starting material **19a–i** affords dihydrobenzo[c]phenanthridines **12** and the radical anion **23a** (step 3), thus continuing the catalytic cycle (Scheme 4.9).

The hexahydropyrroloindole alkaloids (also known as cyclotryptamine alkaloids) constitute a large class of natural compounds that are formally derived from tryptophan [31]. A fascinating array of dimeric and oligomeric derivatives exist, a large subset of which contain vicinal quaternary stereocenters at C3a and C3a′ of the hexahydropyrroloindole substructure (**24a–e**; Figure 4.5). The complex architecture of these alkaloids combined with their biological activity has led to significant interest in their total syntheses. Important contributions in this area have established a biosynthetically inspired oxidative dimerization [32] of oxindoles and tryptamines for accessing derivatives in racemic and meso form [33].

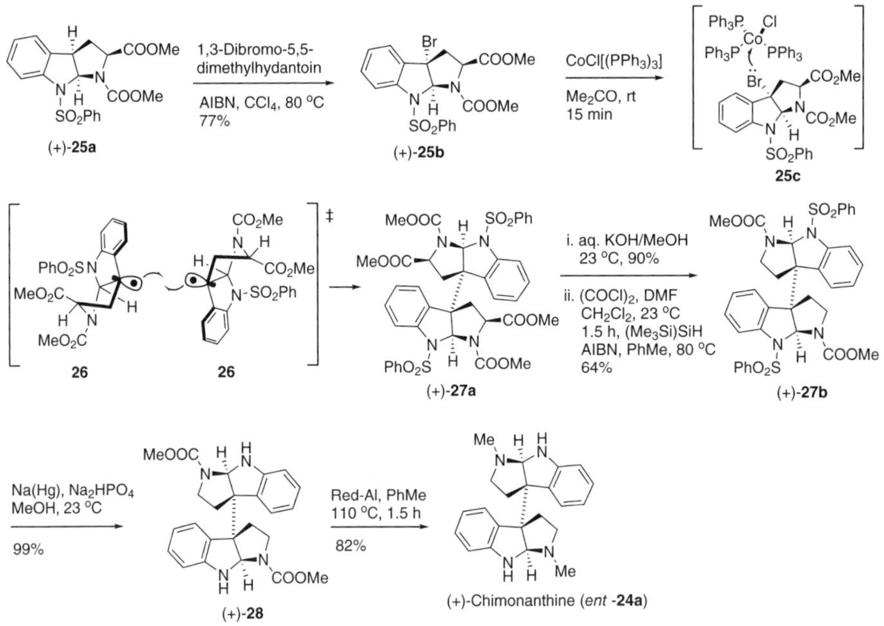

Scheme 4.10 Total synthesis of (±)-chimonanthine (*ent*-**24a**).

The concise and efficient total synthesis of (+)-chimonanthine (*ent*-**24a**), (+)-folicanthine (*ent*-**24b**), and (−)-calycanthine (*ent*-**24e**) by Movassaghi et al. [34] utilizes a single protecting-group consideration (benzenesulfonyl group on indole nitrogen). Here the treatment of the commercially available N_a-methoxycarbonyl-L-tryptophan methyl ester with neat phosphoric acid followed by N-sulfonylation provided the desired tricyclic hexahydropyrroloindole (+)-**25a** (>99% de, >99% ee) on multigram scale (Scheme 4.10). Benzylic bromination of (+)-**25a** was optimally achieved using 1,3-dibromo-5,5-dimethylhydantoin on gram scale to afford the bromide (+)-**25b** as a single diastereomer in 77% yield and >99% ee. Treatment of the tricyclic bromide (+)-**25b** with [CoCl(PPh₃)₃] in acetone at 23 °C for 15 min under optimum reaction conditions provided the desired dimer (+)-**27a** in 60% yield and >99% ee on gram scale. This dimerization process may involve a bromide abstraction from (+)-**25b**, thus allowing the union of two free radicals (+)-**26** [35], or alternatively, this may involve a net oxidative insertion of the cobalt(I) reagent to afford a tertiary benzylic cobalt(III) complex that could afford (+)-**27a** by homodimerization followed by fragmentation of the C—Co bonds and formation of a C—C bond [34]. Later, treatment of the diester (+)-**27a** with a mixture of MeOH and aqueous KOH provided the corresponding dicarboxylic acid in 90% yield. Decarboxylation at the C2 and C2′ position was achieved by sequential conversion into the dicarboxylic acid chloride followed by treatment with tris(trimethylsilyl)silane and AIBN in toluene at 80 °C to afford the hexacycle (+)-**27b** in 64% yield. Finally, treatment of a methanolic solution of hexacycle (+)-**27b** and sodium phosphate dibasic with freshly prepared sodium amalgam afforded the desired diamine (+)-**28** in 99% yield and >99% ee. Reduction of the two *N*-methoxycarbonyl groups of **28** with sodium bis(2-methoxyethoxy)aluminum hydride (Red-Al) gave (+)-chimonanthine (*ent*-**24a**) in 82% yield. N-methylation of (+)-*ent*-**24a**

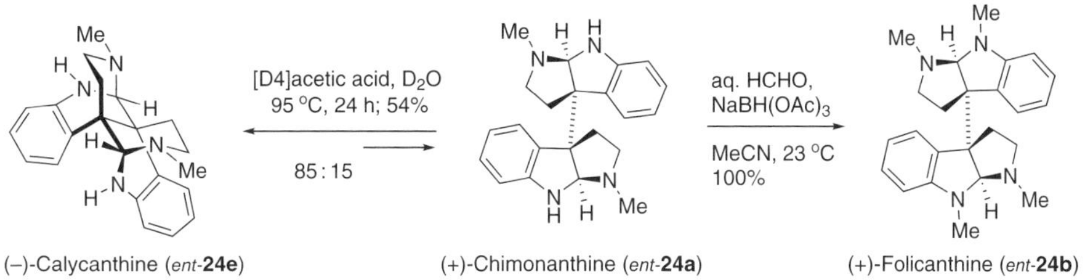

(–)-Calycanthine (*ent*-**24e**) (+)-Chimonanthine (*ent*-**24a**) (+)-Folicanthine (*ent*-**24b**)

Scheme 4.11 Syntheses of (+)-folicanthine (*ent*-**24b**) and (–)-calycanthine (*ent*-**24e**).

using formalin and sodium triacetoxyborohydride provided the first synthetic sample of (+)-folicanthine (*ent*-**24b**) in quantitative yield (Scheme 4.11). Dissolution of (+)-*ent*-**24a** in a mixture of [D$_4$]-acetic acid and deuterium oxide followed by heating to 95 °C led to the formation of the isomeric calycanthine (**24e**) [36].

It is worth mentioning that the chimonanthines are found in nature in both forms of enantiomers, such as (–)-chimonanthine (**24a**; Figure 4.5) [37a] isolated from plant sources, whereas (+)-chimonanthine (*ent*-**24a**) from the skin of a Colombian frog [37b] and from plant sources [37c]. In this regard, catalytic enantioselective processes in the presence of enantioenriched ligands are welcome, so as to obtain chiral nonracemic compounds in both enantiomeric forms. To this end, construction of vicinal all-carbon quaternary stereocenters in an enantioselective fashion is one of the most challenging aspects in the synthesis of complex molecules [38]. It is usually the second quaternary center that complicates the asymmetric assembly. To date, only limited methods exist in the literature that can assemble vicinal quaternary all-carbon stereocenters in an asymmetric fashion in a single-step operation [39].

In search for unified total syntheses of either enantiomer of hexahydropyrroloindole alkaloids (Figure 4.5), Bisai et al. envisioned an efficient enantioselective Pd(0)-catalyzed double stereoablative allylation [40a] of racemic and *meso* diastereomers of bisesters **29** to furnish a mixture of enantioenriched **30** and *meso*-**30** (Figure 4.6).

The major issues of this transformation include the presence of preexisting stereocenters in the substrates that may interfere with inherent catalyst selectivity [40b–d]. In this process, the mixture of diastereomers **29** must undergo decarboxylative deallylation to produce racemic enolates **31**, which upon subsequent monoallylation may form intermediate **32**, in order to achieve the desired outcome (Figure 4.6). It is obvious that if enantiopure ligand **34** [41] controls the formation of the new stereocenter, a pair of ester diastereomers **32**, i.e. (*S,S*)-**32** and (*R,S*)-**32**, will predominantly form over the other pair, i.e. (*S,R*)-**32** and (*R,R*)-**32**. Therefore, the influence of the remaining substrate stereocenter may either reinforce or conflict with catalyst control during this enantioselective methodology. From this diastereomeric mixture of monoesters, a second stereoablative process would then generate a mixture of enolates (*S*)-**33** and (*R*)-**33**. A second facially selective allylation would afford the doubly allylated product as a diastereomeric mixture of enantioenriched (*S,S*)-**30** and *meso*-**30** (Figure 4.6).

Treatment of **29** in the presence of 2.5 mol% of Pd$_2$(dba)$_3$ and 7.5 mol% of Trost ligand **34** in Et$_2$O at –10 °C afforded products (*S,S*)-**30** and *meso*-**30** as a 17 : 1 mixture in 88% yield (Scheme 4.12). Gratifyingly, it was found that the major diastereomer **30** is formed in 97% ee and the reaction proceeds without any additives. Interestingly, the high degree

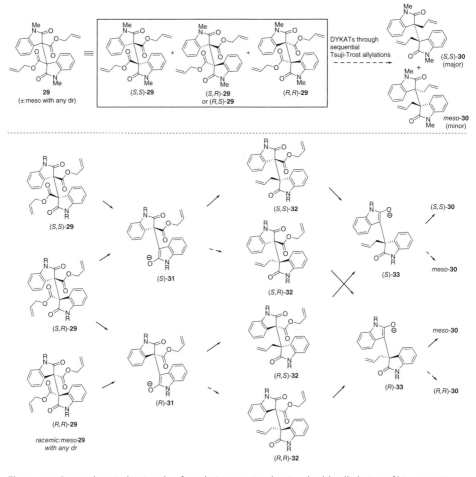

Figure 4.6 Stereochemical rationale of catalytic enantioselective double alkylation of bis-ester **29**.

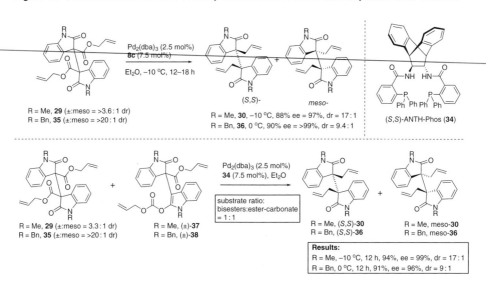

Scheme 4.12 Catalytic enantioselective double allylation of bis-esters (±)- and *meso*-**29** and **35**.

of stereoconvergence and stereochemical control predominantly favors formation of the C2-symmetric product (*S,S*)-**30** over the *meso*-product **30**. Therefore, in a single step, two sterically congested vicinal quaternary stereogenic carbon centers are constructed (Scheme 4.12). Under optimized conditions, *N*-benzyl-protected substrate **35** (dr >20 : 1) afforded products (*S,S*)-**36** and *meso*-**36** as a 9.4 : 1 mixture in 90% yield and >99% ee (Scheme 4.12) [42a].

Further, it was also hypothesized that ester carbonates **37**–**38** would also undergo decarboxylative deallylation to produce a racemic ester enolate, which would then follow the same pathway as predicted for the diastereomeric mixture of **29** (Figure 4.6). Thus, they applied the double enantioselective decarboxylative allylations to a 1 : 1 mixture of bis-esters and ester carbonates (Scheme 4.12). The reaction proceeded in providing C_2-symmetric products (*S,S*)-**30** and (*S,S*)-**36** predominantly over *meso*-**30** and **36**. The Pd-catalyzed double stereoablative alkylations of ester carbonates **37** and **38** also independently afforded similar results (Scheme 4.12).

For synthesis of (–)-folicanthine (**24b**), the oxidative degradation of bis-allyl groups of (*S,S*)-**30** afforded bis-aldehyde (*S,S*)-**39a** in 95% yield (Scheme 4.13), which was reduced with NaBH$_4$ to afford the diol intermediate (*S,S*)-**39b**. Mitsunobu reaction of (*S,S*)-**39b** with diphenylphosphoryl azide afforded the di-azide (*S,S*)-**39c** in 96% yield, which was then reduced by using triphenylphosphine in water, and subsequent protection with chloromethyl formate provided the bis-Moc-protected compound (*S,S*)-**39d** in 85% yield over two steps. Finally, compound (*S,S*)-**39d** was reduced with Red-Al in toluene at 110 °C to complete the total synthesis of (–)-folicanthine (**24b**). Later, catalytic enantioselective total synthesis of (+)-folicanthine (*ent*-**24a**) was reported via a Pd-catalyzed DYKAT of **29** in the presence of Pd(0)-(*R,R*)-**34**, which afforded (*S*)-**30** in similar efficiency. Thus, total synthesis of (+)-folicanthine (*ent*-**24b**) was also reported following a similar strategy as shown in Scheme 4.13 [42b].

Further, it is interesting to note that C_1-symmetric dimeric tryptamine alkaloids such as *meso*-chimonanthine (**40a**) [43] and *meso*-calycanthine (**40c**) were also isolated from various sources from nature (Figure 4.7) [44]. *meso*-Calycanthine (**40c**) is biosynthetically obtained by acid-catalyzed rearrangement of the bis(pyrroloindoline) alkaloid *meso*-chimonanthine (**40a**).

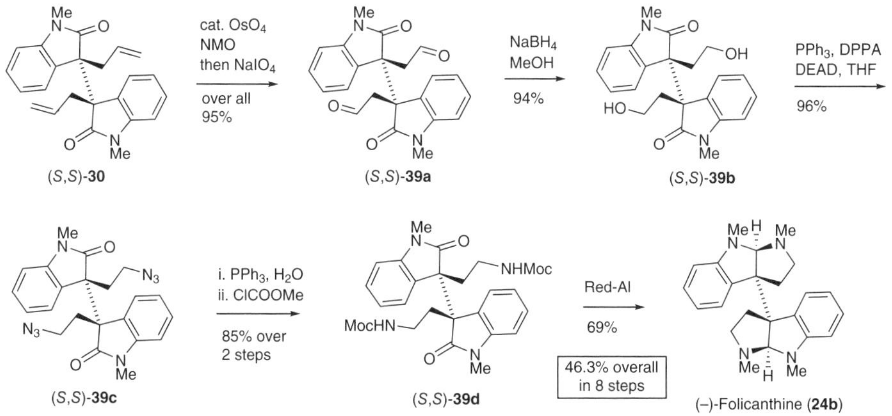

Scheme 4.13 Catalytic enantioselective total synthesis of (–)-folicanthine (**24b**).

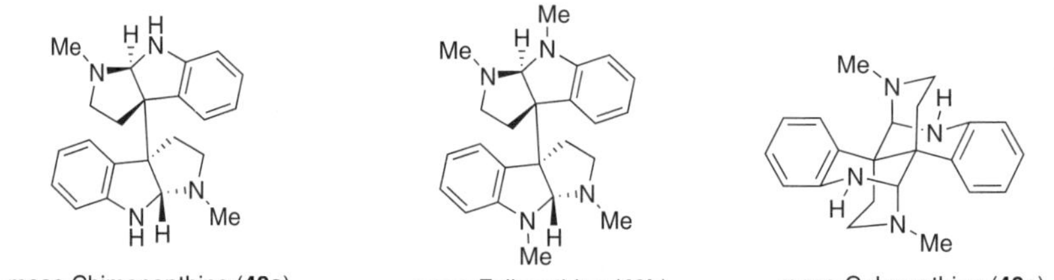

Figure 4.7 Representative of *meso*-dimeric pyrrolidinoindoline alkaloids.

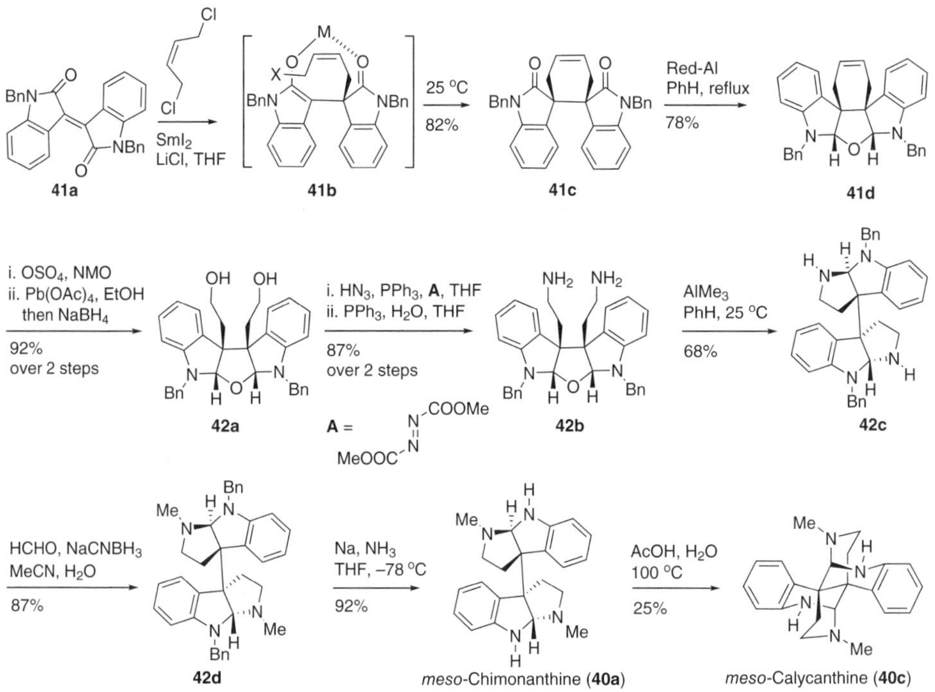

Scheme 4.14 Total synthesis of *meso*-chimonanthine (**40a**) and *meso*-calycanthine (**40c**).

Overman et al. [44a] have reported a concise stereocontrolled route to *meso*-chimonanthine (**40a**) and *meso*-calycanthine (**40c**) relying on only *N*-benzyl protecting group (Scheme 4.14). This synthesis features an unusual samarium-mediated reductive dialkylation to control the relative stereochemistry of the two critical quaternary centers. Isoindigo **41a** was reduced at room temperature with 2 equiv of SmI$_2$ in the presence of 10 equiv of LiCl and then alkylated at this temperature with *cis*-1,4-dichloro-2-butene for 8 h to furnish cyclohexene **41c** in 82% yield. Next, **41c** was treated with Red-Al [sodium bis(2-methoxyethoxy)aluminum hydride] in refluxing benzene to provide hexacycle **41d** in high yield. The unusual structure of this intermediate was verified by X-ray crystallography. The cyclohexene ring then was cleaved to provide diol **42a** through sequential reaction of **41d** with OsO$_4$-*N*-methylmorpholine *N*-oxide (NMO), lead tetraacetate (LTA), and NaBH$_4$. This unstable diol was immediately converted into diazide via Mitsunobu reaction using hydrazoic acid (HN_3), which was further reduced to the corresponding diamine **42b**. The latter was then reacted with excess Me_3Al at room temperature to provide the desired bis(pyrroloindoline) **42c** in 68% yield. Reductive amination of **42c** with formaldehyde in the presence of sodium cyanoborohydride yielded **42d**, which upon treatment with Na/NH_3 delivered *meso*-chimonanthine (**40a**) in 80% yield over two steps. Finally, exposure of **40a** to hot dilute acetic acid provided *meso*-calycanthine (**40c**) as per biosynthetic proposal.

In 2017, Bisai et al. [45] anticipated that achiral hexacyclic unit comprising the skeleton of *meso*-chimonanthine (**40a**), *meso*-folicanthine (**40b**), and *meso*-calycanthine (**40c**) could be constructed from dimeric 2-oxindole of type *meso*-**45** (Scheme 4.15), formed via sequential allylation through *retro*-Claisen activation [46]. These intermediates contain key vicinal quaternary carbon centers required for total synthesis of **40a–c** (Figure 4.7). Toward this goal, dimeric 2-oxindole **43** (*meso*: ± ~ 2.5 : 1) was used as pronucleophile and subjected to allyl alcohol as proelectrophile. Delightfully, when the sequential allylations via *retro*-Claisen activation were carried out on substrates **43a** and **b** with allyl alcohol, it afforded **45a** and **b** in up to 90% yield, favoring the formation of *meso*-bisoxindole **45** with dr up to 2.2 : 1 (Scheme 4.15). Fortunately, ester carbonate **46** and bis-carbonate **47** also enabled the formation of bisallylated dimeric 2-oxindoles **45a** in excellent yields with moderate dr in favor of *meso*-**45a**. It is noteworthy that ester carbonate **46** with different carbonyl functionalities having two different reaction rates also afforded **45a** in similar efficiencies under optimized condition [47].

For total syntheses of *meso*-chimonanthine (**40a**) and *meso*-folicanthine (**40b**) (Figure 4.7), compounds *meso*-**45a–b** were converted to bis-Moc compounds *meso*-**49a–b** in six steps through the formation of diol intermediates *meso*-**48a–b** (Scheme 4.16). Reduction of *meso*-**49a** with Red-Al completed the first total synthesis of *meso*-folicanthine (**40b**). On the other hand, reduction of *meso*-**49b** afforded *meso*-**50**, from where the total synthesis of *meso*-chimonanthine (**40a**) was accomplished by hydrogenolysis in the presence of Pd/C under H_2 (1 atm). Therefore, this synthesis enlightens usage of minimum protecting group (benzyl group). Since the total synthesis of *meso*-calycanthine (**40c**) is known from *meso*-chimonanthine (**40a**) [44a], these efforts culminated in formal total synthesis of **40c** (Scheme 4.16).

Secondary metabolites of fungal origin that feature a common bicyclo[2.2.2]diazaoctane core constitute a large number of architecturally complex indole-based alkaloids, such as versicolamide B (**51**) [48]. It has been postulated that this ring system is

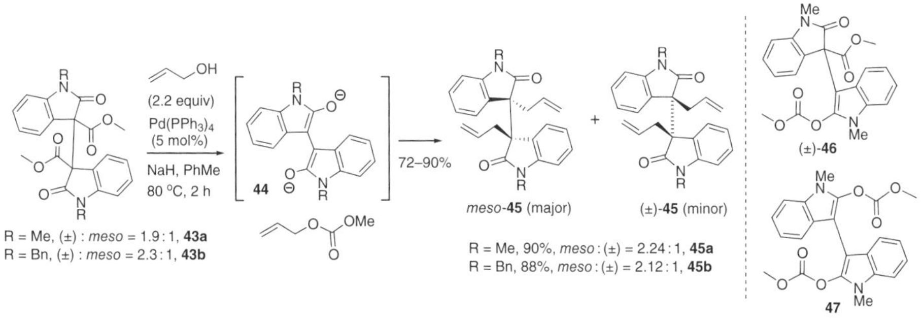

Scheme 4.15 Sequential deacylative allylations (DaA) of bis-esters **43a** and **b**.

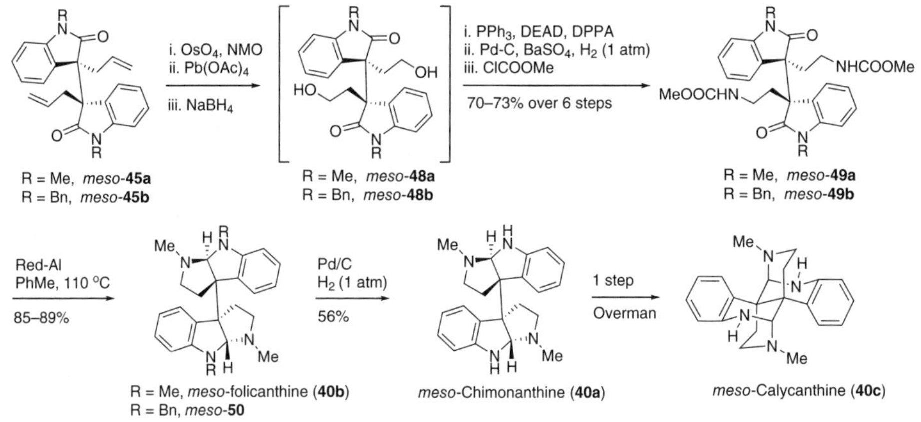

Scheme 4.16 Total synthesis of *meso*-chimonanthine (**40a**) and *meso*-calycanthine (**40c**).

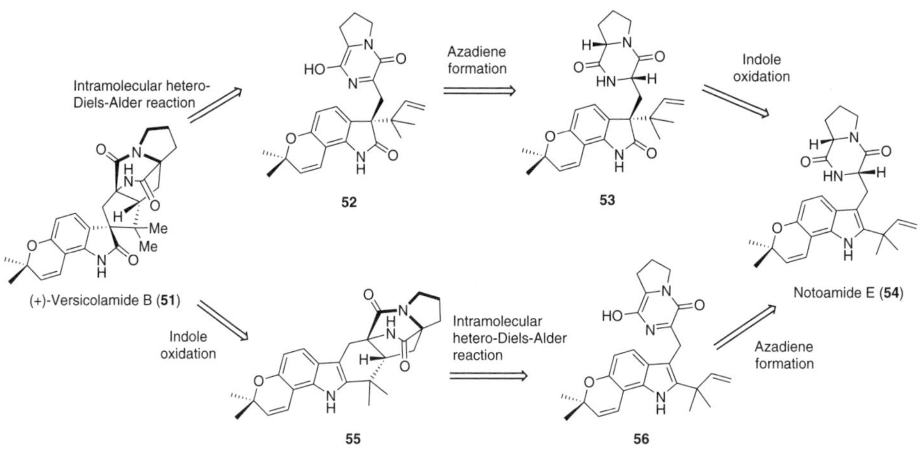

Scheme 4.17 Proposed biosynthetic route to (+)-versicolamide B (**51**) from notoamide E (**54**).

generated through an intramolecular Diels–Alder cycloaddition of the C5 moiety across the α-carbons of the amino acid subunits, as depicted in Scheme 4.17 [49]. Williams' group addressed the asymmetric total synthesis of (+)- and (−)-versicolamide B (**51**) by employing a hetero-Diels–Alder reaction for construction of the bicyclo[2.2.2]diazaoctane core in line with their biosynthetic hypothesis [50].

Based on structural resemblance of related congeners, they identified notoamide E (**54**) as a potential precursor to their target compound via a highly effective hetero-Diels–Alder strategy. They envisioned two plausible sequences of events for conversion of notoamide E (**54**) to (+)-versicolamide B (**51**). If oxidation to the azadiene **56** takes place first followed by hetero-Diels–Alder and indole oxidation, the stereochemical information in notoamide E (**54**) is lost. Alternatively, an initial stereoselective indole oxidation and rearrangement to the oxindole **53** followed by azadiene formation **52** set the stage for an intramolecular hetero-Diels–Alder reaction to give the desired product **51** in a diastereoselective manner (Scheme 4.17).

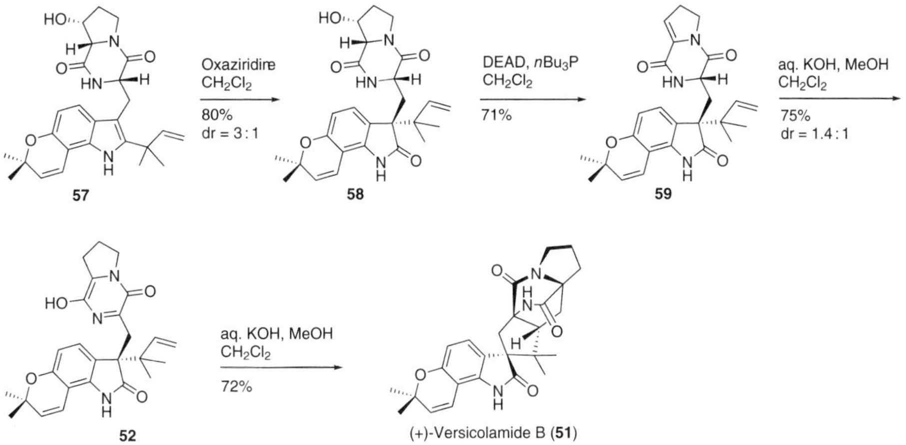

Scheme 4.18 Total synthesis of (+)-versicolamide B (**51**) by Williams et al.

In a synthetic sequence, prenylated indole derivative **57** was subjected to oxidative rearrangement upon treatment with Davis' oxaziridine to give the oxindole **58** (Scheme 4.18) as major product in 80% yield along with a minor C3 diastereomer (dr = 3 : 1) [51]. Dehydration gave rise to the dehydroproline derivative **59** that upon tautomerization and intramolecular hetero-Diels–Alder reaction gave (+)-versicolamide B (**51**) along with a minor diastereomer (dr = 1.4 : 1) in 75% yield. In fact, for the first time, an experimental support for the biosynthetic hypothesis that (+)-versicolamide B (**51**) is likely the result of an intramolecular hetero-Diels–Alder reaction of an oxindole derivative such as **59** through **52** was established (Scheme 4.18).

Psychotrimine (**60**) is a polymeric pyrroloindoline alkaloid that contains the less common N1–C3 union (compared with C3-C3, indole numbering) and is formally derived from tryptamine (Figure 4.8) [52]. Owing to the rarity of this structural motif, novel oxidative synthesis of pyrroloindoline having an aromatic amine was developed by Newhouse and Baran [53].

Treatment of tryptamine derivative **63a** with *N*-iodosuccinimide (NIS) and 2-iodoaniline yielded pyrroloindoline **63c** in good yield (61-67%) via intermediacy of **63b**, generating the key N1–C3 bond in a single step (Scheme 4.19) [53]. One might think that the carbamate functionality may be considered to be a blocking group for the primary amine, but because a carbon from the carbamate is actually incorporated into the natural product, it should be considered a latent methylamine. Next, an iodide-selective Larock annulation of **63c** with alkyne afforded 3-(1-indolyl)pyrroloindoline **63d**, where palladium did not affect the bromoarene [54]. This was followed by a chemoselective Buchwald–Goldberg–Ullmann reaction of **63d** with tryptamine with carbamate functionality **63e** to afford *N*-arylated product **63f**. Finally, Red-Al reduction of **63f** yielded (±)-psychotrimine (**60**) in a short five-step sequence. Over 2.5 g of this natural product has been prepared using this route [55].

Psychotetramine (**71**) was isolated from a rubiaceous plant *Psychotria rostrata* (Scheme 4.20) [56a]. It represents the curious case of a tryptamine tetramer containing both types of connectivity: a C3–C3 linked dimer bonded to a C3–N1 dimer through an unusual C7–N1 linkage. The stereochemical determination and constitutional

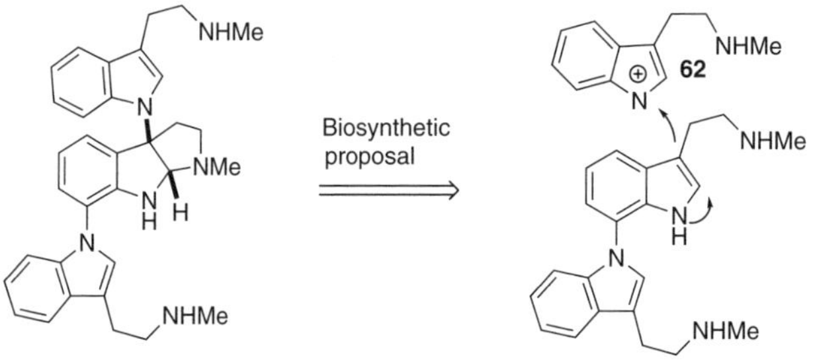

Figure 4.8 Psychotrimine (**60**) and plausible biogenetic pathway.

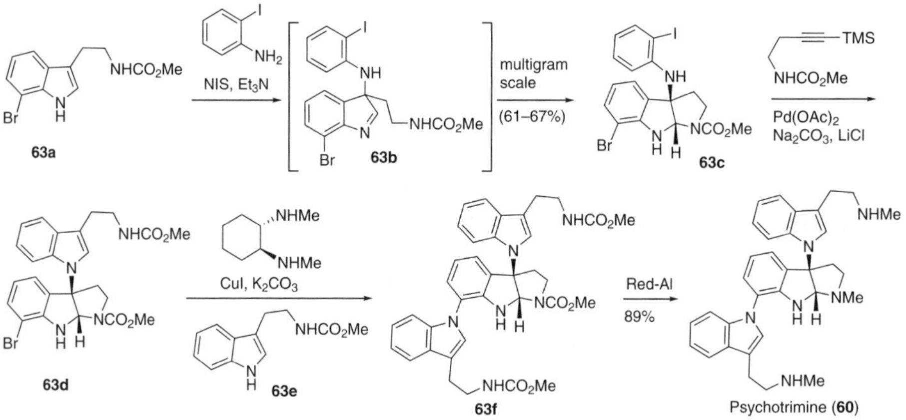

Scheme 4.19 Gram-scale total synthesis of (±)-psychotrimine (**60**).

assignment of **71** was systematically accomplished by Baran [56a]. Since the three pyrrolidine subunits of **71** must be *cis*-fused, there are eight possible stereoisomers (four pairs of enantiomers).

The synthesis commenced with oxidative coupling of *o*-iodoaniline and 7-bromo-D-tryptophan derivative (−)-**64** in the presence of NIS and pyridinium *p*-toluenesulfonate (PPTS) at −45 °C (Scheme 4.20). This process furnished the required pyrroloindoline (+)-**65** in 66% yields, which sets the key N1−C3 bond in a single step. Subsequent Larock annulation [54] of (+)-**65** with known alkyne led to tryptamine–tryptophan dimer (+)-**66** in 46% isolated yield (debromination not observed). Following saponification with aqueous KOH followed by Barton decarboxylation [57], enantiopure (+)-**68** was obtained in 72% yield over two steps. At this stage, amination between two sterically congested coupling partners, in this case between an indoline nitrogen (of chimonanthine (+)-**69**) and an ortho-substituted aryl halide (+)-**68**, in the presence of a total of four free N−H's is a challenging task (Scheme 4.20). This was accomplished with 1 equiv of (+)-**69** with 2.6 equiv of (+)-**68** using 5 mol% Pd$_2$(dba)$_3$ (with respect to (+)-**68**) and 20 mol% RuPhos (with respect to (+)-**68**) and in the presence of *t*-BuONa via an unusually complex Buchwald–Hartwig amination to provide (+)-**70** in 41% yield. Finally, reduction of this enantiopure compound was carried out with allane (AlH$_3$) to accomplish the total synthesis of (+)-psychotetramine (**71**).

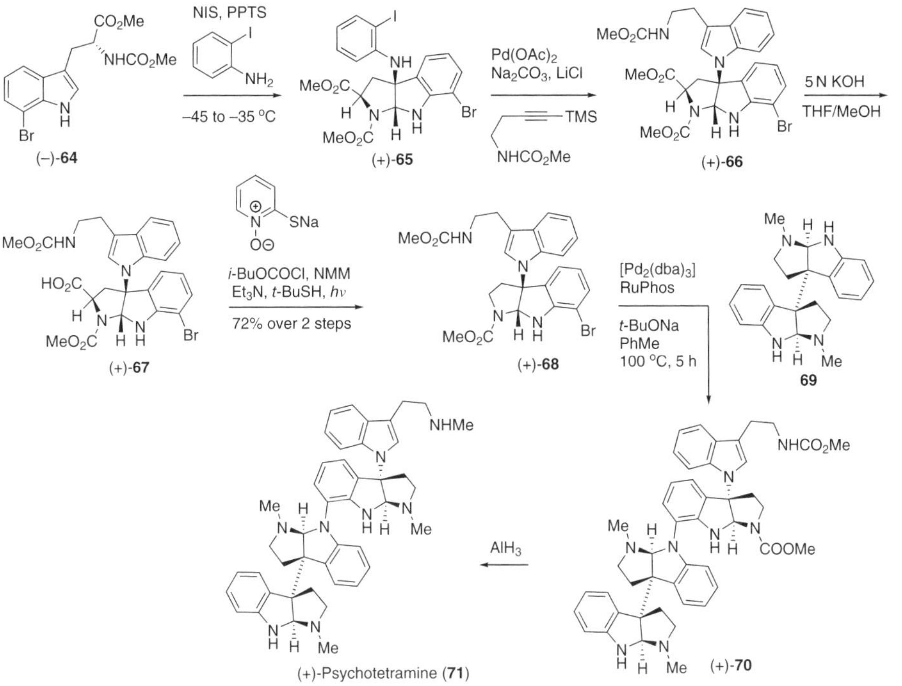

Scheme 4.20 Synthesis of enantiopure pyrroloindoline **68** and (+)-psychotetramine (**71**).

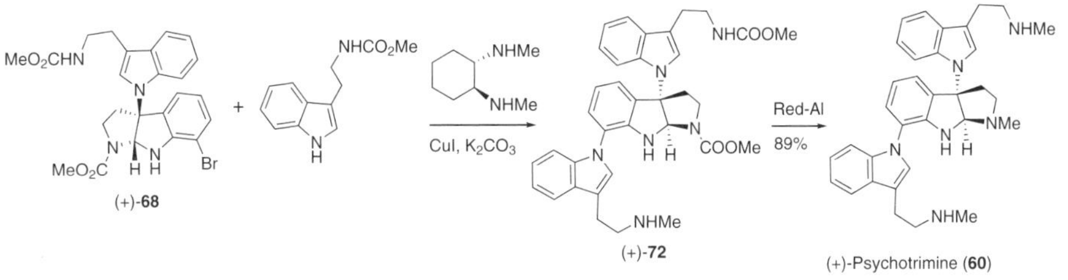

Scheme 4.21 Asymmetric total synthesis of (+)-psychotrimine (**60**).

With a scalable route to enantiopure (+)-**68** in hand, an enantioselective total synthesis of psychotrimine (**60**) was also pursued (Scheme 4.21). Thus, copper-mediated amination of (+)-**68** with the known tryptamine derivative, followed by reduction of the methyl carbamate groups using Red-Al, furnished psychotrimine (+)-**60** rapidly, $[\alpha]_D^{20} = +193$ ($c = 1.0$, CHCl$_3$) [natural $[\alpha]_D^{18} = +179$ ($c = 0.2$, CHCl$_3$)] [52]. This represents the shortest and highest-yielding route to enantiopure psychotrimine (**60**, 9 steps, 7% overall from 7-bromoindole).

4.3 Syntheses of Naturally Occurring Terpenoids

Naturally occurring terpenes and their derivatives have profoundly impacted the human experience. As flavors, fragrances, poisons, and medicines, nearly every human on Earth has experienced their effects. In this context, linear triquinane natural products

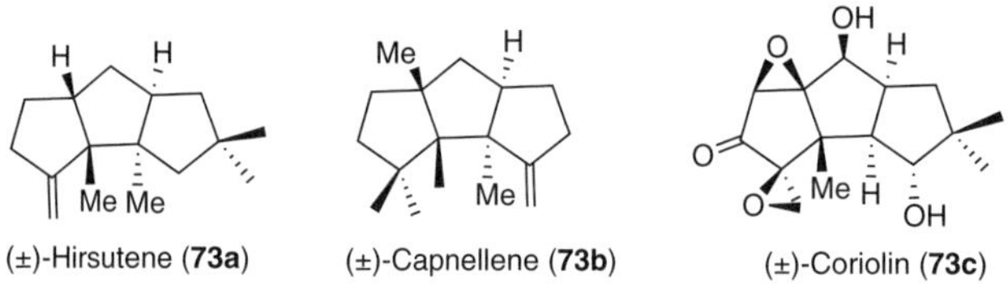

Figure 4.9 Naturally occurring triquinane natural products **73a–c**.

are a subset of polyquinanes that constitute an important class of sesquiterpenoids [58]. Linear triquinane natural products such as **73a–c** (Figure 4.9) are isolated from plants, microbes, and marine organisms, and they have been attracting continuous attention from synthetic chemists due to their promising biological activity and novel architecture [58]. The isolation and characterization of this linearly fused triquinanes attracted early attention as a consequence of their reputed antibacterial and antitumor properties [59, 60].

One of the classical total syntheses of hirsutene (**73a**) was carried out under PGF strategy by Mehta et al. in 1986 [61], although the concept of PGF synthesis was not extensively explored until the twenty-first century. In their synthetic sequence, Diels–Alder cycloaddition between 1,3-cyclopentadiene and 2,4-dimethyl-*p*-benzoquinone furnished **74a** in 90% yield (Scheme 4.22). This was irradiated from a 450 W UV lamp or sunlight through Pyrex filter, which resulted in a smooth intramolecular cycloaddition to the crystalline pentacyclic dione **74b** in 85% yield. The key step involving the uncaging of the caged dione was effected by employing the flash vacuum pyrolysis (FVP) technique. Sublimation of **74b** through a quartz column filled with quartz chip at 500 °C (0.1 torr) led to the triquinane bis-enone **74c** via a regioselective cyclobutane fragmentation. When a solution of *cis-,syn-,cis*-**74c** was refluxed in benzyl benzoate, the required *cis-,anti-,cis*-bisenone **74d** was obtained in 37% yield. Catalytic hydrogenation

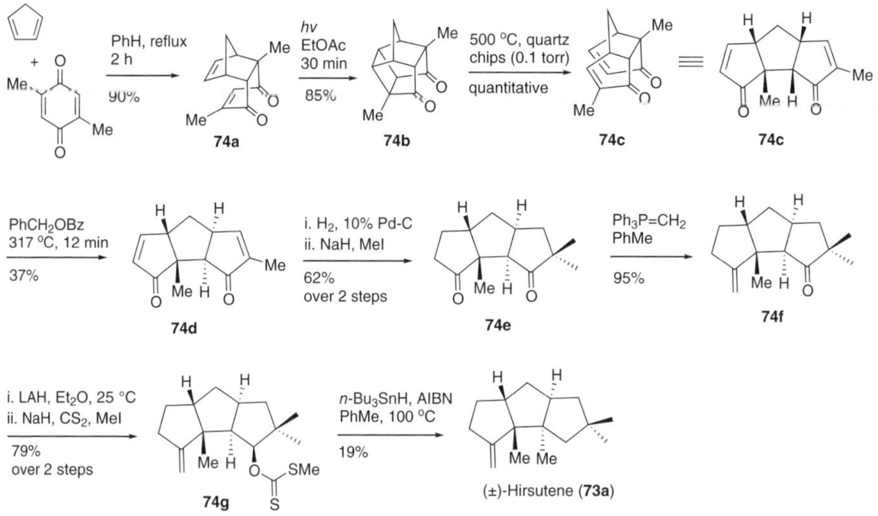

Scheme 4.22 Total synthesis of (±)-hirsutene (**73a**) by Mehta et al.

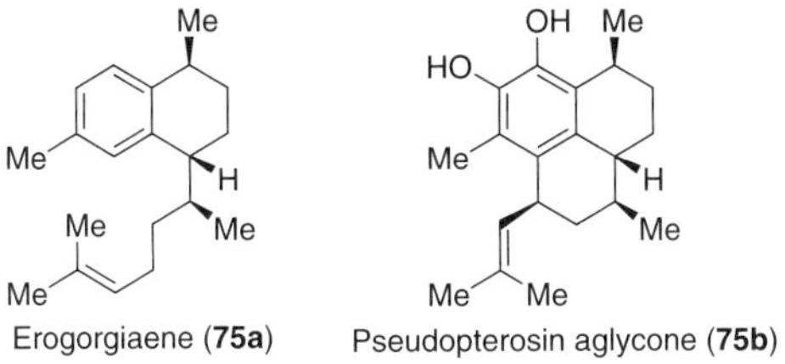

Erogorgiaene (**75a**)   Pseudopterosin aglycone (**75b**)

Figure 4.10 Erogorgiaene (**75a**) and pseudopterosin aglycon (**75b**).

of **74d** over Pd/C furnished the tetrahydro derivative, which was methylated in the presence of NaH to form crystalline dione **74e** in 62% yield over two steps. Wittig olefination of **74e** proceeded selectively at the less hindered keto-carbonyl group to afford **74f** in 84% yield. The latter was reduced with LiAlH$_4$ followed by reaction with CS$_2$/MeI to form *S*-methyl dithiocarbonate derivative **74g** in 79% over two steps, which was deoxygenated under Barton deoxygenation [62] to complete the total synthesis of hirsutene (**73a**) [61].

(+)-Erogorgiaene (**75a**) is a member of the marine diterpenes isolated from the West Indian sea whip *Pseudopterogorgia elisabethae* and displays promising activity against *Mycobacterium tuberculosis* H37Rv (Figure 4.10) [63a]. Several important biologically active secondary metabolites have been isolated from the *Pseudopterogorgia* corals as well [63b–d]. Consequently, there has been extensive activity directed toward the synthesis of these natural products [64].

In 2004, the first enantioselective PGF synthesis of erogorgiaene (**75a**) was reported by Hoveyda and coworkers [65] by utilizing a key asymmetric conjugate addition of alkylzinc onto acyclic α,β-unsaturated ketones catalyzed by Cu(I)-peptidic phosphine ligands (Scheme 4.23). The synthesis commenced with Heck coupling of methyl vinyl ketone (MVK) with commercially available 2-bromo-3-methyl iodobenzene to afford α,β-unsaturated ketone **76a** in 88% yield, which was treated with Me$_2$Zn in the presence of Cu(I) complex of phosphine ligand **A** to deliver conjugate addition product **76b** in 94% yield with >98% ee. Next, a Pd-catalyzed Stille coupling with 5 mol% Pd(PPh$_3$)$_4$ and 1.2 equiv of Bu$_3$SnCCSiMe$_3$ at 100 °C afforded alkyne **76c** in 96% yield. The latter under kinetic enolization with TMSCl in the presence of LiTMP furnished desired silyl enol ether **76d** with 9 : 1 regioselectivity and 97% overall yield. The latter was converted to enyne **76e** in 60% overall yield in three steps, viz. triflation using MeLi and *N*-phenyl triflamate followed by reduction using *n*-Bu$_3$SnH in the presence of 5 mol% Pd(PPh$_3$)$_4$ and then by treatment of K$_2$CO$_3$ in methanol to remove TMS group (Scheme 4.23). Further, treatment of enyne **76e** with 5 mol% Grubb's catalyst **B** generated diene **76f** via enyne metathesis in 84% yield, which was further treated with 2 equiv of MVK in the presence of 10 mol% **B**, resulting in the formation of unsaturated ketone **76g** in 74% isolated yield in >95 : 5 *E/Z* ratio. Diastereoselective conjugate addition of Me$_2$Zn onto enone **76g** was carried out with Me$_2$Zn in the presence of 5 mol% (CuOTf)$_2$·C$_6$H$_6$ and 12 mol% chiral phosphine **C** in toluene at 4 °C to afford desired ketone **76h** in 50% yield with 97 : 3 diastereoselectivity and 9 : 1 regioselectivity (1,4-addition/1,6-addition). Later, **76h** was reduced with NaBH$_4$, and treatment of the resulting secondary alcohol to dissolving metal conditions (Li/NH$_3$) and oxidation of secondary alcohol with Dess–Martin periodinane (DMP) afforded **76i** in 73% overall yield and with 85 : 15 diastereoselectivity. The major diastereomer of **76i** was converted into primary alcohol **76j** in

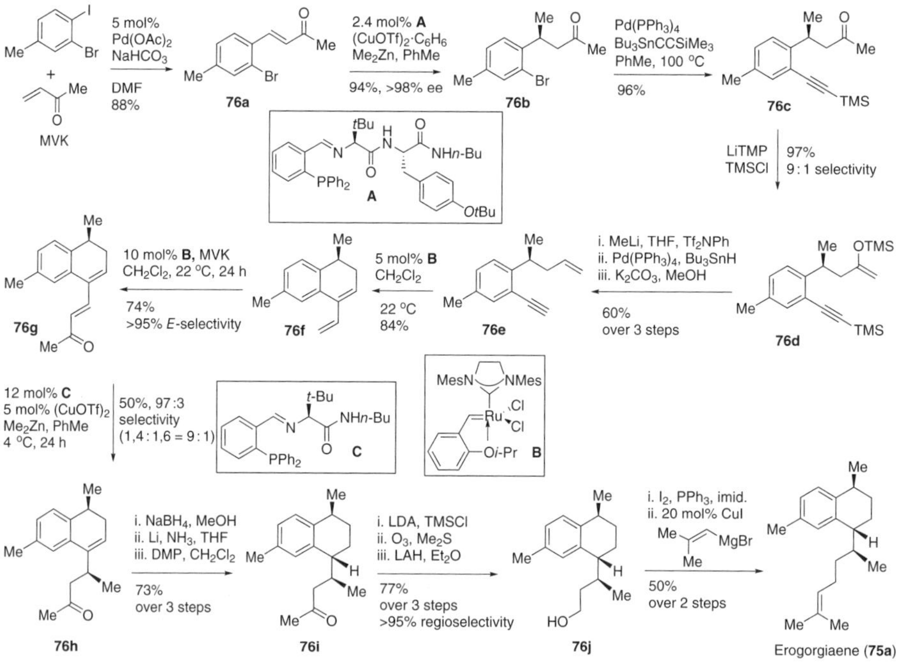

Scheme 4.23 Catalytic enantioselective synthesis of (+)-erogorgiaene (**75a**) by Hoveyda.

77% overall yield in three steps: kinetic enolization using TMSCl (>95% selectivity) and ozonolysis, followed by reduction using lithium aluminum hydride. Finally, transformation of the primary alcohol **76j** to the terminal iodide followed by Cu-catalyzed alkylation with 2-propenylmagnesium bromide completed the total synthesis of erogorgiaene (**75a**) (Scheme 4.23).

In 2005, a much improved PGF synthesis of (+)-erogorgiaene (**75a**) was reported by Davies and Walji (Scheme 4.24) [66]. They have utilized C–H activation/Cope rearrangement catalyzed by dirhodium tetraprolinate [Rh$_2$(dosp)$_4$] (dosp = (*N*-dodecylbenzenesulfonyl)prolinate) [67] for total synthesis of (+)-erogorgiaene (**75a**). In their synthetic design, it was revealed that only (*S*)-**77** would be capable of a matched combined C–H activation/Cope rearrangement to form **79** via intermediate **78**, whereas (*R*)-**77** would be matched for a cyclopropanation to form **80a** (Scheme 4.24). It was observed that reaction catalyzed by [Rh$_2$(*R*-dosp)$_4$] is truly exceptional and results in a 1 : 1 mixture of the combined C–H activation product **79** with 90% ee and the cyclopropane **80a** with just a trace of the diastereomer **80b** (Scheme 4.24). Owing to the tendency of **79** to undergo a retro-Cope rearrangement, the combined mixture of **79** and **80** was globally hydrogenated, and the ester was reduced to the alcohol **81**, which was isolated in 31% overall yield from the dihydronaphthalene (±)-**77**. Oxidation of **81** to the aldehyde with pyridinium chlorochromate (PCC) followed by a Wittig reaction completed the total synthesis of (+)-erogorgiaene (**75a**) without taking help of any protecting group.

Abietanes are a family of naturally occurring diterpenoids that have been isolated from a variety of terrestrial plant sources. These compounds exhibit a wide variety of

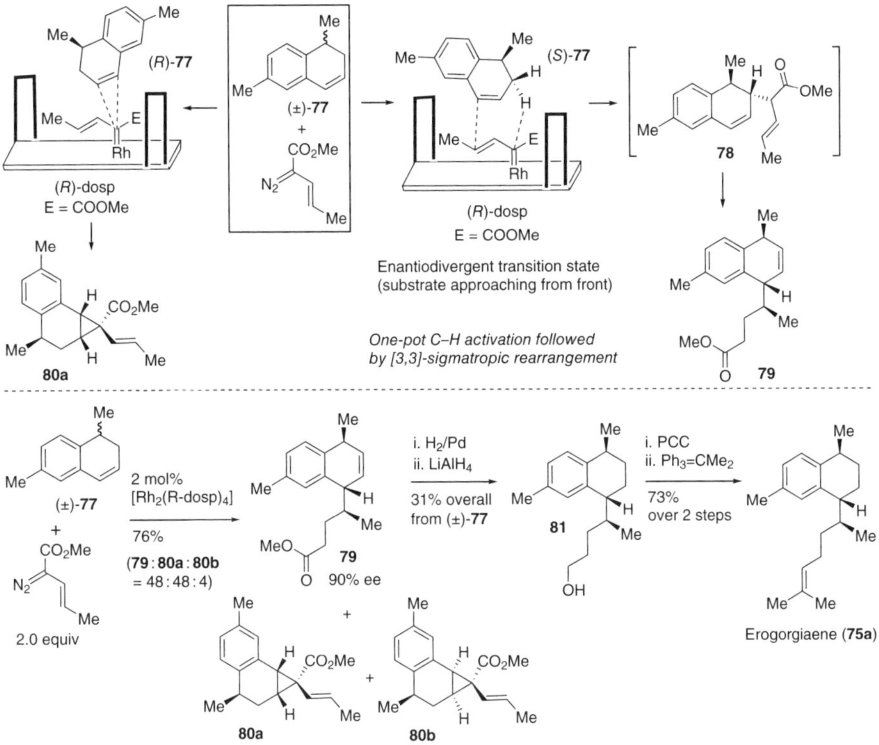

Scheme 4.24 Total synthesis of erogorgiaene (**75a**) by Davies.

interesting biological activities, which have generated significant interest from the medicinal and pharmacological communities. In fact, the aromatic abietanes have displayed a wide spectrum of interesting biological properties including antimicrobial, antileishmanial, antiplasmodial, antifungal, antitumor, cytotoxicity, antiviral, antiulcer, cardiovascular, antioxidant, and anti-inflammatory activity [68]. For example, (±)-nimbiol (**82a**) possesses antimicrobial activity [69a] (Figure 4.11). (±)-Nimbidiol (**82b**), which had been isolated from the root bark of *Azadirachta indica A. Juss* [69b] (Indian neem), is used in the indigenous system of medicine in India. Ferruginol (**82c**) is known to have antihepatomic, antitumor, antibacterial, and fungicidal [69c] properties. In addition, there are a number of highly oxygenated abietane diterpenoids such as royleanone (**82d**) and inuroyleanone (**82e**) that have been isolated from various species [69d].

Ramana and Bhar [70] disclosed PGF total synthesis of several naturally occurring diterpenoids, such as (±)-nimbiol (**82a**), (±)-nimbidiol (**82b**), and (±)-ferruginol (**82c**), via a concerted domino acylation–cycloalkylation/alkylation–cycloacylation as the principal step to construct the basic carbocyclic framework required for the *trans*-fused octahydrophenanthrene nucleus, starting from the readily available acyclic monoterpene citral (Scheme 4.25). Initially citral was cyclized to 2,6,6-trimethyl-1-cyclohexene-1-carboxaldehyde (β-cyclocitral) **83a** [71] in the presence of aniline and 95% H_2SO_4. β-Cyclocitral (**83a**) was converted into (2,6,6-trimethylcyclohex-1-enyl)acetic acid **83b**

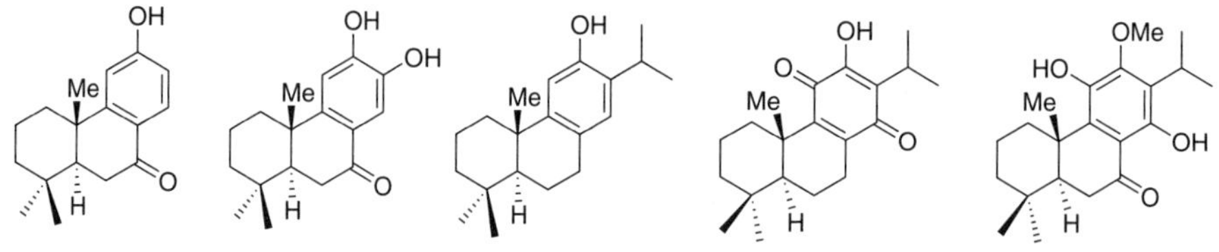

Nimbiol (**82a**) Nimbidiol (**82b**) Ferruginol (**82c**) Royleanone (**82d**) Inuroyleanol (**82e**)

Figure 4.11 Selected diterpenoids (**82a–e**) of biological relevance.

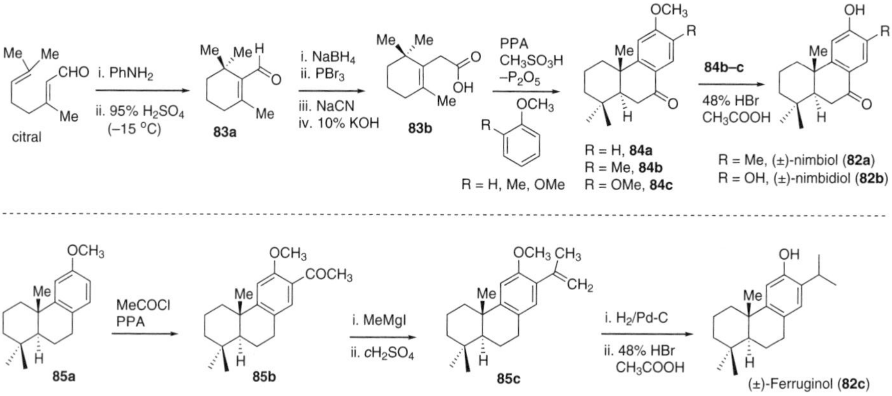

Scheme 4.25 Total syntheses of (±)-nimbiol (**82a**), (±)-nimbidiol (**82b**), and (±)-ferruginol (**82c**).

following a series of reactions, viz. reduction with $NaBH_4$ and reaction with PBr_3 followed by NaCN, and hydrolyzed with dilute KOH. Later, acid **83b** was subjected to $CH_3SO_3H–P_2O_5$ (10 : 1) to promote domino acylation–cycloalkylation with *o*-methylanisole to yield the tricyclic ketone **84b**, which was subsequently transformed into the diterpene (±)-nimbiol (**82a**) via a demethylation reaction in the presence of 48% HBr in AcOH. Under similar condition, acid **83b** was reacted with veratrole to yield the tricyclic ketone **84c**, which was reacted with 48% HBr in AcOH to afford (±)-nimbidiol (**82b**). Along similar line, acid **83b** was reacted with anisole to yield the tricyclic ketone **84a**, which was subsequently reduced under Wolff–Kishner condition to furnish tricycle **85a**. The latter was converted to (±)-ferruginol (**82c**) following simple synthetic elaborations, viz. Friedel–Crafts acylation, to give **85b**, followed by reaction with MeMgI and subsequent elimination of hydroxyl group to furnish **85c** and finally reduction and treatment with 48% HBr in AcOH.

Taiwaniaquinoids (**86a–h**) [72] are a family of unusual diterpenoids possessing a [6,5,6]-*abeo*-abietane skeleton (see ABC ring systems; Figure 4.12) sharing an all-carbon quaternary stereocenter at the pseudobenzylic position. Most of these diterpenoids have been isolated since 1995 from *Taiwania cryptomerioides* Hayata (Taxodiaceae) of the central mountains of Taiwan independently by Cheng [73] and Kuo [74]; from *Salvia dichroantha* Stapf (Lamiaceae), a Turkish flowering sage by Kawazoe [75a]; and from *Thuja standishii* (Cupressaceae), a Japanese conifer by Tanaka [75b]. It is also interesting to note that there could be oxidized A-ring present in taiwanaquinoids, such as taiwaniaquinol F (**86h**). Reportedly, a few members of taiwaniaquinoids are found to

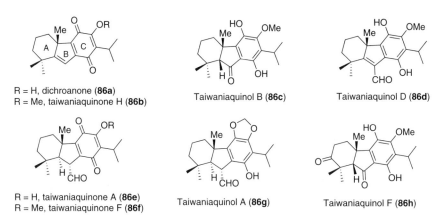

R = H, dichroanone (**86a**)
R = Me, taiwaniaquinone H (**86b**)

Taiwaniaquinol B (**86c**)

Taiwaniaquinol D (**86d**)

R = H, taiwaniaquinone A (**86e**)
R = Me, taiwaniaquinone F (**86f**)

Taiwaniaquinol A (**86g**)

Taiwaniaquinol F (**86h**)

Figure 4.12 Selected taiwaniaquinoids sharing [6,5,6]-structural scaffolds.

exhibit potent cytotoxic activity against KB epidermoid carcinoma cancer cells [74b] and aromatase inhibitory activity [76]. Because of their diverse biological profiles and uncommon structural features, taiwaniaquinoids have gained extensive attention from the synthetic community, leading to numerous efficient approaches. The first catalytic enantioselective total synthesis of (+)-dichroanone (**86a**) was reported by McFadden and Stoltz [77]. This synthesis takes advantage of a Pd(0)-catalyzed asymmetric allylation methodology to generate all-carbon quaternary centers adjacent to carbonyls. Treatment of compound **87b** with a palladium catalyst (2.5 mol% Pd_2dba_3) and chiral ligand (6.25 mol% of (*S*)-*t*Bu-PHOX) installed the quaternary center in **87c** with 91% ee (Scheme 4.26). Wacker oxidation [78a] of keto-olefin **87c** followed by condensation using KOH in xylenes with azeotropic water removal provided bicyclic enone **87e** in excellent yield [78b]. Michael addition of the lithium enolate of **87e** to MVK formed the keto-enone **87f** with high diastereoselectivity. The latter was reacted under Robinson annulation strategy to form the final ring of dichroanone that was trapped with *N*-phenyltriflimide to give enol triflate **87g**. This was immediately subjected to Kumada coupling with isopropenylmagnesium bromide, giving mixture of isomeric products, which converted irreversibly to compound **87h** upon exposure to acid. Later, a formylation of **87h**, a nonoxygenated skeleton, afforded aldehyde **87i** in 79% yield, which under Baeyer–Villiger sequence installed the first oxygen in **87j** in a straightforward manner (74% yield). Finally, a one-pot oxidative reaction sequence was followed to complete the synthesis of (+)-dichroanone (**86a**) under PGF manner.

In 2006, Trauner et al. reported the total synthesis of (±)-taiwaniaquinol B (**86c**), (±)-taiwaniaquinone H (**86b**), and (±)-dichroanone (**86a**) via a key Nazarov cyclization [79]. Lithiation of bromoarene **88a**, followed by addition of the commercially available β-cyclocitral **83a**, afforded arylvinylcarbinol **88b**, which was oxidized immediately to yield aryl vinyl ketone **88c** (Scheme 4.27). It is reported that **88c** could be cyclized in the presence of trimethylsilyl triflate (TMSOTf) in nitromethane to afford the highly unstable silyl enol ether **88e**, presumably through the intermediacy of cation **88d**. Upon aqueous work-up, this procedure afforded the thermodynamically more favorable *cis*-indane product **89** as the only stereoisomer observed. From this, (±)-taiwaniaquinol B (**86c**) was synthesized in three steps, viz. selective demethylation by BCl_3, oxidation in

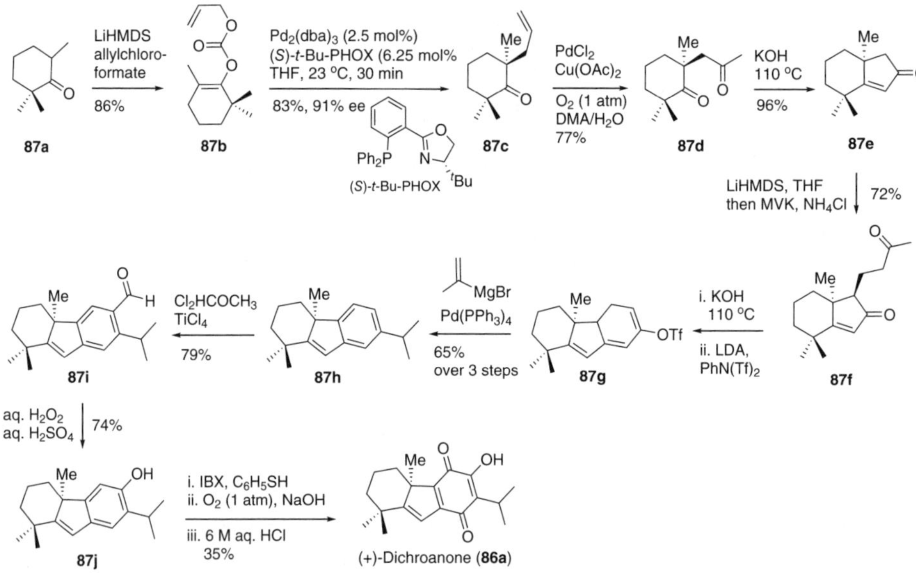

Scheme 4.26 Total synthesis of (+)-dichroanone (**86a**) by Stoltz.

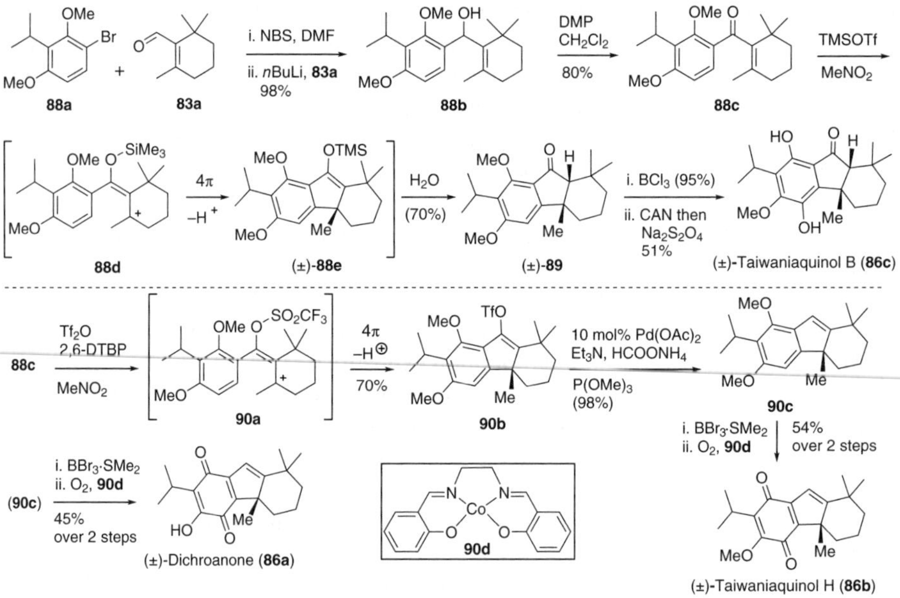

Scheme 4.27 Collective total syntheses of taiwaniaquinoids (**86a–c**) by Trauner.

the presence of ceric(IV) ammonium nitrate (CAN), and reduction of *p*-benzoquinone with $Na_2S_2O_4$ to taiwaniaquinol B (**86c**). More interestingly, heating of aryl vinyl ketone **88c** with triflic anhydride in the presence of a hindered base (2,6-di-*tert*-butylpyridine (2,6-DTBP)) cleanly afforded trifloxy indene **90b** through the intermediacy of cation **90a**.

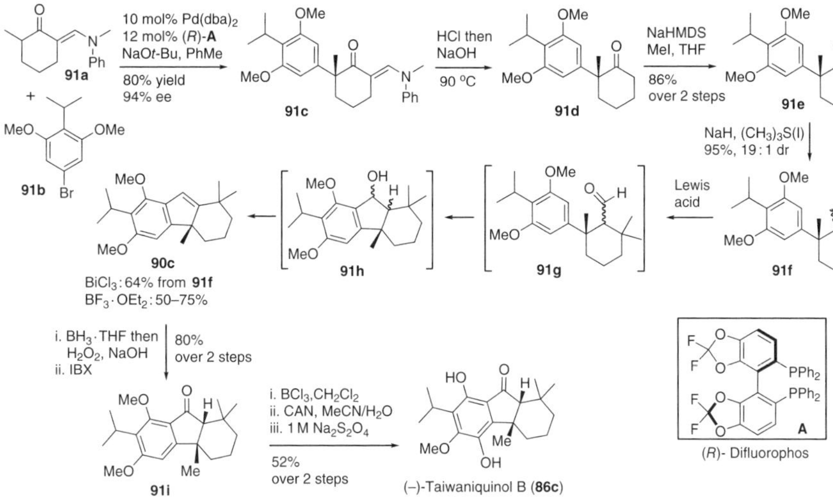

Scheme 4.28 Total syntheses of (−)-taiwaniaquinol B (**86c**) by Hartwig.

This reaction is proposed to proceed through 4π electrocyclization of intermediate cation **90a** followed by deprotonation to yield **90b**. Later, Pd(0)-catalyzed reduction of **90b** gave indene **90c** in excellent yield. Global demethylation, followed by oxidation catalyzed by salcomine **90d**, furnished (±)-dichroanone (**86a**). Alternatively, a more selective demethylation and oxidation led to (±)-taiwaniaquinol H (**86b**).

The first enantioselective total synthesis of (−)-taiwaniaquinol B (**86c**) was reported by Hartwig et al. [80]. The reaction of aryl bromide **91b** with the 2-methylcyclohexanone derivative **91a** [81] in the presence of Pd(dba)$_2$ and (R)-difluorophos **A** (10 mol%) formed **91c** in 80% yield with 94% ee (Scheme 4.28). Acid-promoted hydrolysis of ketone **91c** followed by deformylation in the presence of aqueous NaOH provided ketone **91d** (91% yield over two steps). Dimethylation of **91d** with MeI using NaHMDS afforded tetrasubstituted cyclohexanone **91e** in 86% yield. Corey–Chaykovsky epoxidation of **91e** using more reactive sulfur ylide (dimethylsulfonium methylide) [82] furnished epoxide **91f** in 95% yield with excellent diastereoselectivity. Later, Lewis acid (BF$_3$·OEt$_2$ or 20 mol% BiCl$_3$)-catalyzed rearrangement of the epoxide **91f** to the aldehyde **91g** followed by Friedel–Crafts alkylation with electron-rich arene part of **91g** afforded corresponding alcohol **91h**, which underwent elimination to form the target tetrahydrofluorene **90c** in good yield. Further, hydroboration of **90c** followed by oxidation in the presence of basic hydrogen peroxide and further oxidation with 2-iodoxybenzoic acid (IBX) gave **91i** as a single diastereomer in 80% yield over two steps. Finally, selective monodemethylation of **91i** with BCl$_3$, oxidation with CAN, and reductive work-up with sodium dithionite gave (−)-taiwaniaquinol B (**86c**) in 52% yield over two steps.

In 2010, Gademann reported a biogenetic hypothesis for the transformation of an abietane-type diterpene into the 6–5–6 skeleton of the taiwaniaquinoids by a ring contraction of an oxidized precursor (Scheme 4.29) [83]. This strategy has been successfully utilized for the total synthesis of (+)-taiwaniaquinone H (**86b**) under PGF conditions. The starting material **92e** was obtained from commercially available methyl

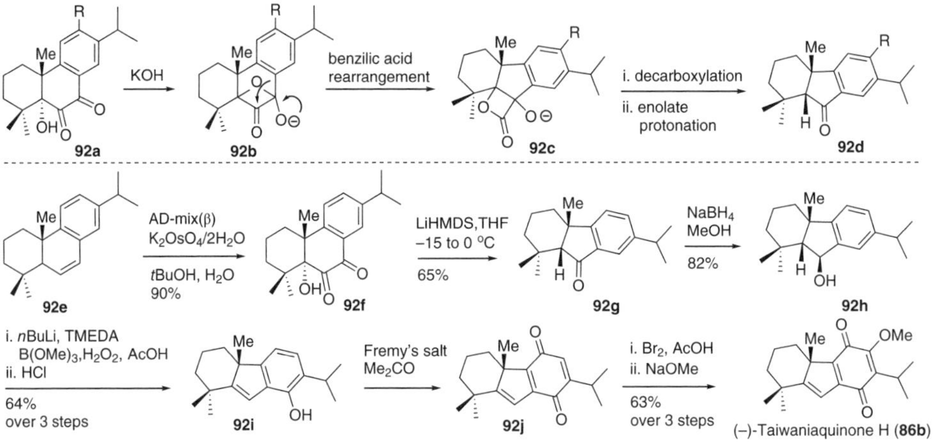

Scheme 4.29 Hypothesis of conversion of abietane (**92a**) to *abeo*-abietane (**92d**) and synthesis of (+)-taiwaniaquinone H (**86b**) by Gademann.

dehydroabietate in five steps according to literature procedures [84]. The hydroxydiketone functionality of the key intermediate **92f** was installed through a Sharpless asymmetric dihydroxylation reaction [85a] (Scheme 4.29). Treatment of the hydroxydione **92f** with LiHMDS gave the hydrofluorene derivative **92g** in 65% at −15 °C, which was reduced with NaBH$_4$ to afford **92h**. Later, a one-pot sequence of hydroxy-group-directed *ortho* lithiation of the benzylic alcohol **92h** [85b] and subsequent borylation of the aryllithium compound and oxidation of the corresponding aryl boronate gave the hydroxylated phenol derivative, which was dehydrated under acidic conditions to furnish phenol **92i** in 64% yield over three steps. The latter was then cleanly oxidized using Fremy's salt to the corresponding *p*-quinone **92j**. Electrophilic bromination of the quinone **92j** followed by substitution with a methoxy group provided (+)-taiwaniaquinone H (**86b**) in 63% yield over three steps.

Later, in 2013, Gademan et al. have reported PGF syntheses of (−)-taiwaniaquinone F (**86f**) and (+)-taiwaniaquinol A (**86g**) via a Wolff ring contraction of the diazo derivative of sugiol methyl ether **93b** [86a]. The latter was synthesized from commercially available abietic acid **93a** in 36% overall yield in nine steps (Scheme 4.30). In the synthetic sequence, sugiol methyl ether **93b** was converted to α-diazoketone **93c** (by a reaction with *p*-acetamidobenzenesulfonic acid (*p*-ABSA)), which was reacted with MeOH in the presence of light to afford methyl ester **93d**. This was epimerized and reduced with lithium aluminum hydride to furnish primary alcohol **93g** via methyl ester **93f** in almost quantitative yield. Carbotricycle containing primary alcohol **93g** was then reacted with Br$_2$ to afford bromoarene, which was reacted with *n*BuLi in the presence of TMEDA, and subsequent reaction with molecular oxygen afforded **93h** in 47% yield over two steps. Later, the synthesis of (−)-taiwaniaquinone F (**86f**) was accomplished from **93h** via two steps, viz. oxidation of aromatic ring using molecular oxygen in the presence of catalytic Co(Salen) complex followed by oxidation of primary alcohol with DMP. Finally, an unanticipated formation of a methylene catechol moiety was envisioned by photolysis of (−)-taiwaniaquinone F (**86f**) to (+)-taiwaniaquinol A (**86g**) via a remote C−H functionalization (Scheme 4.30). This study clearly demonstrates the hypothesis for the

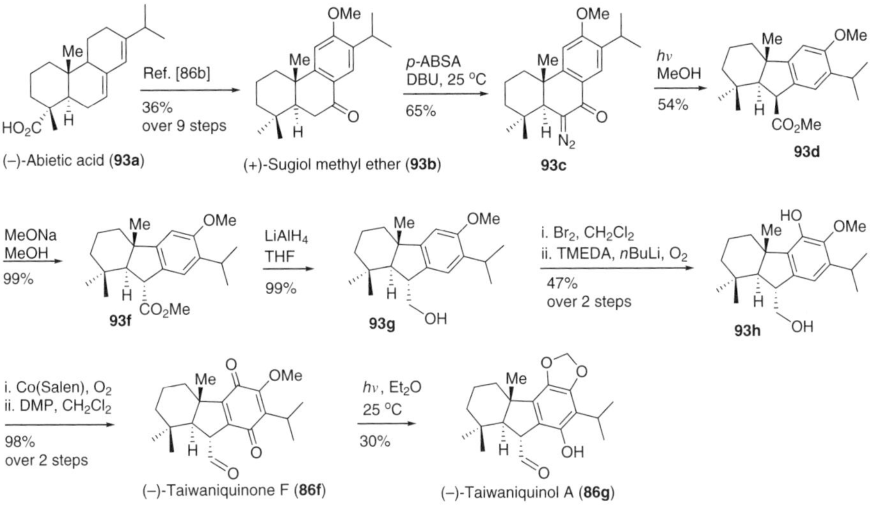

Scheme 4.30 Total syntheses of (−)-taiwaniaquinone F (**86f**) and (−)-taiwaniaquinol A (**86g**) by Gademann.

biogenesis of the frequently encountered methylenedioxy aromatic motifs in natural products.

In the same year, independently, Li et al. have developed an efficient route for the divergent total synthesis of (±)-taiwaniaquinones A (**86e**) and F (**86f**) and (±)-taiwani-aquinols B (**86c**) and D (**86d**) via a Wolff-type ring contraction reaction of a *trans*-fused [6,5,6]-ring system [87]. Treatment of **94a** with methylenetriphenylphosphorane fur-nished styrene, which was hydroborated with 9-BBN to afford the corresponding alky-lborane (Scheme 4.31) [88b, c]. This *in situ* generated borane was subjected to Suzuki–Miyaura coupling [88c] in the presence of alkenyl iodide **94b** to give homogera-nyl arene **94c** in 83% overall yield from **94a**. Treatment of **94c** with 10 mol% Bi(OTf)$_3$ in MeNO$_2$ at 80 °C provided tricyclic compound in 71% yield as a single diastereomer with *trans*-A/B ring junction, which was oxidized at benzylic position with freshly prepared CrO$_3$/3,5-dimethylpyrazole at −10 °C to provide **94d** in 89% yield. Ketone **94d** was then converted to α-diazoketone **94e** in 78% yield by treatment of 2,4,6-triisopropyl benze-nesulfonylazide (TrisN$_3$) and Bu$_4$NOH. Wolff rearrangement of **94e** under thermal conditions (BnOH, 2,4,6-collidine, 160 °C) [89] afforded the corresponding benzyl ester **94f** as a single detectable diastereomer in 56% yield. Reduction of **94f** with LiAlH$_4$ fol-lowed by oxidation with DMP furnished aldehyde **94g** (75% over two steps). The latter was treated with K$_2$CO$_3$/EtOH at 80 °C to epimerize to **94h** in 93% yield, which was reacted with excess BBr$_3$ (global demethylation) followed by spontaneous aerobic oxi-dation furnishing (±)-taiwaniaquinone A (**86e**) in 76% yield. Intermediate **94h** was oxidized with CAN to afford (±)-taiwaniaquinone F (**86f**) in 76% yield. In another direc-tion, treatment of **94g** with TMSOTf and Et$_3$N gave silyl enol ether **94i** (ca. 1.2 : 1 *cis/trans* isomers) in 92% yield, which underwent a sequence of Saegusa–Ito oxidation [90], monodemethylation, and one-pot oxidation/reduction to afford (±)-taiwaniaquinol D (**86d**) in 80% yield over three steps. Further, dihydroxylation of **94i** using K$_2$OsO$_2$(OH)$_4$ and NMO afforded ketone **94j**, which underwent the same sequence used for

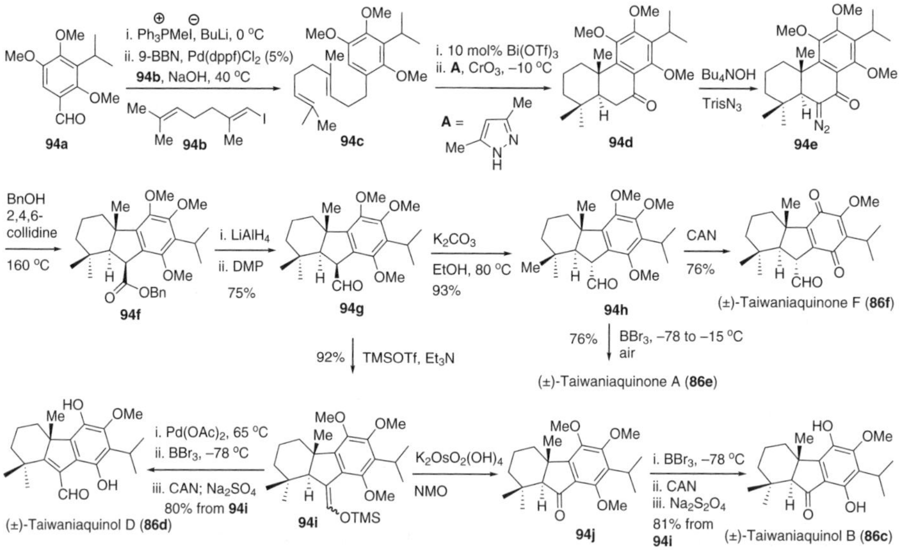

Scheme 4.31 Collective total synthesis of taiwaniaquinoids by Li.

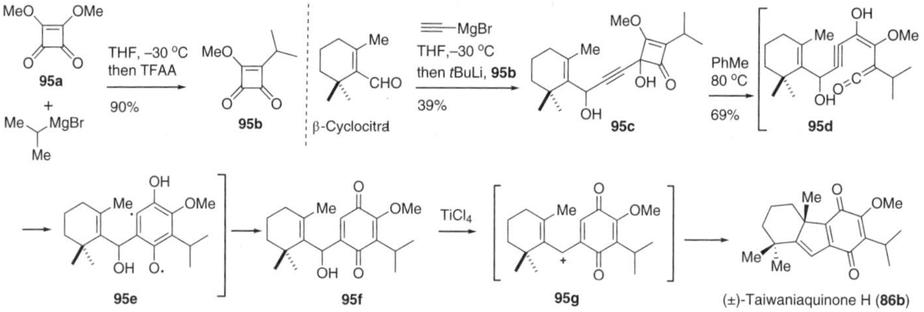

Scheme 4.32 Total synthesis of (±)-taiwaniaquinone H (**86b**) by Hu and Yan.

(±)-taiwaniaquinol D (**86d**) synthesis to complete (±)-taiwaniaquinol B (**86c**) in 81% overall yield (Scheme 4.31).

In 2014, Yan and Hu reported PGF total synthesis of (±)-taiwaniaquinone H (**86b**) by thermal ring expansion/4π electrocyclization of 2-hydroxy-cyclobutenone derivative **95c** (Scheme 4.32) [91]. In their synthesis, cyclobutenedione **95b** was prepared in 90% yield from dimethyl squarate **95a** and isopropylmagnesium bromide following Moore's method [92a], thereby completing the introduction of the methoxy group and the iso-propyl group of the final target at the early synthetic stage. Next, ethynylmagnesium bromide was added to commercially available β-cyclocitral at −30 °C, which was subse-quently treated with *t*-BuLi and cyclobutenedione **95b** sequentially, affording **95c** in 39% yield in a one-pot fashion [92b]. The expected thermal ring expansion/4π electro-cyclization process of **95c** occurred smoothly in toluene at 80 °C and gave the desired ring expansion product **95f** in 69% yield. However, interestingly, when the reaction was carried out in the presence of TiCl₄, the expected thermal ring expansion/4π

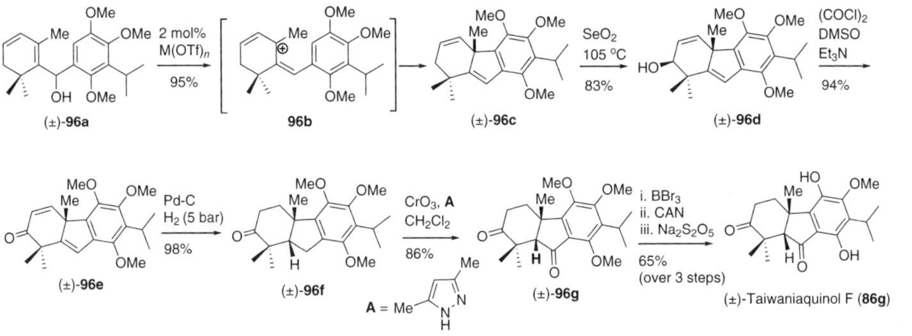

Scheme 4.33 Total synthesis of (±)-taiwaniaquinol F (**86g**) by Bisai.

electrocyclization process afforded (±)-taiwaniaquinone H (**86b**) in 41% yield from **95c** (Scheme 4.32).

Bisai et al. have reported the first total synthesis of (±)-taiwaniaquinol F (**86g**), sharing an oxidized A-ring in *abeo*-abietane scaffold [93]. This synthesis began with Lewis acid-catalyzed cyclization of aryl divinyl carbinol **96a** to carbotricyclic core **96c** via the intermediacy of **96b** (Scheme 4.33) [94]. Compound **96a** was prepared via 1,2-addition of known bromoarene onto commercially available safranal [93]. Allylic oxidation of (±)-**96c** in presence of SeO_2 afforded allyl alcohol (±)-**96d** as single diastereomer in 83% yield. Excellent diastereoselectivity of allylic oxidation was attributed to the oxidation taken place from less hindered convex face. Next, (±)-**96d** was oxidized to α,β-unsaturated ketone (±)-**96e** in 94% yield under Swern oxidation and then hydrogenated to afford tricyclic ketone (±)-**96f** in 98% yield. Later, (±)-**96f** was reacted with CrO_3/3,5-dimethylpyrazole to affect benzylic oxidation to give diketone (±)-**96g** in 86% yield, which was confirmed by X-ray crystal structure. Total synthesis of (±)-taiwaniaquinol F **86g** was completed in three steps in 65% overall yield by treatment with BBr_3 followed by CAN oxidation and $Na_2S_2O_5$ reduction. This strategy was successfully utilized for concise and straightforward total synthesis of (±)-dichroanone (**86a**), (±)-taiwaniaquinone H (**86b**), and (±)-taiwaniaquinol B (**86b**) as well [93].

Hongoquercin A (**97a**) and a number of related meroterpenoids were isolated from extracts of an unidentified terrestrial fungus (Figure 4.13) [95]. They belong to a group of natural products known as the meroterpenoids – hybrid natural products composed of both terpenoid and polyketide-derived substructures [95b, c]. This class of natural products generally consists of a drimane or rearranged drimane sesquiterpenoid attached to a quinone or hydroquinone moiety. A large number of meroterpenoids with related structures are known, and synthetic pathways to these molecules have been investigated extensively. Hongoquercin A (**97a**) exhibits antibiotic activity against vancomycin-resistant *Enterococcus faecium* and methicillin-resistant *Staphylococcus aureus* [95]. Because of their interesting structural scaffold, total synthesis of hongoquercins has been reported [96].

In 2006, Kurdyumov and Hsung [97] have disclosed an unusual polyene cyclization and cationic [2+2]-cycloaddition strategy that led to a divergent total synthesis of hongoquercin A (**97a**) and rhododaurichromanic acid A (**97e**) (Scheme 4.34). The reaction (+)-daurichromenic ester (**97d**) with trifluoroacetic acid (TFA) in CH_2Cl_2 afforded two

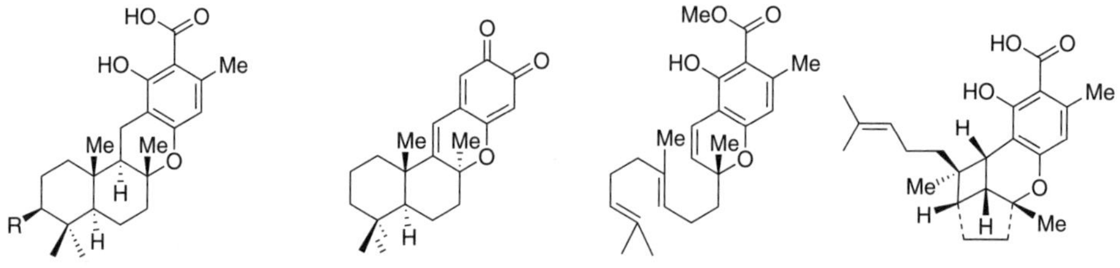

R = H, (+)-hongoquercin A (**97a**)   Puupehedione (**97c**) (+)-Daurichromenic ester (**97d**) Rhododaurichromanic
R = OAc, (+)-hongoquercin B (**97b**)   acid A (**97e**)

Figure 4.13 Hongoquercins A and B (**97a–b**) and related meroterpenoids.

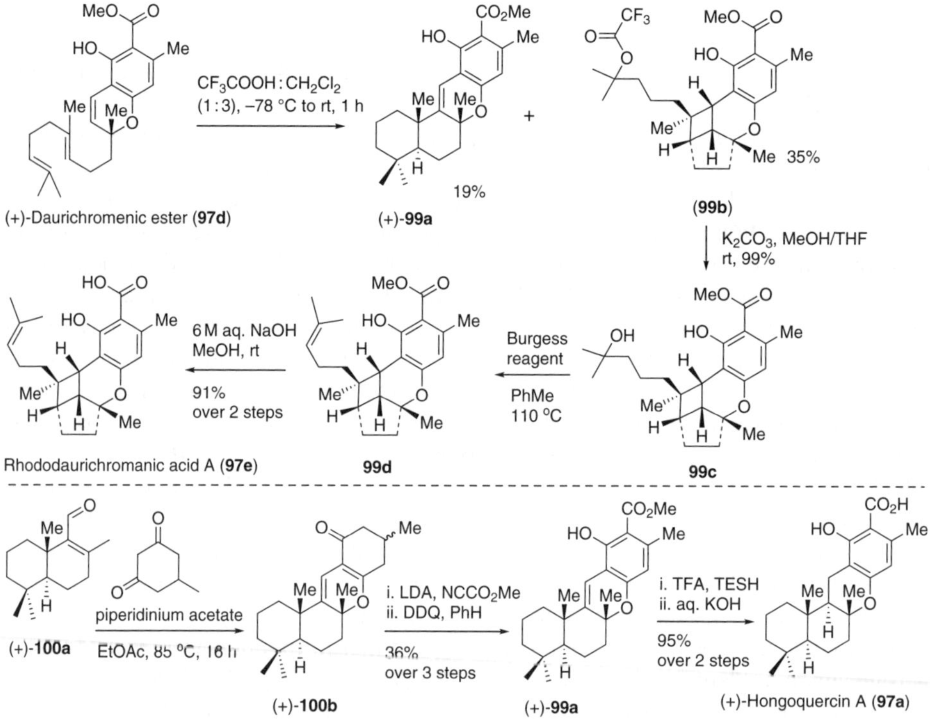

Scheme 4.34 Total synthesis of hongoquercin A (**97a**) and rhododaurichromanic acid A (**97e**).

products **99a** in 19% yield and cyclobutane **99b** in 35% yield with each as a single diastereomer. In this condition no other relevant products or related stereoisomers are being observed. Later, cyclobutane **99b** was reacted with K_2CO_3 in MeOH to furnish **99c** in 99% yield, which was treated with Burgess reagent to give olefin **99d**, followed by saponification to afford rhododaurichromanic acid A (**97e**) in 91% over two steps. Further, (+)-**99a** was synthesized independently from enal (+)-**100a** following three steps, viz. reaction of 5-methylcyclohexanedione in the presence of piperidinium acetate to form (+)-**100b** followed by reaction with Mander's reagent in presence of LDA and aromatization with DDQ. Finally, reduction of (+)-**99a** with triethylsilane (TESH)

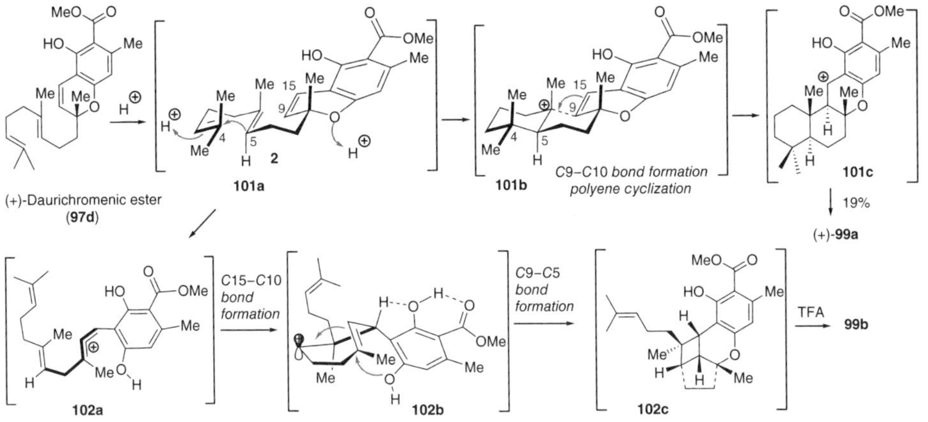

Scheme 4.35 Proposed mechanism for the formation of (+)-**99a** and **99b**.

in the presence of TFA and saponification afforded hongoquercin A (**97a**) as a single diastereomer (Scheme 4.34).

Mechanistically, the formation of **99a** likely proceeds through the classical polyene cyclization via the intermediacy of **101a–c** (Scheme 4.35). The observed high diastereoselectivity is remarkable given that the existing stereocenter at C8 is rather remote from C3 to C4 olefin, which is the initiating site. This stereochemical control is likely a result of geometric and conformational arrangement of the triene motif required in (+)-daurichromenic ester **97d** during the cyclization. The cyclobutane ring formation in **99** is more intriguing in this cyclization process (Scheme 4.35). Daurichromenic ester **97d** could be ionized first through protonation of the chromene oxygen to give allyl cation intermediate **102a**, which is stabilized by the aromatic ring. The C15–10 bond formation could then occur to generate intermediate **102b**. It is important to note that the other possibility could have been the formation of the C15–C5 bond from **102b**, which would have been a six-membered ring formation instead of the seven-membered ring shown in **102c**. Overall, the proposed pathway appears to be reasonable and resembles a Gassman-like cationic [2 + 2]-cycloaddition pathway [98].

In 2012, Baran and coworkers have reported unified approach to meroterpenoids from a common source, borono-sclareolide **104a** (Scheme 4.36) [99]. In the synthetic sequence, DIBAL-H reduction of (3a*R*)-(+)-sclareolide (**103a**) afforded scleral [100a], which upon hypoiodite-mediated C–C bond cleavage conditions of Suárez (PIDA/I$_2$/hν) delivered iodoformate **103b**. Later, a two-step dehydroiodination/hydrolysis in the presence of AgF in pyridine followed by K$_2$CO$_3$ in methanol afforded the sponge metabolite (+)-drim-9(11)-en-8α-ol (**103c**) [100b–d] in 84% overall yield from (3a*R*)-(+)-sclareolide **103a**, without the use of chromatographic separation and on 10 gram scale for each transformation. Hydroboration of **103c** resulted in a 3 : 1 mixture of diastereomers **104a** and **104b** in 96% combined yield, which were readily separated by column chromatography and characterized by X-ray crystallographic analysis. The reaction of borono-sclareolide **104a** with 2 equiv of 1,4-benzoquinone and 5 equiv of K$_2$S$_2$O$_8$ afforded concise synthesis of (+)-chromazonarol (**105a**) in 60% yield.

Zonarol (**105c**) [101a], isozonarol (**105b**) [101b, c], zonarone (**108a**) [101d], and isozonarone (**108b**) [101e] were isolated by Fenical and coworkers in 1973 from the brown

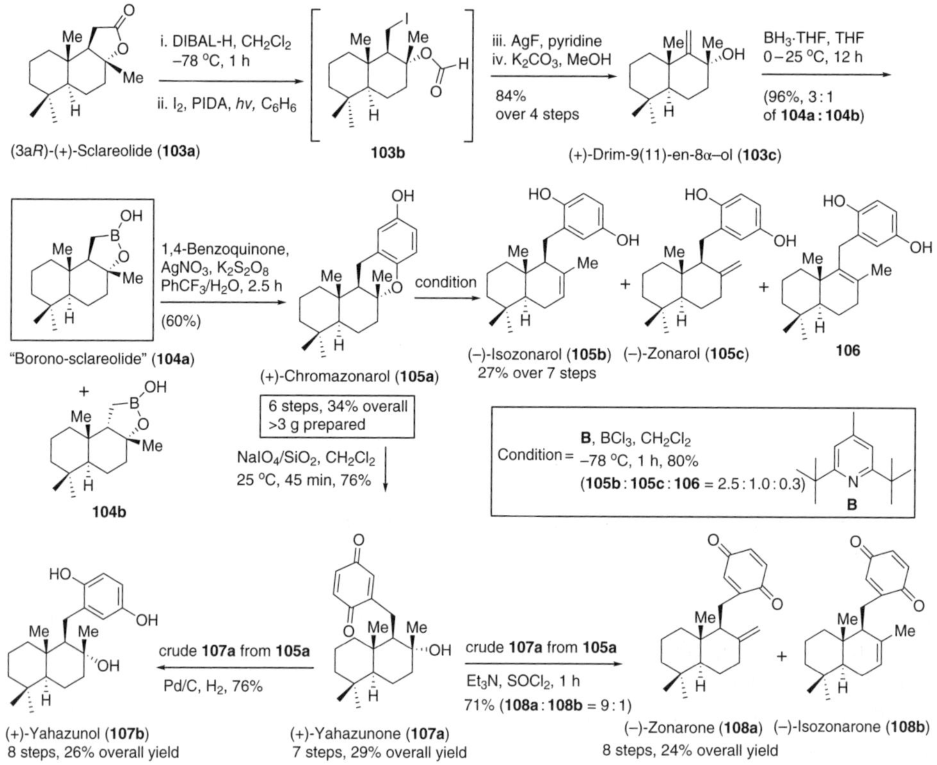

Scheme 4.36 Divergent synthesis of meroterpenoid natural products.

seaweed *Dictyopteris zonarioides*. Treatment of (+)-chromazonarol (**105a**) with BCl₃ [102a] in the presence of 2,6-di-*tert*-butyl-4-methylpyridine **B** at −78 °C afforded a 2.5 : 1.0 : 0.3 mixture of (−)-isozonarol (**105b**), (−)-zonarol (**105c**), and tetrasubstituted olefin **106** [102b, c] in a combined 80% yield (Scheme 4.36). (+)-Chromazonarol (**105a**) with silica gel-supported sodium periodate yielded yahazunone (**107a**) in 76% yield. When crude **107a** (obtained from **105a**) was hydrogenated, it led to (+)-yahazunol (**107b**), which was then purified via trituration with CH₂Cl₂ [103a]. Similarly, crude **107a** was treated with SOCl₂ in the presence of Et₃N at −78 °C to yield a 9 : 1 mixture of (−)-zonarone (**108a**) and (−)-isozonarone (**108b**) in a combined yield of 71% over two steps (Scheme 4.36).

The antitumor and highly potent angiogenesis inhibitor (+)-8-*epi*-puupehedione (**110b**), a synthetic analog of puupehedione, was also targeted (Scheme 4.37) [103b–d]. After exhaustive screening of various oxidants, it was found that the oxidation of (+)-chromazonarol (**105a**) (condition as per report by Pettus) [104] using IBX in DMF resulted in 50% yield of a ∼1 : 1 mixture of **109a** and **109b**, which was readily separable by column chromatography. Later, reduction of **109b** with NaBH₄ and subsequent oxidation of crude puupehenol (**110a**) with chloranil in refluxing *tert*-butanol afforded (+)-8-*epi*-puupehedione (**110b**) in 24% overall yield from (+)-chromazonarol (**105a**).

To further establish the utility of borono-sclareolide **104a** as a versatile terpene donor, it was exposed to a variety of aryl bromides **111a–b** under Suzuki conditions developed

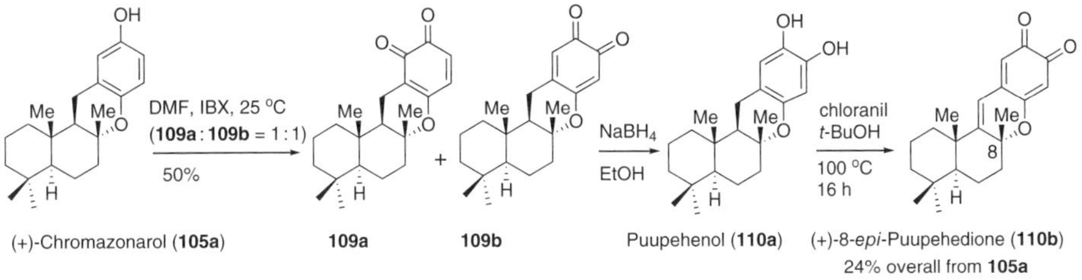

Scheme 4.37 Synthesis of puupehenol (**110a**) and (+)-8-*epi*-puupehedione (**110b**).

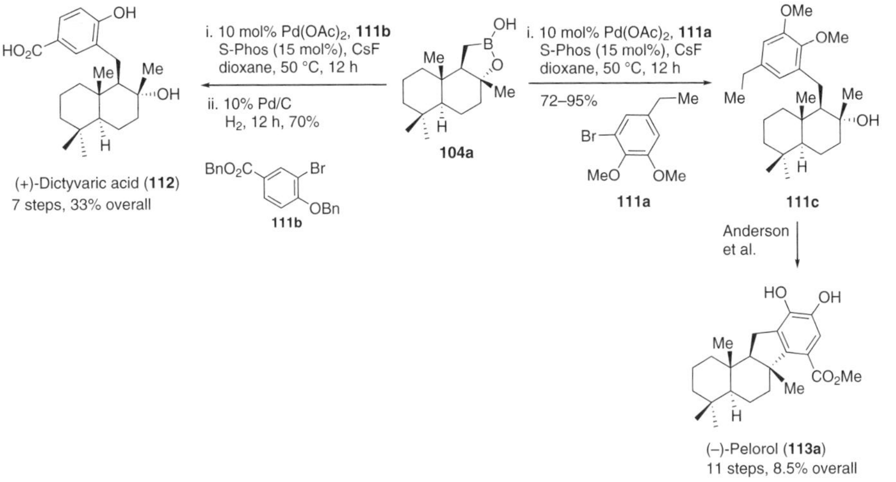

Scheme 4.38 Synthesis of (−)-pelorol (**113a**) and (+)-dictyvaric acid (**112**).

by Buchwald [105a], resulting in excellent yields of the coupled products (Scheme 4.38). The formal synthesis of the potent SHIP2 inhibitor (−)-pelorol (**113a**) was accomplished by intercepting Andersen's intermediate **111c** in 87% yield [105b]. Coupling between **104a** and benzyl ester **111b** followed by hydrogenation with 10% Pd/C under 200 psi H_2 yielded meroterpenoid (+)-dictyvaric acid (**112**) in 55% yield from **104a** (Scheme 4.38) [105c].

Merosesquiterpenes, sharing a tetracarbocyclic core, are natural products of mixed biosynthetic origin (arising from polyketide–terpenoid) containing a sesquiterpene unit joined to a phenolic (or sometimes as quinone moiety) [106]. Compounds bearing a bicyclic terpene (especially drimane) moiety are the most important group of this family, owing to their wide variety of structural types and potent biological activities [107a]. Among various merosesquiterpenes [107b], pelorol (**113a**) was isolated from *Dactylospongia elegans* in 2000 [108], and akaol A (**113b**) was isolated from a *Micronesian sponge* of the genus *Aka* in 2003 (Figure 4.14) [109]. Other merosesquiterpenes such as dasyscyphins A, B and D, and E (**113d–e** and **113f–g**) were isolated from the ascomycete *Dasyscyphus niveus* [110a–c].

In 2008, Yue and coworkers also isolated two unprecedented skeletons with C24 nortriterpenoid structural motifs, walsucochins A (**113h**) and B (**113i**), from *Walsura cochinchinensis* (Figure 4.14) [111a]. These novel C24 nortriterpenoids exhibit

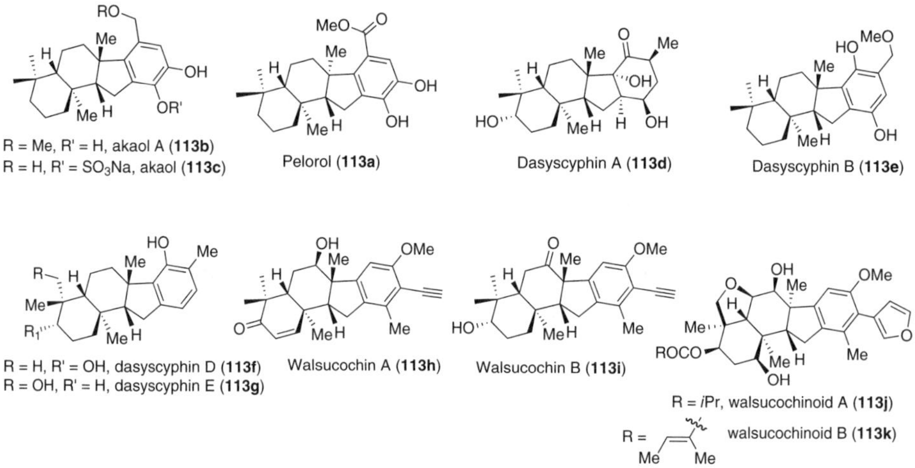

R = Me, R' = H, akaol A (**113b**)
R = H, R' = SO₃Na, akaol (**113c**)

Pelorol (**113a**)

Dasyscyphin A (**113d**)

Dasyscyphin B (**113e**)

R = H, R' = OH, dasyscyphin D (**113f**)
R = OH, R' = H, dasyscyphin E (**113g**)

Walsucochin A (**113h**)

Walsucochin B (**113i**)

R = *i*Pr, walsucochinoid A (**113j**)

R = walsucochinoid B (**113k**)

Figure 4.14 Selected naturally occurring merosesquiterpenoids (**113a–k**).

significant cell protecting activity against H_2O_2-induced PC12 cell damage [111a]. Recently, in 2012, the same group isolated two more new C24 nortriterpenoids walsucochinoids A (**113j**) and B (**113k**) from *W. cochinchinensis* [111b]. Even though the bioactivities of this family of compounds have yet to be examined comprehensively, preliminary studies have revealed that pelorol (**113a**) is an activator of the Src homology 2 domain-containing inositol 5-phosphatase (SHIP) [112], whereas dasyscyphin B (**113e**) shows potent cytotoxic activities in several human cell lines [110a] and dasyscyphin D (**113f**) and E (**113g**) exhibit antifungal properties [110c].

In 2005, Andersen et al. reported total synthesis of pelorol (**113a**) starting from (+)-sclareolide via a key intramolecular Friedel–Crafts alkylation to create the cyclopentane C-ring (Scheme 4.39) [113]. In their synthesis, (+)-sclareolide was converted to the diol **114a** in excellent yield (90%) using a one-pot three-step sequence by modification of the literature procedure [114]. Swern oxidation cleanly oxidized the diol **114a** to aldehyde **114b**, which on subsequent reaction with aryllithium **114c** gave the diastereomeric benzyl alcohols **114d** in good yield. Hydrogenolysis of the mixture **114d** cleanly removed the benzylic alcohol to give **111c** (also see Scheme 4.38). Cyclization of **111c** using SnCl₄ as a catalyst gave the desired tetracyclic intermediate **114e** in high yield, which was reacted with PCC to afford methyl ketone **114f**. The latter on reaction with Br₂ in aqueous NaOH led to the formation of desired benzoic acid, which was esterified with MeI to give **114g**. Finally, selective cleavage of aryl methyl ethers with BI₃ at −78 °C afforded pelorol (**113a**) in 50% yield.

In 2011, She et al. [115] reported the total synthesis of dasyscyphin D (**113f**) via a key PtCl₂-catalyzed pentannulation reaction and acid-catalyzed Robinson annulations (Scheme 4.40). Methylation and addition of ethynylmagnesium bromide onto acetophenone **115a** afforded the corresponding alcohol, which was further protected as acetyl to obtain **115b** in 95% yield over three steps. Treatment of **115b** with 5 mol% PtCl₂ in toluene underwent the desired PtCl₂-catalyzed pentannulation reaction followed by acidic hydrolysis to give the desired 2-indanone **115c** in 72% yield over two steps. The reaction between 2-indanone **115c** and 1.5 equiv of **115d** in the presence of

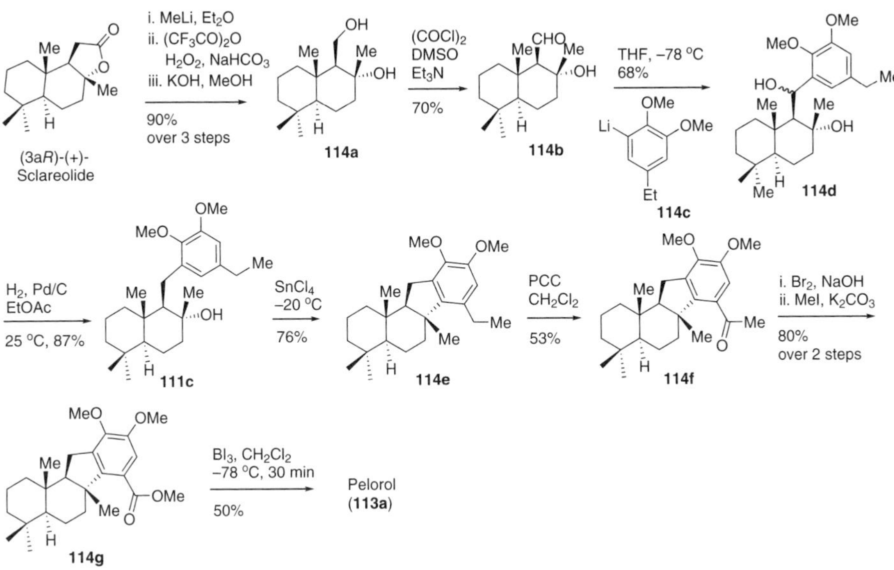

Scheme 4.39 Total synthesis of pelorol (**113a**) by Anderson.

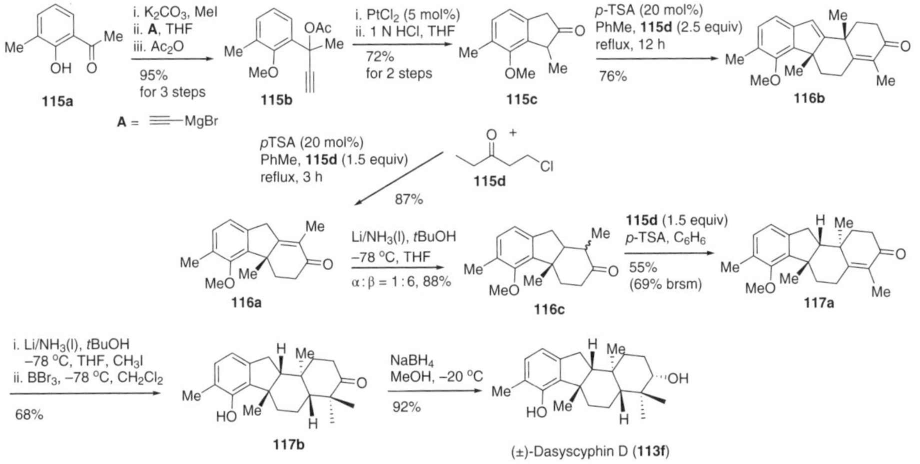

Scheme 4.40 Total synthesis of (±)-dasyscyphin D (**113f**) by She.

catalytic *p*-toluenesulfonic acid (*p*TSA) in refluxing toluene afforded tricycle **116a** in 87% yield. Interestingly, when this reaction was carried out with 2.5 equiv of **115d**, it furnished tetracycle **116b** in 76% yield in 12 h (Scheme 4.40). Later, Birch reduction of enone **116a** afforded tricycle ketone **116c** in 6 : 1 dr in favor of β-methyl in 88% yield. Next, tricyclic ketone **116c** was treated with 1.5 equiv of **115d** catalyzed by *p*TSA in refluxing benzene to afford tetracycle **117a** in 55% yield (69% based on recovered starting material) [116]. The sequential Birch reduction–alkylation reaction (with methyl iodide) of **117a** [117], followed by demethylation, furnished tetracyclic ketone **117b** in

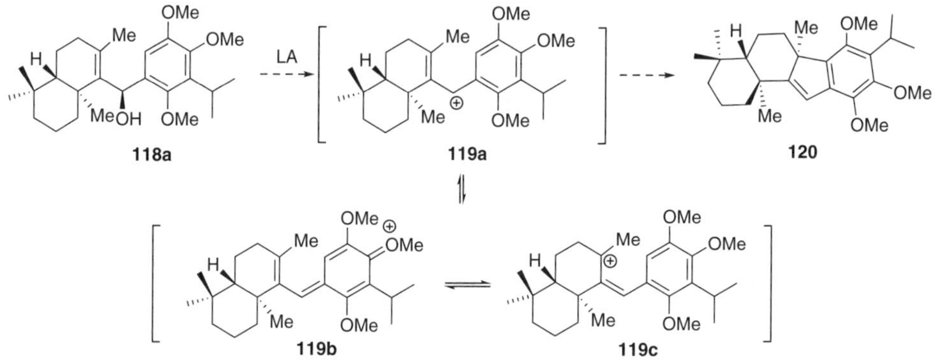

Scheme 4.41 Proposed hypothesis for synthesis of merosesquiterpenoid by Bisai.

68% yield over three steps. Finally, reduction of ketone **117b** with NaBH$_4$ in MeOH completed the total synthesis of (±)-dasyscyphin D (**113f**).

The first enantiospecific synthesis of akaol A (**113b**) has been reported from commercial (−)-sclareol [118] by Alvarez-Manzaneda via a key Diels–Alder approach. The similar approach was utilized for the synthesis of dasyscyphin B (**113e**) from commercially available abietic acid [119]. Recently, Bisai et al. [120] reported the synthesis of akaol A (**113b**) through a Lewis acid-catalyzed Nazarov-type cyclization of arylvinylcarbinols of type **118a** (Scheme 4.41). It was hypothesized that arylvinylcarbinol **118a** would generate allyl benzyl carbocation species **119a** in the presence of Lewis acid, which would further be stabilized via oxocarbenium **119b** and carbocation **119c**. This would finally react by electron-rich aromatics to generate tetracyclic core of merosesquiterpene of type **120**. In fact, cyclization of **118a** in the presence of 2 mol% Bi(OTf)$_3$ or Sn(OTf)$_2$ in refluxing dichloroethane for 2 h afforded tetracycle **120** in 92–93% yields.

The reaction of arylvinylcabinols **121** having a hydroxymethyl group led to the mixture of diastereomers **122a** and **122b** in 90% yield with dr ~1 : 1 (Scheme 4.42) [121]. For synthesis of akaol A (**113b**), the mixture of **122a** and **122b** was hydrogenated (Pd/C in MeOH, 1 atm I$_2$) to afford chromatographically separable carbotetracycles **123a** and **123b** in 98% yield with dr = 1 : 1. Next, **123b** was oxidized using PCC to furnish aldehyde **124** in 97% yield, which was then demethylated with PhSH in the presence of K$_2$CO$_3$ to afford aldehyde **125** in 83% yield. The latter was then reduced with LiAlH$_4$ in THF followed by a chemoselective methylation of benzyl alcohol with conc. HCl in MeOH completed the synthesis of akaol A (**113b**) in 88% yields over two steps (Scheme 4.42).

4.4 Conclusions

PGF synthesis is a promising area for further study with regard to the total synthesis of complex molecules following atom economy, step economy, and redox economy. Last 10 years have witnessed the development of a number of total syntheses that attempt to minimize protecting groups at the planning stage. To this end, several methodologies

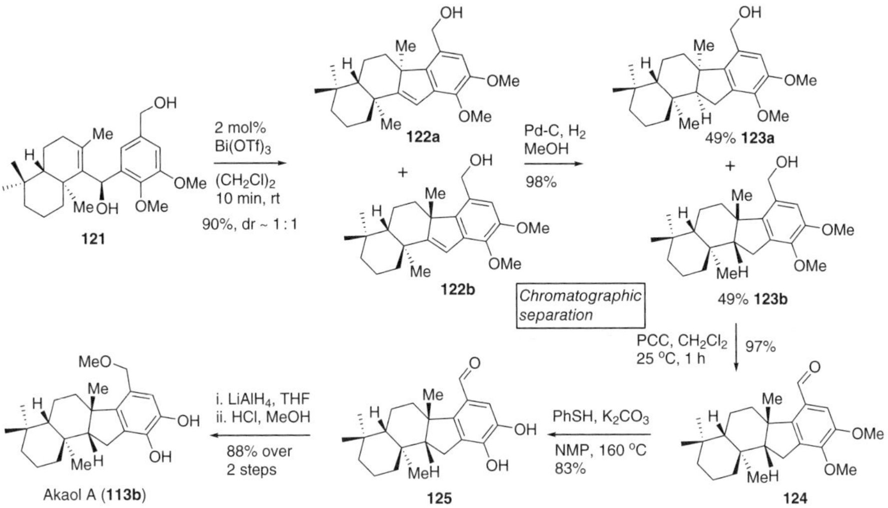

Scheme 4.42 Synthesis of merosesquiterpene quinol akaol A (**113a**).

have also arisen with this general goal in mind. There are examples where biomimetic approaches are considered while developing PGF total synthesis on natural products of biological potential.

References

1 Trost, B.M. (1991). *Science* 254: 1471.
2 Wender, P.A., Verma, V.A., Paxton, T.J., and Pillow, T.H. (2008). *Acc. Chem. Res.* 41: 40.
3 Bruggink, A., Schoevaart, R., and Kieboom, T. (2003). *Org. Process Res. Dev.* 7: 622.
4 (a) Corey, E.J. and Venkateswarlu, A. (1972). *J. Am. Chem. Soc.* 94: 6190; (b) Corey, E.J., Cho, H., Rucker, C., and Hua, D.H. (1981). *Tetrahedron Lett.* 22: 3455; (c) Merrifield, R.B. (1963). *J. Am. Chem. Soc.* 85: 2149; (d) Hoffmann, R.W. (2006). *Synthesis* 3531.
5 Humphrey, A.J. and O'Hagan, D. (2001). *Nat. Prod. Rep.* 18: 494.
6 Grynkiewicz, G. and Gadzikowska, M. (2008). *Pharmacol. Rep.* 60: 439.
7 Medley, J.W. and Movassaghi, M. (2013). *Chem. Commun.* 49: 10775.
8 (a) Robinson, R.A. (1917). *J. Chem. Soc. Trans.* 111: 762. (b) Mondal, N., Mandal, S.C., Das, G.K., and Mukherjee, S. (2003). *J. Chem. Res. (S)* 580.
9 Tandem events in total syntheses of natural products, see:Nicolaou, K.C., Edmonds, D.J., and Bulger, P.G. (2006). *Angew. Chem. Int. Ed.* 45: 7134.
10 Stoermer, D. and Heathcock, C.H. (1993). *J. Org. Chem.* 58: 564.
11 Kyburz, R., Schopp, E., Bick, I.R.C., and Hess, M. (1981). *Helv. Chim. Acta* 64: 2555.
12 Stevens, R.V. and Kenney, P.M. (1983). *J. Chem. Soc. Chem. Commun.* 384.
13 (a) Kobayashi, J. and Kubota, T. (2009). *Nat. Prod. Rep.* 26: 936; (b) Kobayashi, J. and Morita, H. (2003). The Alkaloids, vol. 60 (ed. G.A. Cordell), 165. New York: Academic Press; (c) Xu, J.-B., Zhang, H., Gan, L.-S. et al. (2014). *Am. Chem. Soc.* 136: 7631 and references cited therein.

14 (a) Weiss, M.E. and Carreira, E.M. (2011). *Angew. Chem. Int. Ed.* 50: 11501; (b) Lu, Z.Y., Li, Y., Deng, J., and Li, A. (2013). *Nat. Chem.* 5: 679; (c) Shvartsbart, A. and Smith, A.B. (2014). *J. Am. Chem. Soc.* 136: 870; (d) Shvartsbart, A. and Smith, A.B. (2015). *J. Am. Chem. Soc.* 137: 3510.

15 (a) Ruggeri, R.B. and Heathcock, C.H. (1989). *Pure Appl. Chem.* 61: 289; (b) Heathcock, C.H., Hansen, M.M., Ruggeri, R.B., and Kath, J.C. (1992). *J. Org. Chem.* 57: 2544; (c) Heathcock, C.H., Kath, J.C., and Ruggeri, R.B. (1995). *J. Org. Chem.* 60: 1120; (d) Ruggeri, R.B. and Heathcock, C.H. (1990). *J. Org. Chem.* 55: 3714;(e) Stafford, J.A. and Heathcock, C.H. (1990). *J. Org. Chem.* 55: 5433.

16 (a) Heathcock, C.H. (1996). *Proc. Natl. Acad. Sci. U.S.A.* 93: 14323; (b) Kim, J. and Movassaghi, M. (2009). *Chem. Soc. Rev.* 38: 3035.

17 (a) Kokayashi, J., Sekiguchi, M., Shimamoto, S. et al. (2002). *J. Org. Chem.* 67: 6449; (b) Ishiyama, H., Matsumoto, M., Sekiguchi, M. et al. (2005). *Heterocycles* 66: 651; (c) Cheng, X., Duhaime, C.M., and Waters, S.P. (2010). *J. Org. Chem.* 75: 7026; (d) Gao, P., Liu, Y., Zhang, L. et al. (2006). *J. Org. Chem.* 71: 9495.

18 Chen, P., Cao, L., and Li, C. (2009). *J. Org. Chem.* 74: 7533.

19 (a) De Risi, C., Pollini, G.P., Trapella, C. et al. (2001). *Bioorg. Med. Chem.* 9: 1871; (b) Showell, G.A., Gibbons, T.L., Kneen, C.O. et al. (1991). *J. Med. Chem.* 34: 1086.

20 (a) Davis, F.A., Chao, B., and Rao, A. (2001). *Org. Lett.* 3: 3169; (b) Davis, F.A., Zhang, Y., and Anilkumar, G. (2003). *J. Org. Chem.* 68: 8061; (c) Mahrwald, R. (1999). *Chem. Rev.* 99: 1095; (d) Yan, T.-H., Hung, A.-W., Lee, H.-C. et al. (1995). *J. Org. Chem.* 60: 3301; (e) Crimmins, M.T. and Chaudhary, K. (2000). *Org. Lett.* 2: 775.

21 Yu, F., Cheng, B., and Zhai, H. (2011). *Org. Lett.* 13: 5782.

22 (a) Bentley, K.W. (1992). *Nat. Prod. Rep.* 9: 365; (b) Bentley, K.W. (1984). *Nat. Prod. Rep.* 1: 355; (c) Fotie, J., Bohle, D.S., Olivier, M. et al. (2007). *J. Nat. Prod.* 70: 1650.

23 (a) Arthur, H.R., Hui, W.H., and Ng, Y.L. (1959). *J. Chem. Soc.* 4007; (b) Wang, S.-Q., Wang, G.-Q., Huang, B.-K. et al. (2007). *Chem. Nat. Compd.* 43: 100; (c) Hu, J., Zhang, W.-D., Liu, R.-H. et al. (2006). *Chem. Biodiversity* 3: 990.

24 (a) Moura, N.F.D., Ribeiro, H.B., Machado, E.C.S. et al. (1997). *Phytochemistry* 46: 1443; (b) Arthur, H.R., Hui, W.H., and Ng, Y.L. (1959). *J. Chem. Soc.* 1840; (c) Wall, M.E., Wani, M.C., and Taylor, H. (1987). *J. Nat. Prod.* 50: 1095; (d) Chang, C.-T., Doong, S.-C., Tsai, I.-L., and Chen, I.-S. (1997). *Phytochemistry* 45: 1419.

25 Korivi, R.P. and Cheng, C.-H. (2010). *Chem. Eur. J.* 16: 282.

26 (a) Vardamides, J.C. (2006). *Chem. Pharm. Bull.* 54: 1034; (b) Mattson, A.E. and Scheidt, K.A. (2007). *J. Am. Chem. Soc.* 129: 4508.

27 (a) Bunnett, J.F. (1978). *Acc. Chem. Res.* 11: 413; For reviews of HAS with aryl radicals, see: (b) Bolton, R. and Williams, G.H. (1986). *Chem. Soc. Rev.* 15: 261; (c) Fossey, J., Lefort, D., and Sorba, J. (1995). Chapter 14. In: Free Radicals in Organic Chemistry, 166–180. Chichester: Wiley.

28 (a) Studer, A. and Bossart, M. (2001). Chapter 1.4. In: Radicals in Organic Synthesis, vol. 2 (ed. P. Renaud and M.P. Sibi), 62–80. Weinheim: Wiley-VCH. (b) Rossi, R.A., Pierini, A.B., and Peñéñory, A.B. (2003). *Chem. Rev.* 103: 71.

29 (a) De, S., Ghosh, S., Bhunia, S. et al. (2012). *Org. Lett.* 14: 4466. (b) De, S., Mishra, S., Kakde, B.N. et al. (2013). *J. Org. Chem.* 78: 7823.

30 For strategies using Pd-catalyzed regioselective C-H activation, see: (a) Harayama, T., Hori, A., Nakano, Y. et al. (2002). *Heterocycles* 58: 159; (b) Harayama, T., Sato, T., Hori,

A. et al. (2003). *Synlett* 1441;(c) Harayama, T., Sato, T., Hori, A. et al. (2004). *Synthesis* 1446.

31 (a) Cordell, G.A. and Saxton, J.E. (1981). Bisindole alkaloids. In: The Alkaloids: Chemistry and Physiology, vol. 20 (ed. R.H.F. Manske and R.G.A. Rodrigo), 3–294. New York: Academic Press. (b) Crich, D. and Banerjee, A. (2007). *Acc. Chem. Res.* 40: 151.

32 (a) Woodward, R.B., Yang, N.C., Katz, T.J. et al. (1960). *Proc. Chem. Soc.* 76; (b) Robinson, R. and Teuber, H.J. (1954). *Chem. Ind.* 783.

33 (a) Hendrickson, J.B., Rees, R., and Göschke, R. (1962). *Proc. Chem. Soc.* 383; (b) Hino, T. and Yamada, S. (1963). *Tetrahedron Lett.* 4: 1757; (c) Scott, A.I., McCapra, F., and Hall, E.S. (1964). *J. Am. Chem. Soc.* 86: 302; (d) Nakagawa, M., Sugumi, H., Kodato, S., and Hino, T. (1981). *Tetrahedron Lett.* 22: 5323; (e) Fang, C.-L., Horne, S., Taylor, N., and Rodrigo, R. (1994). *J. Am. Chem. Soc.* 116: 9480; (f) Somei, M., Osikiri, N., Hasegawa, M., and Yamada, F. (1999). *Heterocycles* 51: 1237; (g) Ishikawa, H., Takayama, H., and Aimi, N. (2002). *Tetrahedron Lett.* 43: 5637; (h) Matsudz, Y., Kitajima, M., and Takayama, H. (2005). *Heterocycles* 65: 1031.

34 Movassaghi, M. and Schmidt, M.A. (2007). *Angew. Chem. Int. Ed.* 46: 3725.

35 (a) Aresta, M., Rossi, M., and Sacco, A. (1969). *Inorg. Chim. Acta* 3: 227; (b) Yamada, Y. and Momose, D.-I. (1981). *Chem. Lett.* 1277; (c) Bagal, S.K., Adlington, R.M., Baldwin, J.E., and Marquez, R. (2004). *J. Org. Chem.* 69: 9100.

36 (a) Hendrickson, J.B., Göschke, R., and Rees, R. (1964). *Tetrahedron* 20: 565; (b) Hall, E.S., McCarpa, F., and Scott, A.I. (1967). *Tetrahedron* 23: 4131; (c) Guéritte-Voegelein, F., Sévenet, T., Pusset, J. et al. (1992). *J. Nat. Prod.* 55: 923; (d) Lebsack, A.D., Link, J.T., Overman, L.E., and Stearns, B.A. (2002). *J. Am. Chem. Soc.* 124: 9008.

37 (a) Hodson, H.F., Robinson, B., and Smith, G.F. (1961). *Proc. Chem. Soc. Lond.* 465; (b) Tokuyama, T. and Daly, J.W. (1983). *Tetrahedron* 39: 41; (c) Verotta, L., Pilati, T., Tatò, M. et al. (1998). *J. Nat. Prod.* 61: 392.

38 For a general method for the construction of quaternary carbon stereocentres, see:(a)Hong, A.Y. and Stoltz, B.M. (2013). *Eur. J. Org. Chem.* 2745; (b) Douglas, C.J. and Overman, L.E. (2004). *Proc. Natl. Acad. Sci. U.S.A.* 101: 5363; (c) Christoffers, J. and Baro, A. (2003). *Angew. Chem. Int. Ed.* 42: 1688; (d) Corey, E.J. and Guzman-Perez, A. (1998). *Angew. Chem. Int. Ed.* 37: 388.

39 (a) Corey, E.J. and Lin, S. (1996). *J. Am. Chem. Soc.* 118: 8765. (b) Peterson, E.A. and Overman, L.E. (2004). *Proc. Natl. Acad. Sci. U.S.A.* 101: 11943; (c) Steven, A. and Overman, L.E. (2007). *Angew. Chem. Int. Ed.* 119: 5584; (d) Schmidt, M.A. and Movassaghi, M. (2008). *Synlett* 313; (e) Trost, B.M., Malhotra, S., and Chan, W.H. (2011). *J. Am. Chem. Soc.* 133: 7328.

40 (a) Weaver, J.D., Recio, A. III, Grenning, A.J., and Tunge, J.A. (2011). *Chem. Rev.* 111: 1846; (b) Enquist, J.A. Jr. and Stoltz, B.M. (2008). *Nature* 453: 1228; (c) Enquist, J.A. Jr., Virgil, S.C., and Stoltz, B.M. (2011). *Chem. -Eur. J.* 17: 9957; (d) Shibuya, G.M., Enquist, J.A. Jr., and Stoltz, B.M. (2013). *Org. Lett.* 15: 3480.

41 (a) Trost, B.M. and Van Vranken, D.L. (1992). *Angew. Chem. Int. Ed.* 31: 228; (b) Trost, B.M., Thaisrivongs, D.A., and Hartwig, J. (2011). *J. Am. Chem. Soc.* 133: 12439.

42 (a) Ghosh, S., Bhunia, S., Kakde, B.N. et al. (2014). *Chem. Commun.* 50: 2434; (b) Ghosh, S., Chaudhuri, S., and Bisai, A. (2015). *Chem. -Eur. J.* 21: 17479.

43 Adjibadé, Y., Weniger, B., Quirion, J.C. et al. (1992). *Phytochemistry* 31: 317.

44 (a) Link, J.T. and Overman, L.E. (1996). *J. Am. Chem. Soc.* 118: 8166; (b) Menozzi, C., Dalko, P.I., and Cossy, J. (2006). *Chem. Commun.* 4638.

45 Kumar, N., Das, M.K., Ghosh, S., and Bisai, A. (2017). *Chem. Commun.* 53: 2170.

46 (a) Grenning, A.J. and Tunge, J.A. (2011). *Angew. Chem. Int. Ed.* 50: 1688; (b) Grenning, A.J. and Tunge, J.A. (2011). *J. Am. Chem. Soc.* 133: 14785; (c) Grenning, A.J., Allen, C.K.V., Maji, T. et al. (2013). *J. Org. Chem.* 78: 7281; (d) Deacylative benzylation (DaB):Maji, T., Ramkumar, K., and Tunge, J.A. (2014). *Chem. Commun.* 50: 14045.

47 Lithium prefers to give C_2 symmetric alkylation products, see:Hoyt, S.B. and Overman, L.E. (2000). *Org. Lett.* 2: 3241.

48 (a) Williams, R.M., Sanz-Cervera, J.F., Sancenon, F. et al. (1998). *J. Am. Chem. Soc.* 120: 1090; (b) Williams, R.M., Glinka, T., Kwast, E. et al. (1990). *J. Am. Chem. Soc.* 112: 808.

49 (a) Williams, R.M., Cao, J., Tsujishima, H., and Cox, R.J. (2003). *J. Am. Chem. Soc.* 125: 12172; (b) Sanz-Cervera, J.F. and Williams, R.M. (2002). *J. Am. Chem. Soc.* 124: 2556.

50 (a) Porter, A.E.A. and Sammes, P.G.A. (1970). *J. Chem. Soc. D* 1103; (b) Baldas, J., Birch, A.J., and Russell, R.A. (1974). *J. Chem. Soc., Perkin Trans. 1* 50; (c) Williams, R.M. and Cox, R.J. (2003). *Acc. Chem. Res.* 36: 127.

51 Miller, K.A., Tsukamoto, S., and Williams, R.M. (2009). *Nat. Chem.* 1: 63.

52 Takayama, H., Mori, I., Kitajima, M. et al. (2004). *Org. Lett.* 6: 2945.

53 Newhouse, T. and Baran, P.S. (2008). *J. Am. Chem. Soc.* 130: 10886.

54 (a) Larock, R.C. and Yum, E.K. (1991). *J. Am. Chem. Soc.* 113: 6689; (b) Larock, R.C. and Yum, E.K. (1991). *J. Am. Chem. Soc.* 113: 6689; (c) Ma, C., Liu, X., Li, X. et al. (2001). *J. Org. Chem.* 66: 4525.

55 Matsuda, Y., Kitajima, M., and Takayama, H. (2008). *Org. Lett.* 10: 125.

56 Foo, K., Newhouse, T., Mori, I. et al. (2011). *Angew. Chem. Int. Ed.* 50: 2716.

57 (a) Barton, D.H.R. and Serebryakov, E.P. (1962). *Proc. Chem. Soc.* 309; (b) Martin, S.F., Clark, C.W., and Corbett, J.W. (1995). *J. Org. Chem.* 60: 3236.

58 Reviews:(a)Singh, V. and Thomas, B. (1998). *Tetrahedron* 54: 3647; (b) Mehta, G. and Srikrishna, A. (1997). *Chem. Rev.* 97: 671.

59 (a) Kupka, J., Anke, T., Giannetti, B.M., and Steglich, W. (1981). *Arch. Microbiol.* 130: 223; (b) Giannetti, B.M., Steffan, B., Steglich, W. et al. (1986). *Tetrahedron* 42: 3587; (c) Steglich, W. (1981). *Pure Appl. Chem.* 53: 1233; (d) Padwa, A., Curtis, E.A., and Sandanayaka, V.P. (1996). *J. Org. Chem.* 61: 73.

60 (a) Paquette, L.A. and Doyon, J. (1995). *J. Am. Chem. Soc.* 117: 6799; (b) Paquette, L.A. and Doyon, J. (1997). *J. Org. Chem.* 62: 1723; (c) Shibasaki, M., Iseki, K., and Ikegami, S. (1981). *Tetrahedron* 37: 4411.

61 Mehta, G., Murthy, A.N., Reddy, D.S., and Reddy, A.V. (1986). *J. Am. Chem. Soc.* 108: 3443.

62 Barton, D.H.R. and McCombie, S.W. (1975). *J. Chem. Soc. Perkin Trans. 1* 1574.

63 (a) Rodriguez, A.D. and Ramirez, C. (2001). *J. Nat. Prod.* 64: 100; (b) Look, S.A., Fenical, W., Jacobs, R.S., and Clardy, J. (1986). *Proc. Natl. Acad. Sci. U.S.A.* 83: 6238; (c) Rodriguez, A.D. and Shi, Y.-P. (2000). *Tetrahedron* 56: 9015; (d) Rodriguez, A.D. (1995). *Tetrahedron* 51: 4571.

64 (a) Pseudopterosins:Corey, E.J. and Lazerwith, S.E. (1998). *J. Am. Chem. Soc.* 120: 12777; (b) Pseudopteroxazole:Davidson, J.P. and Corey, E.J. (2003). *J. Am. Chem. Soc.* 125: 13486; (c) Elisabethins:Heckrodt, T.J. and Mulzer, J. (2003). *J. Am. Chem. Soc.* 125: 4680.

65 Cesati III, R.R., de Armas, J., and Hoveyda, A.H. (2004). *J. Am. Chem. Soc.* 126: 96.

66 Davies, H.M.L. and Walji, A.M. (2005). *Angew. Chem. Int. Ed.* 44: 1733.

67 (a) Davies, H.M.L. and Jin, Q. (2004). *Proc. Natl. Acad. Sci. U.S.A.* 101: 5472; (b) Davies, H.M.L. and Jin, Q. (2004). *J. Am. Chem. Soc.* 126: 10862.

68 (a) Hanson, J.R. (2013). *Nat. Prod. Rep.* 30: 1346. (b) Gonzalez, M.A. (2015). *Nat. Prod. Rep.* 32: 684.

69 (a) Aladesanmi, A.J. and Odediran, S.A. (2000). *Fitoterapia* 71: 179; (b) Nishino, C., Kobayashi, K., Shiobara, Y., and Kodama, M. (1988). *Agric. Biol. Chem.* 52: 77; (c) Majumder, P.L., Maiti, D.C., Kraus, W., and Bokel, M. (1987). *Phytochemistry* 26: 3021; (d) Manuel, J.E., Enrique, P.M., Alfonsa, L.-R. et al. (1988). *J. Nat. Prod.* 51: 243.

70 (a) Bhar, S.S. and Ramana, M.M.V. (2004). *J. Org. Chem.* 69: 8935; (b) Bhar, S.S. and Ramana, M.M.V. (2006). *Tetrahedron Lett.* 47: 7805.

71 (a) Heather, J.B., Mittal, R.S.D., and Sih, C.J. (1976). *J. Am. Chem. Soc.* 98: 3661; (b) Branca, S.J., Lock, R.L., and Smith, A.B. (1977). *J. Org. Chem.* 42: 3165.

72 (a) Fillion, E. and Fishlock, D. (2005). *J. Am. Chem. Soc.* 127: 13144; (b) Banerjee, M., Mukhopadhyay, R., Achari, B., and Banerjee, A.K. (2006). *J. Org. Chem.* 71: 2787; (c) Planas, L., Mogi, M., Takita, H. et al. (2006). *J. Org. Chem.* 71: 2896; (d) For a review, see:Majetich, G. and Shimkus, J.M. (2010). *J. Nat. Prod.* 73: 284.

73 (a) Lin, W.H., Fang, J.M., and Cheng, Y.S. (1995). *Phytochemistry* 40: 871; (b) Lin, W.H., Fang, J.M., and Cheng, Y.S. (1996). *Phytochemistry* 42: 1657; (c) Kawazoe, K., Yamamoto, M., Takaishi, Y. et al. (1999). *Phytochemistry* 50: 493.

74 (a) Chang, C.I., Chien, S.C., Lee, S.M., and Kuo, Y.H. (2003). *Chem. Pharm. Bull.* 51: 1420; (b) Chang, C.I., Chang, J.Y., Kuo, C.C. et al. (2005). *Planta Med.* 71: 72.

75 (a) Kawazoe, K., Yamamoto, M., Takaishi, Y. et al. (1999). *Phytochemistry* 50: 493; (b) Ohtsu, H., Iwamoto, M., Ohishi, H. et al. (1999). *Tetrahedron Lett.* 40: 6419.

76 Katoh, T., Akagi, T., Noguchi, C. et al. (2007). *Bioorg. Med. Chem.* 15: 2736 and references cited therein.

77 McFadden, R.M. and Stoltz, B.M. (2006). *J. Am. Chem. Soc.* 128: 7738.

78 (a) Smith, A.B. III, Cho, Y.S., and Friestad, G.K. (1998). *Tetrahedron Lett.* 39: 8765; (b) Thomas, A.F., Ozainne, M., and Guntz-Dubini, R. (1980). *Can. J. Chem.* 58: 1810.

79 Liang, G., Xu, Y., Seiple, I.B., and Trauner, D. (2006). *J. Am. Chem. Soc.* 128: 11022.

80 Liao, X., Stanley, L.M., and Hartwig, J.F. (2011). *J. Am. Chem. Soc.* 133: 2088.

81 (a) Chieffi, A., Kamikawa, K., Ahman, J. et al. (2001). *Org. Lett.* 3: 1897; (b) Woodward, R.B., Sondheimer, F., Taub, D. et al. (1952). *J. Am. Chem. Soc.* 74: 4223.

82 (a) Corey, E.J. and Chaykovsky, M. (1962). *J. Am. Chem. Soc.* 84: 867; (b) Corey, E.J. and Chaykovsky, M. (1965). *J. Am. Chem. Soc.* 87: 1353; (c) Corey, E.J. and Chaykovsky, M. (1962). *J. Am. Chem. Soc.* 84: 3782.

83 Jana, C.K., Scopelliti, R., and Gademann, K. (2010). *Chem. Eur. J.* 16: 7692.

84 (a) Lin, W.H., Fang, J.M., and Cheng, Y.S. (1995). *Phytochemistry* 40: 871; (b) Geiwiz, J. and Haslinger, E. (1995). *Helv. Chim. Acta* 78: 818.

85 (a) Sharpless, K.B., Amberg, W., Bennani, Y.L. et al. (1992). *J. Org. Chem.* 57: 2768; (b) Meyer, N. and Seebach, D. (1978). *Angew. Chem. Int. Ed.* 17: 521;(c) Loh, T.P. and Hu, Q.Y. (2001). *Org. Lett.* 3: 279.

86 (a) Thommen, C., Jana, C.K., Neuburger, M., and Gademann, K. (2013). *Org. Lett.* 15: 1390; (b) Portmann, C., Prestinari, C., Myers, T. et al. (2009). *ChemBioChem* 10: 889.

87 Deng, J., Li, R., Luo, Y. et al. (2013). *Org. Lett.* 15: 2022.

88 (a) Padwa, A., Chughtai, M.J., Boonsombat, J., and Rashatasakhon, P. (2008). *Tetrahedron* 64: 4758; (b) Ishiyama, T., Miyaura, N., and Suzuki, A. (1993). *Org. Synth.* 71: 89; (c) Winne, J.M., Guang, B., D'herde, J., and De Clercq, P.J. (2006). *Org. Lett.* 8: 4815.

89 (a) Wilds, A.L. and Meader, A.L. Jr. (1948). *J. Org. Chem.* 13: 763; (b) Sudrik, S.G., Chavan, S.P., Chandrakumar, K.R.S. et al. (2002). *J. Org. Chem.* 67: 1574.

90 Ito, Y., Hirao, T., and Saegusa, T. (1978). *J. Org. Chem.* 43: 1011.

91 Yan, X. and Hu, X. (2014). *J. Org. Chem.* 79: 5282.

92 (a) Tomooka, C.S., Liu, H., and Moore, H.W. (1996). *J. Org. Chem.* 61: 6009. (b) Huang, X., Song, L., Xu, J. et al. (2013). *Angew. Chem. Int. Ed.* 52: 952.

93 Kakde, B.N., Kumari, P., and Bisai, A. (2015). *J. Org. Chem.* 80: 9889.

94 (a) Alvarez-Manzaneda, E., Chahboun, R., Cabrera, E. et al. (2009). *J. Org. Chem.* 74: 3384; (b) Majetich, G. and Shimkus, J.H. (2009). *Tetrahedron Lett.* 50: 3311; (c) Kakde, B.N., De, S., Dey, D., and Bisai, A. (2013). *RSC Adv.* 3: 8176; (d) Kakde, B.N., Parida, A., Kumari, P., and Bisai, A. (2016). *Tetrahedron Lett.* 57: 3179; (e) Kakde, B.N., Parida, A., Kumar, N. et al. (2016). *ChemistrySelect* 1: 3357.

95 (a) Roll, D.M., Manning, J.K., and Carter, G.T. (1998). *J. Antibiot.* 51: 635; (b) Geris, R. and Simpson, T.J. (2009). *Nat. Prod. Rep.* 26: 1063; (c) Shiomi, K. and Tomoda, H. (1999). *Pure Appl. Chem.* 71: 1059.

96 (a) Tsujimori, H., Bando, M., and Mori, K. (2000). *Eur. J. Org. Chem.* 297; (b) Tsujimori, H. and Mori, K. (2000). *Biosci. Biotechnol. Biochem.* 64: 1410; (c) Rosen, B.R., Simke, L.R., Thuy-Boun, P.S. et al. (2013). *Angew. Chem. Int. Ed.* 52: 7317; (d) Barrett, T.N. and Barrett, A.G.M. (2014). *J. Am. Chem. Soc.* 136: 17013.

97 Kurdyumov, A.V. and Hsung, R.P. (2006). *J. Am. Chem. Soc.* 128: 6272.

98 (a) Gassman, P.G., Chavan, S.P., and Fertel, L.B. (1990). *Tetrahedron Lett.* 31: 6489; (b) Gassman, P.G. and Lottes, A.C. (1992). *Tetrahedron Lett.* 33: 157.

99 Dixon, D.D., Locker, J.W., Zhou, Q., and Baran, P.S. (2012). *J. Am. Chem. Soc.* 134: 8432.

100 (a) Margaros, I., Montagnon, T., and Vassilikogiannakis, G. (2007). *Org. Lett.* 9: 5585; (b) Wada, K., Tanaka, S., and Marumo, S. (1983). *Agric. Biol. Chem.* 47: 1075; (c) Vlad, P.F., Aryku, A.N., and Chokyrian, A.G. (2004). *Russ. Chem. Bull.* 53: 443; (d) Vlad, P.F., Kuchkova, K.I., Aryku, A.N., and Deleanu, K. (2005). *Russ. Chem. Bull.* 54: 2656.

101 (a) Villamizar, J., Orcajo, A.L., Fuentes, J. et al. (2002). *J. Chem. Res. Synop.* 8: 395; (b) Welch, S.C. and Rao, A.S.C.P. (1977). *Tetrahedron Lett.* 505; (c) Welch, S.C. and Rao, A.S.C.P. (1978). *J. Org. Chem.* 43: 1957; (d) Schröder, J., Magg, C., and Seifert, K. (2000). *Tetrahedron Lett.* 41: 5469; (e) Fenical, W., Sims, J.J., Squatrito, D. et al. (1973). *J. Org. Chem.* 38: 2383.

102 (a) Snyder, S.A., Treitler, D.S., and Brucks, A.P. (2010). *J. Am. Chem. Soc.* 132: 14303; (b) Djura, P., Stierle, D.B., Sullivan, B. et al. (1980). *J. Org. Chem.* 45: 1435; (c) Villamizar, J., Fuentes, J., Tropper, E. et al. (2003). *Synth. Commun.* 33: 1121.

103 (a) Laube, T., Schröder, J., Stehle, R., and Seifert, K. (2002). *Tetrahedron* 58: 4299; (b) Barrero, A.F., Alvarez-Manzaneda, E.J., Chahboun, R. et al. (1999). *Tetrahedron* 55: 15181; (c) Maiti, S., Sengupta, S., Giri, C. et al. (2001). *Tetrahedron Lett.* 42: 2389; (d) Armstrong, V., Barrero, A.F., Alvarez-Manzaneda, E.J. et al. (2003). *J. Nat. Prod.* 66: 1382.

104 Magdziak, D., Rodriguez, A.A., Van De Water, R.W., and Pettus, T.R.R. (2002). *Org. Lett.* 4: 285.

105 (a) Walker, S.D., Barder, T.E., Martinelli, J.R., and Buchwald, S.L. (2004). *Angew. Chem. Int. Ed.* 43: 1871; (b) Yang, L., Williams, D.E., Mui, A. et al. (2005). *Org. Lett.* 7: 1073; (c) Song, F.H., Fan, X., Xu, X.L. et al. (2004). *Chin. Chem. Lett.* 15: 316.

106 (a) Cornforth, J.W. (1968). *Chem. Br.* 4: 102; (b) Geris, R. and Simpson, T.J. (2009). *Nat. Prod. Rep.* 26: 1063.

107 (a) Marcos, I.S., Conde, A., Moro, R.F. et al. (2010). *Mini-Rev. Org. Chem.* 7: 230; (b) Yong, K.W.L., Jankam, A., Hooper, J.N.A. et al. (2008). *Tetrahedron* 64: 6341.

108 (a) Goclik, E., Koenig, G.M., Wright, A.D., and Kaminsky, R. (2000). *J. Nat. Prod.* 63: 1150; (b) Kwak, J.H., Schmitz, F.J., and Kelly, M. (2000). *J. Nat. Prod.* 63: 1153.

109 Mukku, V.J.R.V., Edrada, R.A., Schmitz, F.J. et al. (2003). *J. Nat. Prod.* 66: 686.

110 (a) de la Rojas, Parra, V., Mierau, V., Anke, T., and Sterner, O. (2006). *Tetrahedron* 62: 1828; (b) Mierau, V., de la Rojas, Parra, V., Sterner, O., and Anke, T. (2006). *J. Antibiot.* 59: 53; (c) Liermann, J.C., Kolshorn, H., Anke, H. et al. (2008). *J. Nat. Prod.* 71: 1654.

111 (a) Zhou, Z.-W., Yin, S., Zhang, H.-Y. et al. (2008). *Org. Lett.* 10: 465; (b) Han, M.-L., Zhang, H., Yang, S.-P., and Yue, J.M. (2012). *Org. Lett.* 14: 486.

112 Kalesnikoff, M., Sly, L.M., Hughes, M.R. et al. (2004). *Rev. Physiol. Biochem. Pharmacol.* 149: 87.

113 Yang, L., Williams, D.E., Mui, A. et al. (2005). *Org. Lett.* 7: 1073.

114 Kuchkova, K.I., Chumakov, Y.M., Simonov, Y.A. et al. (1997). *Synthesis* 1045.

115 Zhang, L., Xie, X., Liu, J. et al. (2011). *Org. Lett.* 13: 2956.

116 (a) Paquette, L.A., Wang, T.Z., and Sivik, M.R. (1994). *J. Am. Chem. Soc.* 116: 2665; (b) Paquette, L.A., Wang, T.Z., and Sivik, M.R. (1994). *J. Am. Chem. Soc.* 116: 11323.

117 (a) Stork, G., Rosen, P., and Goldman, N.L. (1961). *J. Am. Chem. Soc.* 83: 2965; (b) Stork, G., Rosen, P., Goldman, N. et al. (1965). *J. Am. Chem. Soc.* 87: 275; (c) Hiroya, K., Ichihashi, Y., Furutono, A. et al. (2009). *J. Org. Chem.* 74: 6623.

118 Alvarez-Manzaneda, E., Chahboun, R., Alvarez, E. et al. (2012). *Chem. Commun.* 48: 606.

119 Akhaouzan, A., Fernandez, A., Mansour, A.I. et al. (2013). *Org. Biomol. Chem.* 11: 6176.

120 Kakde, B.N., Kumar, N., Mondal, P.K., and Bisai, A. (2016). *Org. Lett.* 18: 1752.

121 For torquoselectivity in Nazarov-type cyclizations, see: (a) Hutson, G.E., Türkmen, Y.E., and Rawal, V.H. (2013). *J. Am. Chem. Soc.* 135: 4988; (b) Denmark, S.E., Wallace, M.A., and Walker, C.B. Jr. (1990). *J. Org. Chem.* 55: 5543.

5

Protecting-Group-Free Synthesis of Heterocycles

Trapti Aggarwal and Akhilesh K. Verma

Department of Chemistry, University of Delhi, India

5.1   Introduction

Heterocyclic compounds play a pivotal role in the metabolism of living cells. If at least one atom other than carbon forms a part of the ring system, then it is designated as a heterocyclic compound. Nitrogen, oxygen, and sulfur are the most common heteroatoms, but heterocyclic rings containing other heteroatoms are also widely known. A heterocycle is enormously distributed in natural products as well as major components of biological molecules such as DNA, RNA, vitamins, and other biomolecules. They have a broad range of applications in pharmaceuticals [1] and have been extensively used in agrochemicals, dyes, biocide formulations, and polymer science [2]. Some of these compounds exhibit a significant solvatochromic, photochromic, and biochemiluminescence properties. Thousands of natural products contain heterocycles as their central building blocks. They also have applications in supramolecular and polymer chemistry. One of the remarkable features of heterocycles is their ability to manifest substituents around a core scaffold, which made them distinctive and important for the drug industry. Moreover, they act as organic conductors, semiconductors, molecular wires, photovoltaic cells, organic light-emitting diodes (OLEDs), light harvesting systems, optical data carriers, chemically controllable switches, and liquid crystalline compounds [3]. The ability of many heterocycles to produce stable complexes with metal ions has great biochemical significance. Therefore, substantial attention has been paid to develop efficient new methods to synthesize heterocycles. Complex heterocyclic compounds are elaborated by microorganisms and are useful as antibiotics in medicines. Marine animals and plants are also a source of complex heterocyclic compounds and are receiving much attention in current research efforts. In the past centuries, synthesis of heterocyclic molecules has been undertaken by many organic chemists. However, the major challenge faced in organic synthesis is the construction of complex molecules without using protecting-group chemistry and thus complete the reaction sequence with less number of steps.

Due to the vast importance of heterocycles, their synthesis has attracted the interest of organic chemists. In this regard, many reports are available in the literature for the

Protecting-Group-Free Organic Synthesis: Improving Economy and Efficiency, First Edition.
Edited by Rodney A. Fernandes.
© 2018 John Wiley & Sons Ltd. Published 2018 by John Wiley & Sons Ltd.

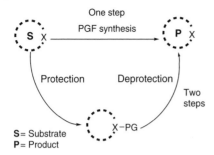

Scheme 5.1 General representation of protecting-group-free strategy.

synthesis of complex heterocycles using protecting-group-free (PGF) strategies that provide step-economical approach to develop new complex molecules or natural products in fewer steps. Incorporation and removal of protecting groups leads to loss of material that further decreases the yield of product (Scheme 5.1). Also, the subsequent atoms and reagents used for the introduction/removal of the blocking group are not present in the final product, so it should be separated from the desired compound and discarded; this is unfavorable due to loss of atoms. For the execution of "PGF" synthesis, the introduction of chemoselective methodologies is proved to be crucial.

The introduction of protecting groups in a reaction sequence increases the overall waste production, leading to violation of the principles of ideal synthesis and the rules of green chemistry [4]. Despite various limitations, the majority of total syntheses use protecting-group strategies [5]. Although shortcomings associated with the use of protecting groups in synthesis have been identified, their incorporation into a synthetic strategy has led to an explosive growth in the complexity of targeted compound synthesis.

The concept of PGF synthesis has previously been reviewed, but recent advances in the design of reagents/processes that induce a high degree of functional group discrimination have led to a surge in the number of complex molecules prepared without protecting groups. During the past decade, PGF strategy for the synthesis of natural products and complex molecules became an emerging area in research. The significance of chemoselective reagents/strategies in accomplishing this task should not be underestimated, and this chapter focuses on various methodologies used for the synthesis of heterocycles.

5.2 Historical Background of Protection-Free Strategy

The concept of PGF synthesis was not extensively explored until the twenty-first century for the synthesis of small heterocyclic compounds, despite there being few reports for the total syntheses that did not use protecting groups. The potential reason for this may be due to the limited functionality in the intermediates, or final product negates competing reactivity that must be masked. PGF syntheses can occur out of necessity and not synthetic planning, as the primary route features a protecting group that either cannot be removed or alters the reactivity of the system in a detrimental way [6]. Another pathway in which the absence of blocking groups has been reported in the past is in biomimetic strategies used for synthesis of complex molecules. Basically protection-free strategy has been employed for the synthesis of natural products. In this chapter we discuss the utility of protection-free synthesis for the development of *N-*, *O-*, and *S-*heterocycles.

For the protection-free synthesis, there is need to use strategic disconnections in order to shield perceived functional group incompatibilities and harness the reactivity of specific functional groups within a complex setting. PGF strategy has emerged as a powerful tool for the synthesis of a variety of natural products containing *N-*, *O-*, and *S-*heterocycles.

5.3 Protecting-Group-Free (PGF) Strategy for the Synthesis of *N*-Heterocycles

5.3.1 Carbazole Substituted Compounds Using PGF Strategy

Previously, the synthesis of complex molecules required multistep procedures involving protection and deprotection of functional groups to achieve the goal. The main interest of organic chemists in PGF strategy has been focused toward the synthesis of natural products as the introduction of protecting groups led to lower yields and efficiency of synthesis.

An important β-carboline analog, borrerine (**4**), employed for the synthesis of various natural products has been developed by PGF strategy in 2012 [7]. May's group has designed the synthesis of borrerine (**4**) from tryptamine (**1**) using Sakai's approach [8] (Scheme 5.2). When borrerine (**4**) was treated with methyl triflate in CHCl$_3$ at 0 °C and then treated with 2.0 equiv of TFA and warmed to 22 °C, the dimethylamine alkaloids **5**, i.e. flinderoles, were formed directly in a 70% combined yield. Flinderoles are important alkaloids isolated from plant genus *Flindersia*, which show excellent antimalarial activity [9]. Earlier reports in the 1970s for the synthesis of flinderoles required more than 15 synthetic steps. Previously, the flinderole skeleton could also arise from a dimerization of borrerine through the reaction of indolodiene with imine. However, this would proceed through a formal [3+2] cycloaddition pathway. This strategy was supported by

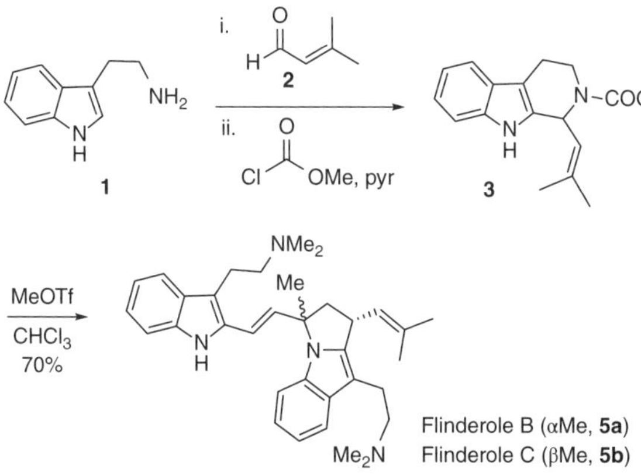

Scheme 5.2 Protection-free synthesis of flinderoles B (**5a**) and C (**5b**).

earlier reported biosynthesis of alkene-appended indoles through [3+2] cycloaddition [10]. Thus the synthesis of antimalarial flindersial alkaloids has been developed by acid-promoted dimerization of borrerine by methylation followed by acid treatment. The sequence allows selective access to alkaloids in only three synthetic steps without using any protection on the NH of indole.

In 2014, Knölker and coworkers reported the Pd(II)-catalyzed synthesis of formyl carbazole analog, murrayaline C (**11**) in a step-economical manner (Scheme 5.3) [11]. Murrayaline C was an important alkaloid isolated in 1986 by Furukawa et al. from the stem bark of *Murraya euchrestifolia* Hayata [12]. The key step in this strategy was a late-stage reduction of cyano group to form the formyl group at C8 position. The reaction proceeded with Buchwald–Hartwig coupling [13] of aniline **6** with benzonitrile **7** in the presence of Pd(II) catalyst to give **8**. Later, the oxidative cyclization of **8** by heating at 120 °C in pivalic acid and catalytic amount of Pd(OAc)$_2$ provided cyano-substituted carbazole **9** in 77% yield. The reduction of the cyano group by treatment with DIBAL-H in CH$_2$Cl$_2$ at low temperature followed by oxidation using MnO$_2$ provided the carbazole-3,8-dicarbaldehyde **10**. Subsequently, cleavage of silyl ether with TBAF at −10 °C afforded the targeted molecule, murrayaline C (**11**), in six steps with overall 20% yield. This methodology described the synthesis of substituted carbazoles through palladium(II)-catalyzed oxidative cyclizations of diarylamines.

An isochromene-fused carbazole core-containing molecule, murrayamine-O (**20**), was isolated from root barks of *M. euchrestifoli* [14]. Dethe's group reported the Lewis acid-catalyzed diastereoselective synthesis of these alkaloids by coupling reaction (Scheme 5.4) [15]. The synthesis of murrayamine-O (**20**) was initiated by coupling substituted aniline **12** with phenylboronic acid **13** in the presence of copper acetate to provide intermediate **14** in 86% yield. The cyclization using Pd(OAc)$_2$ in DMF gave the substituted carbazole **15** in 81% yield, and further demethylation in boron tribromide at 0 °C led to the formation of **16** in 94% yield. The coupling of **16** with (*S*)-*cis*-verbenol

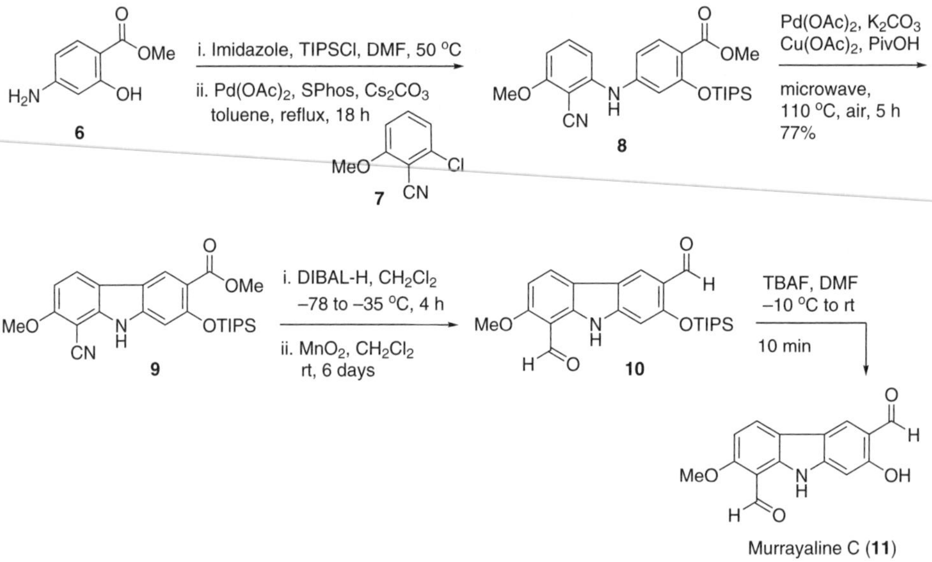

Scheme 5.3 Synthesis of functionalized carbazole murrayaline C (**11**).

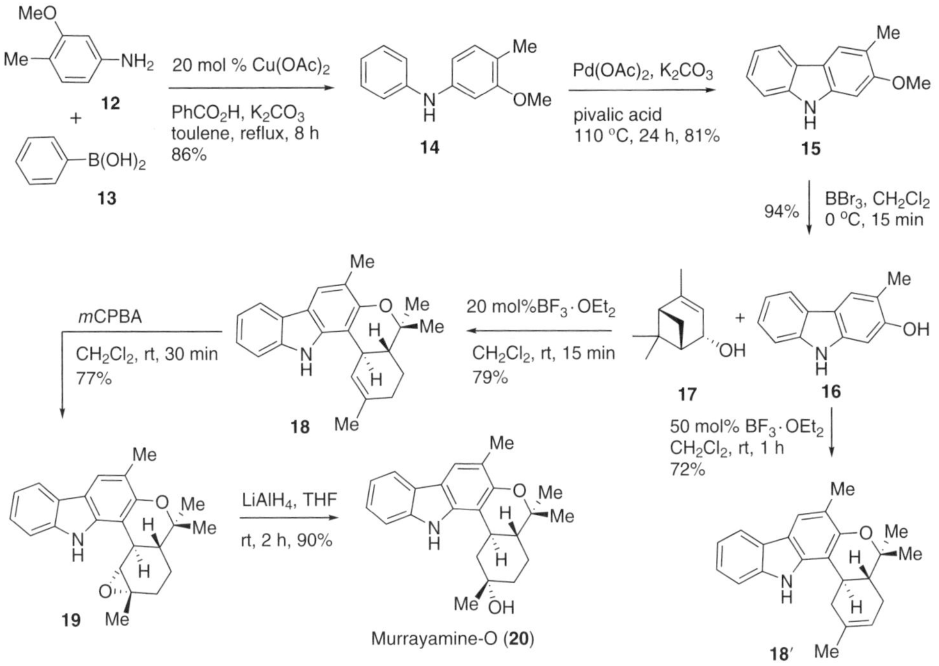

Scheme 5.4 Synthesis of isochromene-fused carbazole murrayamine-O (**20**).

(**17**) in the presence of 20 mol% BF$_3$·OEt$_2$ afforded the single diastereomer **18** in 79% yield. Here, the amount of BF$_3$·OEt$_2$ plays an important role as increasing the quantity of BF$_3$·OEt$_2$ from 20 to 50 mol% afforded another diastereomer **18′** in 72% yield. The compound **18** on treatment with *m*CPBA afforded the epoxide **19** in 77% yield which on reduction with LiAH$_4$ at room temperature led to the formation of murrayamine-O (**20**) in 90% yield.

In 2016, Banwell and coworkers [16] reported the protection-free synthesis of carbazole analog **24**, which further generates alkaloid kopsihainanine A (**25**; Scheme 5.5), which was isolated in 2011 by Gao's group from *Kopsia hainanensis* [17]. The synthesis

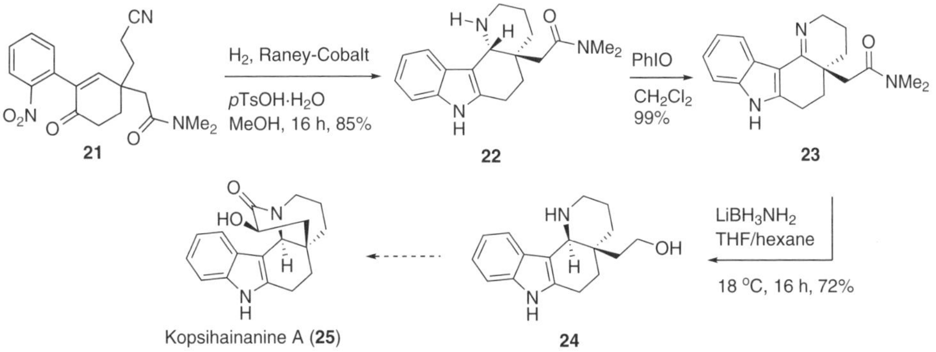

Scheme 5.5 Protection-free synthesis of carbazole analog **24**.

began with hydrogenation followed by tandem cyclization of substrate **21** in one step using Raney cobalt to give fused ring molecule **22** in 85% yield. The oxidation of amine **22** in the presence of iodosylbenzene in CH_2Cl_2 generated corresponding imine **23** in 99% yield. The exhaustive reduction of **23** with lithium amido borohydride ($LiBH_3NH_2$) smoothly provided the corresponding amino alcohol **24** in 72% yield. This reaction sequence provided a direct route for converting a *cis*-ring-fused molecule **22** into the corresponding *trans*-form **24** without using any protecting groups on the free NH. The inverted stereochemistry brought the nonindolic nitrogen into close proximity of the angular β-hydroxyethyl group in the corresponding product **24** that triggers the cyclization for the formation of natural product, kopsihainanine A (**25**), in further steps.

Later, in 2016, Nagarajan and coworkers demonstrated the PGF synthesis of carboxy-substituted β-carboline molecule, metatacarboline A (**33**; Scheme 5.6) [18]. This secondary metabolite was isolated from fruiting bodies of *Mycena metata* in 2013 by Spiteller et al. [19]. The reaction sequence for the synthesis of β-carboline **29** began with the reductive amination of tryptophan **26** and 2,2-dimethoxyacetaldehyde **27** in the presence of trifluoroacetic acid in CH_2Cl_2 at room temperature to give acetal intermediate **28**, which on oxidation followed by hydrolysis gave aldehyde-substituted carboline **29**. Further, the Wittig reaction using phosphorus ylide **30** afforded a mixture of *cis*- and *trans*-alkene products **31**, which were subjected to hydrogenation using 10 mol% Pd/C to give compound **32** in 92% yield (over two steps). The hydrolysis of corresponding compound **32** using aqueous lithium hydroxide afforded metatacarboline A (**33**).

A versatile approach for the synthesis of carbazoles in a protection-free manner has been described by Verma's group using transition metal-catalyzed tandem reactions (Scheme 5.7) [20]. The synthesis began by reacting indole **34** with acrylates in the presence of Pd(II) catalyst to form monoacrylated compounds **35**, which subsequently led to the generation of substituted carbazoles **36** by successive oxidative Heck reactions.

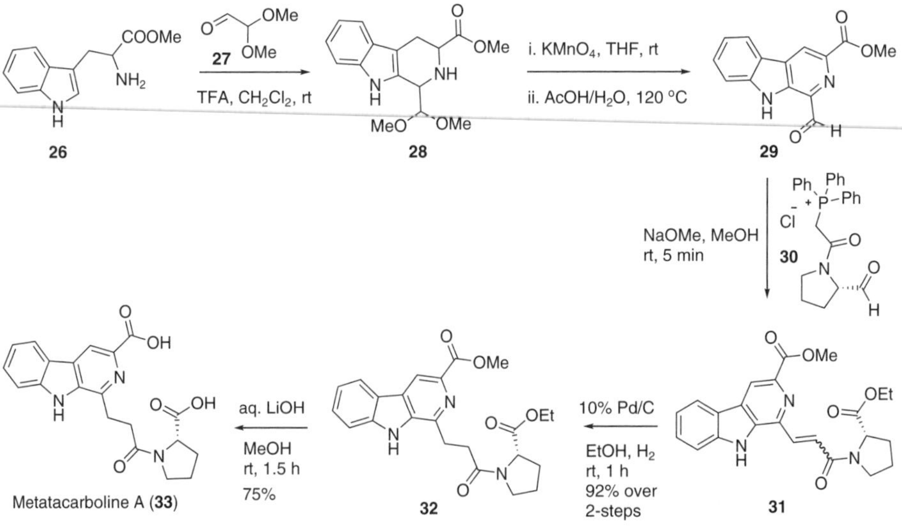

Scheme 5.6 Protecting-group-free synthesis of metatacarboline A (**33**).

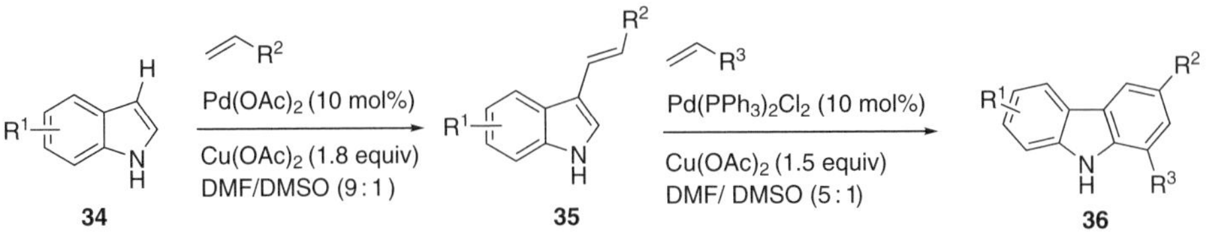

34 35 36

R^1 = H, Br, OMe, NO_2, aryl
R^2 = COOMe, COOEt, COMe, aryl
R^3 = CN, COOMe, COOEt

Scheme 5.7 Free NH-based synthesis of carbazoles **36**.

The designed protocol represents the direct synthesis of functionalized carbazoles from free NH-containing indoles via regioselective triple successive oxidative Heck reactions. Interestingly, the nitrile substitution at R^3 also afforded the C3-functionalized carbazole **36** in good yield via C–H activation instead of reported [4+2] cycloaddition product [21]. The reaction conditions tolerated both electron-deficient and electron-rich alkenes for the incorporation of two different functional groups in the product.

5.3.2 Protection-Free Synthesis of Indole-Substituted Compounds

An bisindolylmaleimide molecule, arcyriarubin A (**40**), was made with a PGF strategy through the coupling between **37** and oxalyl chloride followed by quenching the reaction with methanol at 0 °C to give methyl indole-3-glyoxylate (**38**) in 69% yield (Scheme 5.8) [22]. Further the reaction of **38** with indole-3-acetamide (**39**) in the presence of *t*BuOK formed the desired product arcyriarubin A (**40**) in 96% yield.

Davison and Sperry envisaged the PGF synthesis of C3-substituted indole molecule bufopyramide (**45**), from the coupling of 5-methoxytryptamine **41** with carboxy-substituted pyrrole **42** in the presence of the coupling reagent HATU to give amide-substituted indole **43** in 64% yield (Scheme 5.9) [23]. The selective acetylation of the amide **43** in the presence of the unprotected indole and pyrrole using both acetic anhydride and acetyl chloride led to the formation of product **44** in 38% yield. Due to the partial anionic character of the amide oxygen, the acylation of amides proceeds through an initial O-acylation and rearrangement of the resulting isoimide to the imide to give **44**. The subsequent demethylation in the presence of boron tribromide led to the formation of desired product bufopyramide (**45**). Bufopyramide (**45**) developed by this strategy is a noteworthy alkaloid isolated from the traditional Chinese medicine *ChanSu* used for centuries as an anti-inflammatory, anti-arrhythmic, and anesthetic and in chemotherapy regimens [24].

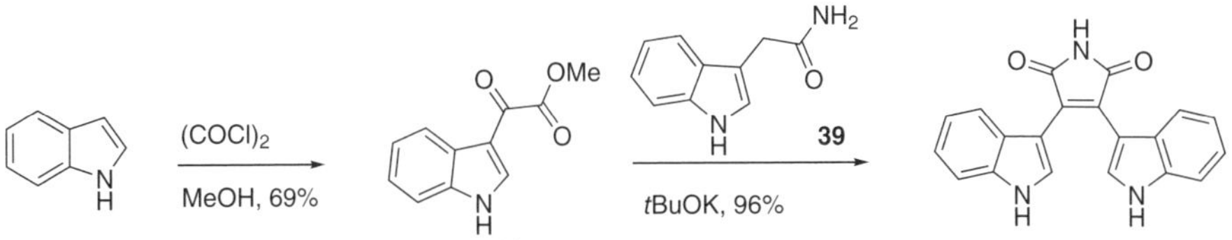

37 38 Arcyriarubin (**40**)

Scheme 5.8 Synthesis of arcyriarubin (**40**) from indole.

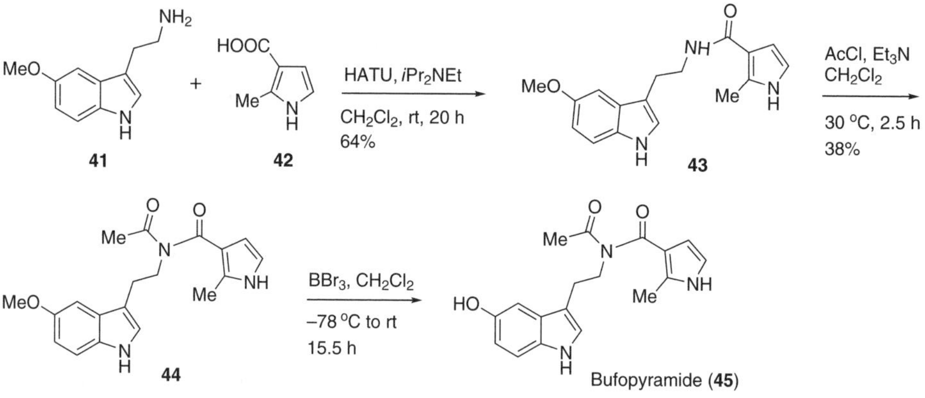

Scheme 5.9 Synthesis of C3-substituted indole compound bufopyramide (**45**).

5.3.3 Synthesis of Pyrrole Analogs Using Protecting-Group-Free Strategy

An interesting synthesis of hydroxypyrrolidine (**50**) from D-arabinose (**46**) has been disclosed by Timmer and Stocker in 2010 (Scheme 5.10) [25]. The reaction was initiated by the formation of iodinated methyl glycoside **47** from **46** in the presence of iodine, triphenylphosphine, and imidazole in two steps with overall 64% yield. Later, the tetrahydrofuran (THF) ring was opened through activated zinc, NaCNBH$_3$, a saturated solution of NH$_4$OAc in ethanol and aqueous NH$_3$ to generate alkenylamine **48** in 93% yield obtained by purification via Dowex H$^+$ resin. Further, the carbamate annulation methodology using iodine and base in aqueous medium gave carbamate **49** in 93% yield. The hydrolysis of carbamate gave corresponding pyrrolidine **50** in 97% yield. This represents an efficient synthesis of pyrrolidine in overall 54% yield in five steps.

An important pyrrole-substituted natural product, marinopyrrole (**58**), having antimicrobial properties was discovered by Clive and Cheng in 2013 [26]. The key intermediate used for the synthesis of marinopyrrole was bipyrrole **53** having a free pyrrole nitrogen atom, which was obtained from the condensation of aminopyrrole **51** and α-ketoester **52** in toluene (Scheme 5.11) [27]. The bipyrrole was converted into Weinreb amides **54** using *N,O*-dimethylhydroxylamine. The corresponding diamide **54** was treated with organolithium reagent **55** to give bisketone **56** in 86% yield. Later, the halogenation of bipyrrole **56** was achieved by using the *N*-chlorosuccinimide along with 10 mol% of additive (2-iodobenzoyl)glycine in CH$_2$Cl$_2$ in 84% yield. The cleavage of the *O*-methyl groups under acidic conditions afforded the marinopyrrole (**58**).

The alkaloid pandamarine (**67**) having diazaspiro unit was isolated by Byrne's group from the leaves of the tropical plant *Pandanus amaryllifolius* [28] (Scheme 5.12). The key intermediate required for the synthesis of pandamarine (**67**) was symmetric difuran **65**. The synthesis was initiated by addition of alkyl lithium reagent on 5-hexynenitrile **59** to form alkynyl lithium anion, which on reaction with 1-hydroxypropan-2-one **60** generated alkynyldiol **61** in 67% yield [29]. The cyclodehydration of **61** using silver nitrate furnished the substituted furan **62**, which on reduction with DIBAL or LiAlH$_4$ afforded the corresponding intermediates **63** and **64**. The reductive amination of substituted aldehyde **63** and amine **64** in the presence of sodium borohydride afforded the symmetric difuran **65** in 81% yield. The photooxygenation of symmetric difuran **65**

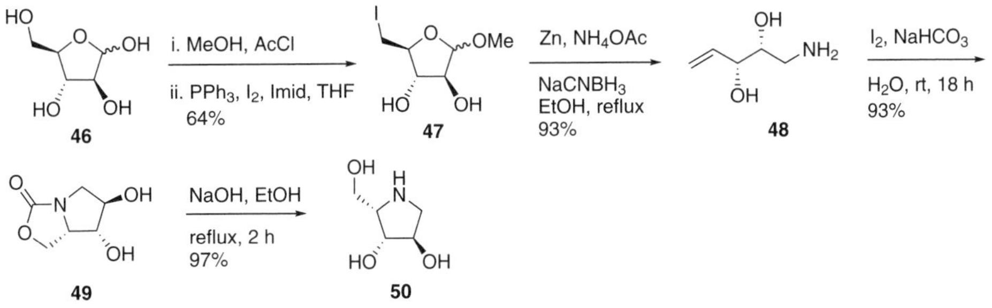

Scheme 5.10 Synthesis of pyrrolidine **50** from unprotected carbohydrate **46**.

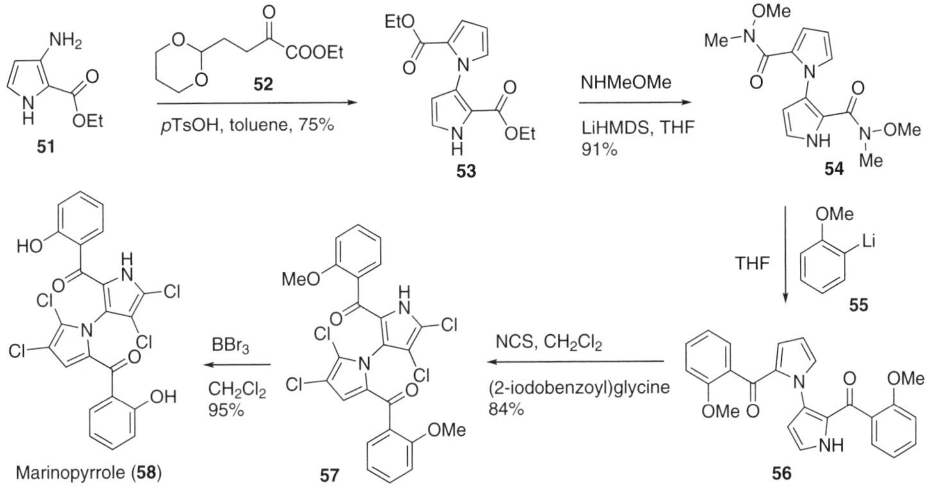

Scheme 5.11 Protecting-group-free synthesis of marinopyrrole (**58**).

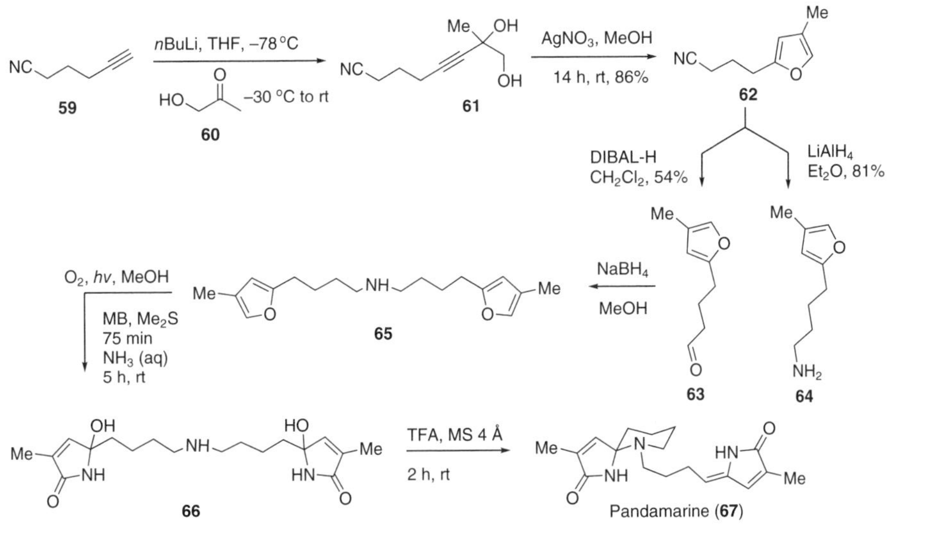

Scheme 5.12 Concise synthesis of pandamarine (**67**) without using protecting groups.

using ammonia as a nitrogen source provided symmetric pyrrolidinone **66**, which was subsequently transformed into pandamarine alkaloid (**67**) in the presence of TFA with overall 30% yield for two steps.

5.4 Protection-Free Synthesis of Quinoline Derivatives

A protection-free synthesis of quinoline derivative actinophenanthroline A (**76**) has been developed by Lindsley and coworkers in 2016 using nitration and reduction strategy (Scheme 5.13) [30]. The regioselective nitrosylation of 2-methylquinolin-8-ol (**68**) to **69** and reduction with tin(II) chloride and concentrated HCl subsequently afforded amine **70**. For obtaining pure amine **70**, various concentrations of HCl were explored, and it was found that HCl at a concentration of 2.2 M led to 98% yield of compound **70**. The pyridine ring annulation using **70** and *trans*-crotonaldehyde **71** in toluene (6.0 M) afforded the tricyclic ring molecule **72** in 90% yield. Again nitrosylation followed by reduction in the presence of Pd/C afforded the aminophenol derivative **74**, which was subsequently coupled with (S)-2-hydroxypropanoic acid (**75**) in the presence of HATU to give the desired actinophenanthroline A (**76**).

In 2014, Selig et al. reported the synthesis of quinolines from protected amine **77** and allene ester **78** via Michael addition followed by aldol condensation in 72% yield (Scheme 5.14) [31]. The *N*-Boc group interestingly got transferred from N to C by 1,3-acyl shift and became part of the products. However, in 2016 Verma and coworkers described the synthesis of quinolines via intermolecular [4+2] cycloaddition of 2-aminobenzyl alcohol in the presence of base without N-protection (Scheme 5.15) [32]. The methodology involves the formation of azadiene intermediate **82** from aminobenzyl alcohol (**81**) in the presence of KOH/DMSO, which on cycloaddition with internal and terminal alkynes **83** afforded the substituted quinolines **84**. The reported strategy has overcome two major regioselectivity challenges as the substrate: amino alcohol has two nucleophilic sites, and similarly the intermediate azadienes also has two nucleophilic sites. Therefore it is interesting that the quinoline products **84** in particular were obtained in good yields without using any protecting group.

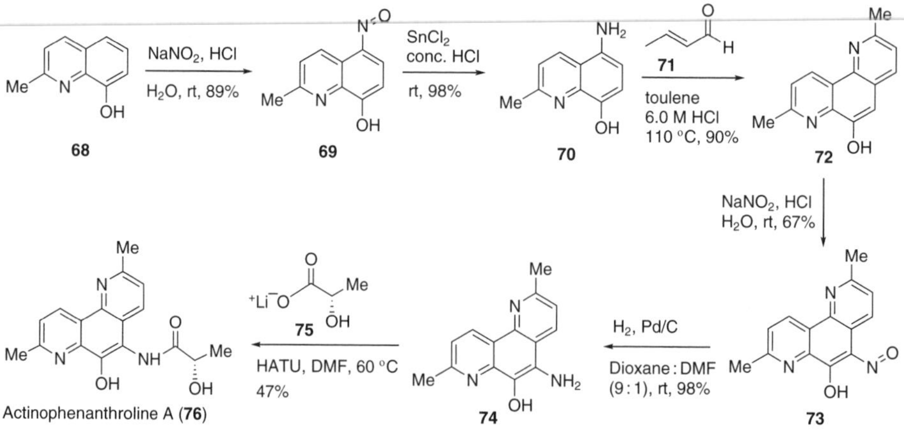

Scheme 5.13 Protection-free synthesis of quinoline derivative **76**.

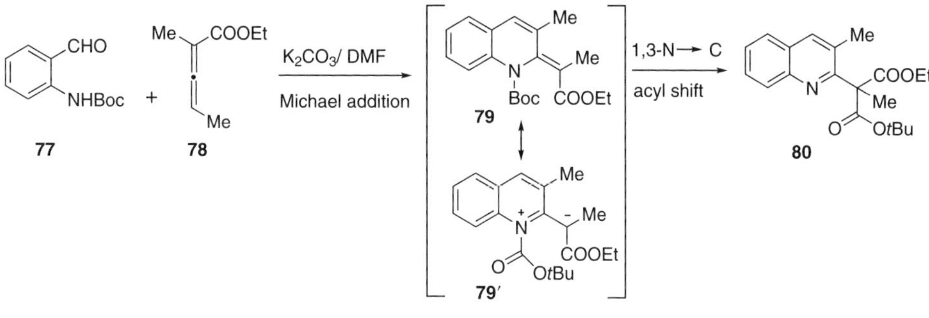

Scheme 5.14 Synthesis of quinoline **80** using protecting group.

R^1 = H, Me, Br, Cl
R^2 = H, Me
R^3 = H, aryl

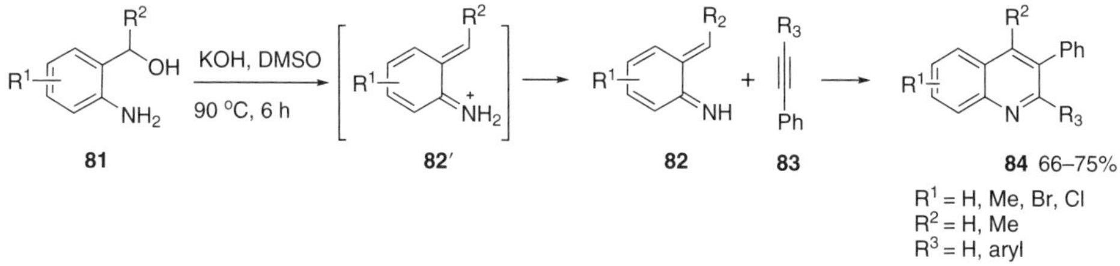

Scheme 5.15 Protection-free approach for the synthesis of substituted quinolines **84**.

5.5 Synthesis of Piperidine-Containing Heterocycles Without Using Protecting Groups

The piperidine-substituted drug (*S*)-lubeluzole (**90**) exhibits potential therapeutic effects for the treatment of acute ischemic stroke (Scheme 5.16). Chakraborti's group has described a water-promoted short sequence for the synthesis of lubeluzole (**90**) [33]. The reaction sequence involved the formation of C–N bond via epoxide aminolysis with subsequent nucleophilic substitution of 2-chlorobenzothiazole (**89**). The water-mediated ring opening of (*S*)-2-(chloromethyl)oxirane (**85**) was followed by N-alkylation of piperidone **86**, which reacted with 3,4-difluorophenol to form *O*-alkylated product **87**. Then, reductive amination triggered by Ti(O*i*Pr)$_4$ followed by treatment with sodium

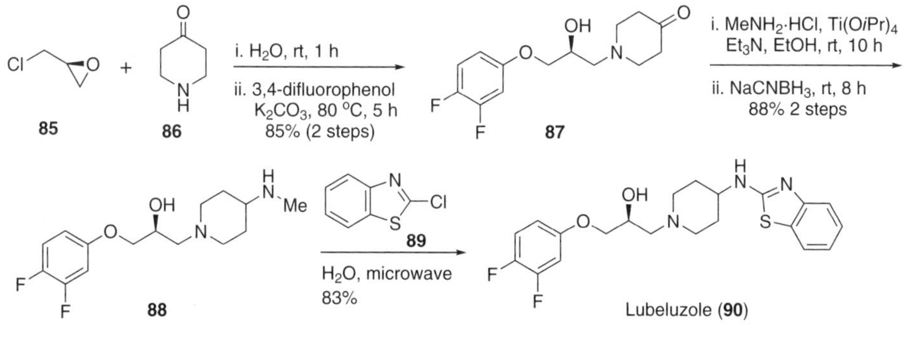

Scheme 5.16 Protection-free synthesis of lubeluzole (**90**).

cyanoborohydride generated intermediate **88**. The target product, (*S*)-lubeluzole (**90**), was then obtained by reacting **88** with 2-chlorobenzothiazole (**79**) under microwave irradiation in water with overall 62% yield.

5.6 Synthesis of Quinazolines Without Using Protecting Groups

Quinazolinone alkaloid asperlicins C (**95**) and **E** [34] are major nonpeptidic antagonists of the gastrointestinal hormone neurotransmitter cholecystokinin (CCK) [35], which were isolated from the fermentation broths produced by *Aspergillus alliaceus* (Scheme 5.17). The synthesis of asperlicin C began with the reaction of isatoic anhydride (**91**) with tryptamine (**1**) in the presence of Et$_3$N followed by cyclization in refluxing glacial AcOH to generate benzodiazepine **92** in one-pot manner [36]. The latter was successively treated with Et$_3$N, DMAP, and *o*-nitrobenzoyl chloride **93** at −20 °C to give the selective *N*-aroylated product **94**. Further the reductive cyclization with titanium (obtained from a mixture of 8.0 equiv of Zn and 4.0 equiv of TiCl$_4$) afforded asperlicin C (**95**).

Another remarkable synthesis of tetrahydroquinazolines **100** was reported by Verma's group in 2015 from 2-aminophenylacrylate with isothiocyanates via amidation and concomitant chemoselective Michael addition reactions (Scheme 5.18) [37]. The reaction proceeds with the coupling of substituted iodo/bromoaniline **96** and acrylate **97**

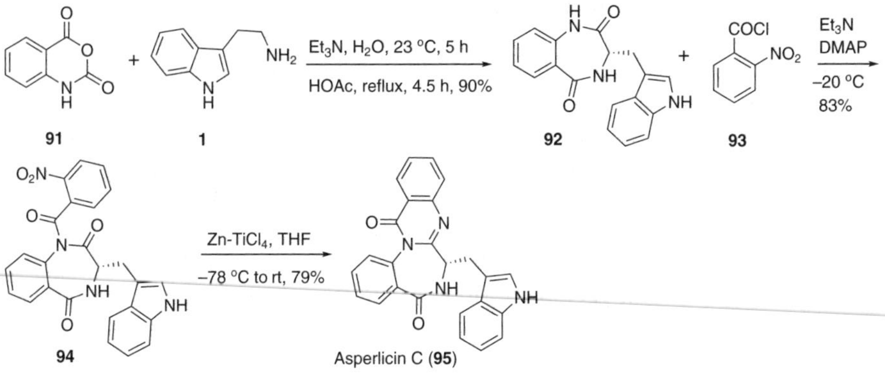

Scheme 5.17 Protection-free synthesis of substituted quinazolinone, asperlicin C (**95**).

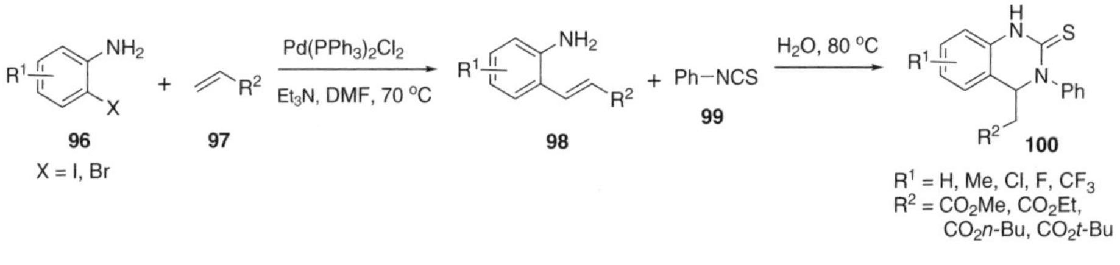

Scheme 5.18 Synthesis of tetrahydroquinazolines **100**.

using Pd catalyst to afford the aminophenyl derivative **98**, which on condensation with aryl isothiocyanate **99** in the presence of water gave the desired tetrahydroquinazoline **100** without using any protecting group on the amine function.

5.7 Protection-Free Synthesis of Pyrrolizine Alkaloid (−)-Rosmarinecine

The first total synthesis of optically pure rosmarinecine (**107**) was reported by Tatsuta et al. and was completed in 18 steps with 7% overall yield [38]. In 2012, Akai's group reported the short and concise reaction sequence for the asymmetric synthesis of rosmarinecine (**107**) in four steps (Scheme 5.19) [39]. The synthesis began with the formation of nitrone intermediate **102** by oxidation of hydroxypyrrolidine **101** in the presence of *tert*-butylamine at room temperature. Then, the lipase CAL-B-catalyzed domino reaction with 1-ethoxyvinyl ethyl fumarate (**103**) resulted in **104**, which immediately cyclized to give intermediate **105** selectively with 99% ee. Later, the reductive cleavage of N—O bond in the presence of Pd(OH)$_2$ to **106** (55%) and treatment with Red-Al afforded (−)-rosmarinecine (**107**) in 90% yield.

5.8 Protecting-Group-Free Synthesis of *O*-Heterocycles

In 2014, Mhaske and coworkers demonstrated an efficient PGF strategy for the synthesis of *O*-heterocycle containing natural product 6-*epi*-cleistenolide **113** (Scheme 5.20) [40]. In 2007, Nkunya et al. [41] reported the isolation of (−)-cleistenolide, which is an analog of the structural motif isolated from *Cleistochlamys kirkii*, a plant species (Annonaceae) found in Tanzania and Mozambique. This plant extract has been used in the treatment of wound infections, rheumatism, and tuberculosis. Cleistenolide has shown *in vitro* antibacterial activity against *Staphylococcus aureus* and *Bacillus anthracis*, as well as antifungal activity against *Candida albicans*. The synthesis began with the benzoylation of furylallyl alcohol (**108**) that was synthesized by Knoevenagel/

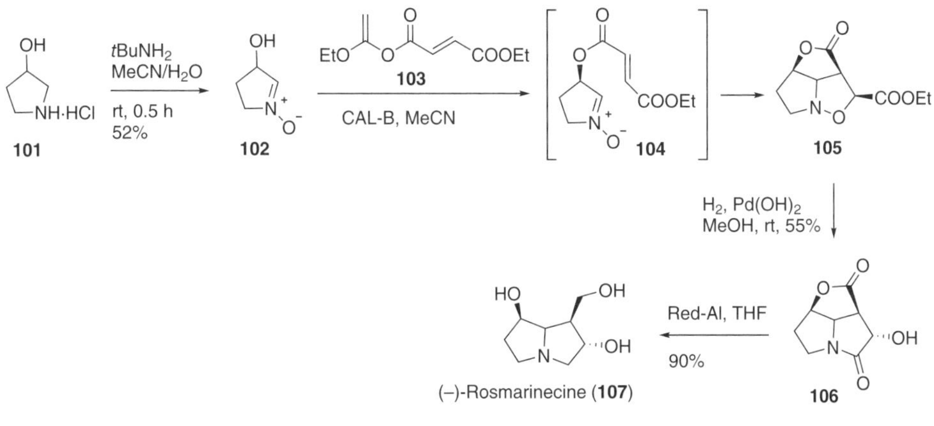

Scheme 5.19 Protecting-group-free synthesis of (−)-rosmarinecine (**107**).

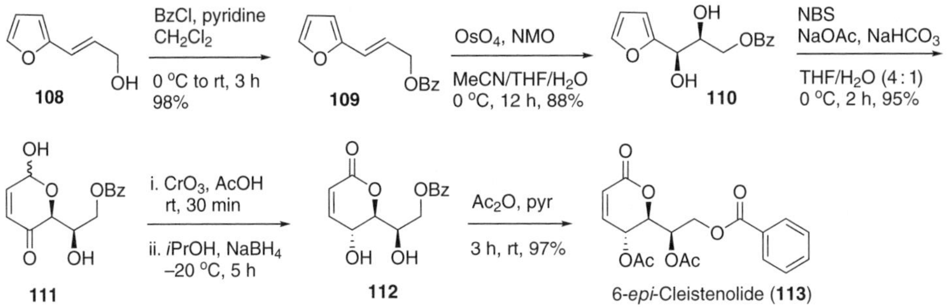

Scheme 5.20 Protection-free synthesis of (±)-6-*epi*-cleistenolide (**113**).

Claisen–Schmidt condensation on furfural followed by reduction [42]. The benzoylated product **109** upon treatment with osmium tetroxide in MeCN/THF/water (4 : 2 : 1) gave furan diol **110** in 88% yield at 0 °C in 12 h. The next synthetic transformation was the conversion of the furan core into a dihydropyran in THF/water (4 : 1) with NBS, NaOAc, and NaHCO₃ through Achmatowicz reaction to give dihydropyranone **111** in 95% yield. Further, the oxidation of pyran using CrO₃ in AcOH and then reaction with sodium borohydride resulted in the formation of intermediate **112**. Subsequent acetylation generated 6-*epi*-cleistenolide (**113**).

Sasai and coworkers reported the concise PGF synthesis of chromanol molecule cordiachromene (**116**) from hydroquinone **114** (Scheme 5.21) [43]. The cordiachromene was isolated from an American tree *Cordia alliodora* and later from *Aplidium constellatum* [44]. The synthesis was accomplished by direct conversion of unprotected hydroquinone and geraniol in the presence of BF₃·Et₂O and dioxane to give **115**. Treating the intermediate **115** using stoichiometric amount of Pd(OCOCF₃)₂ provided the cyclized product cordiachromene (**116**).

Helicascolide C (**123**) isolated from marine alga *Gracilaria* sp. SGR-1 showed potent antifungal activity against phytopathogenic fungus [45]. In 2012, the total synthesis of helicascolides A and C by acid-catalyzed acetonide deprotection followed by lactonization has been disclosed by Krishna et al. in a multistep sequence [46a]. Later in 2016, Breit and coworkers reported a short and concise synthesis of helicascolides A and C without using protecting groups (Scheme 5.22) [46b]. The synthesis was initiated by treatment of the ester **117** with chiral homoallenyl aldehyde using LDA in THF to give **118**. Further treatment with TFA cleaved *tert*-butyl ester and generated the allenyl intermediate acid **119** in excellent yields. The latter on intramolecular hydroxycarbonylation of carboxylic acid to allene in the presence of [Rh(cod)Cl]₂ and DPEphos furnished the substituted pyran-2-one **120**. This was subjected to Pd-catalyzed Wacker oxidation to give **121** and further Wittig reaction followed by treatment with mild

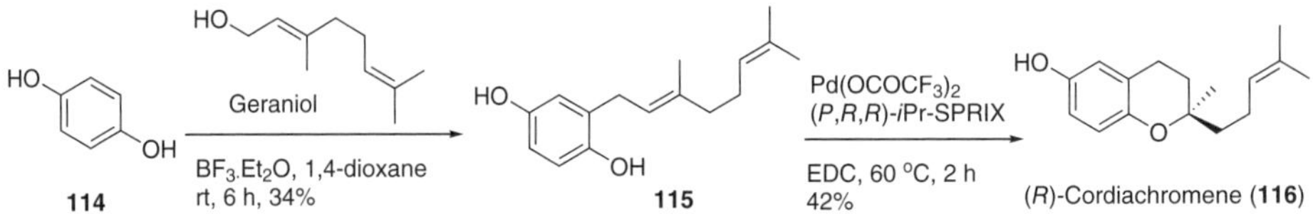

Scheme 5.21 Protection-free synthesis of cordiachromene (**116**).

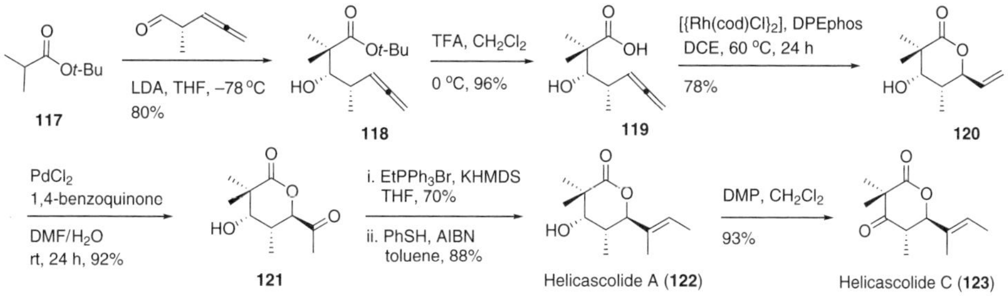

Scheme 5.22 Protection-free approach for the synthesis of helicascolides A (**122**) and C (**123**).

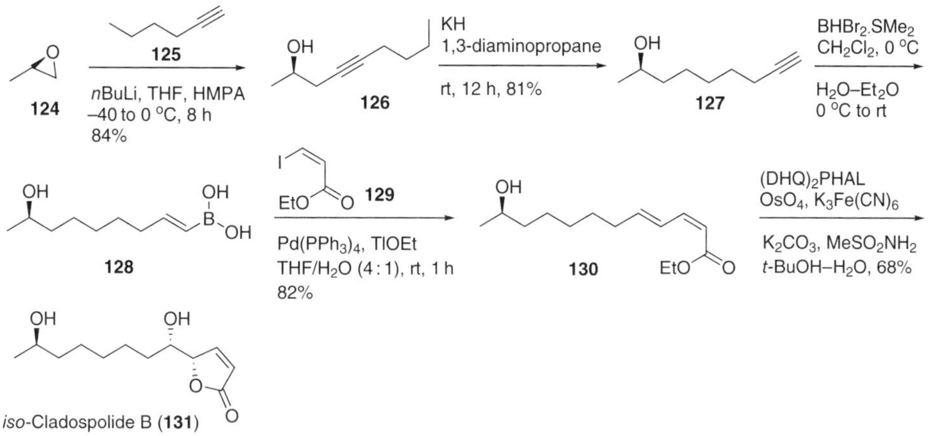

Scheme 5.23 Synthesis of *iso*-cladospolide B (**131**).

radical initiator AIBN in toluene generated helicascolide A (**122**) with good stereoselectivity. The desired lactone, helicascolide C (**123**), was obtained in excellent yield by oxidation of helicascolide A (**122**) with DMP.

The synthesis of furanone analog *iso*-cladospolide B (**131**) proceeded from (2*R*)-2-methyloxirane (**124**) and alkyne **125**, which were transformed into intermediate **126** using *n*BuLi, HMPA, and THF at −40 to 0 °C (Scheme 5.23) [47]. Alkynol **126** on treatment with potassium 3-aminopropylamide (obtained from the reaction of KH and 1,3-diaminopropane) generated **127** through alkyne-zipper reaction, in which the alkyne functionality moved from the internal position to give the terminal alkyne. The alkyne intermediate was subjected to dibromoborane-dimethyl sulfide in dichloromethane to give (*E*)-vinylboronic acid **128** by hydroboration. Then, thallium ethoxide-induced Suzuki coupling of **128** with the *Z*-vinyl iodide **129** gave the (*Z,E*)-dienoate **130** exclusively in 82% yield. Sharpless asymmetric dihydroxylation of dienoate **130** in the presence of osmium tetroxide, hydroquinine 1,4-phthalazinediyl diether, and *tert*-butanol/water gave the desired (4*S*,5*S*,11*R*)-*iso*-cladospolide B (**131**).

Bothwell and Luo reported the protection-free synthesis of Se-adenosyl-L-methionine (SeAM) (**140**) analog from *S*-methyl-L-methionine (**132**; Scheme 5.24) [48]. The latter was treated with methyl iodide and sodium bicarbonate to generate hydroxyl

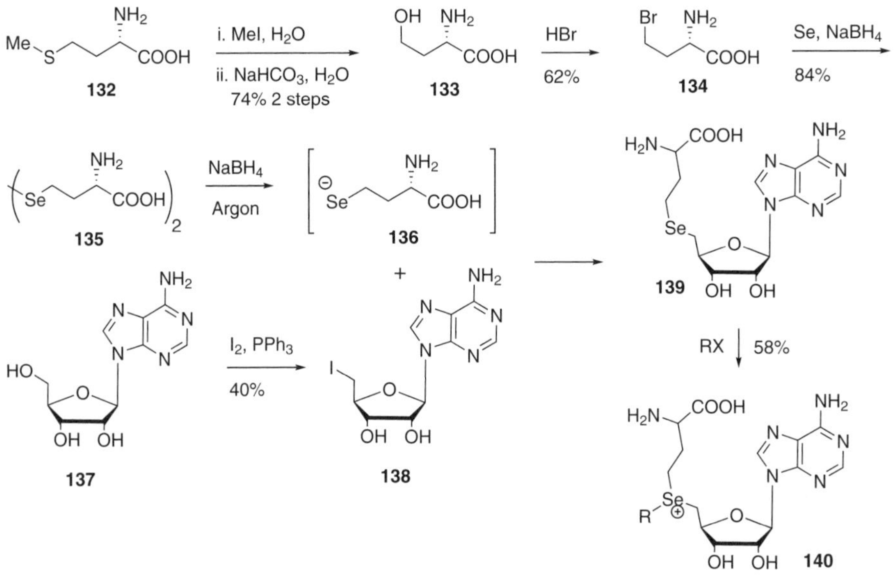

Scheme 5.24 Synthesis of selenium-containing heterocycle **140**.

intermediate **133**, which in the presence of hydrogen bromide was converted into **134** in 62% yield. The intermediate **134** on reaction with Na_2Se_2 (generated *in situ* from selenium powder) gave the key building block selenohomocystine **135** in 84% yield, which was purified by solid-phase extraction with acid-activated Dowex 50WX4 resin. The intermediate **138** was readily prepared from adenosine via Appel reaction [49]. The coupling of intermediate **136** (from **135**) with **138** followed by alkylation afforded the target SeAM (**140**) in moderate yield. This strategy represents the convenient route for selenium-containing nucleotide without using any protection on hydroxyl group present in the substrate.

5.9 Protecting-Group-Free Synthesis of *N,S*-Heterocycles

The specific G protein-coupled receptor subtypes (CCK1 and CCK2) regulate the biological effects of CCK in the gastrointestinal system (Scheme 5.25). CCK is a 33-amino acid peptide predominantly found in the central nervous system [50]. It was found that dual CCK1/CCK2 receptor antagonist **146** provides efficient pharmacologic properties. The PGF synthesis of **146** begins with the formation of amide bond from **142** (obtained from **141** by reaction with thionyl chloride in methanol) and **143** [51]. The coupled intermediate **144** was achieved using DMAP in presence of DMF at room temperature and isolated in 56% yield through column chromatography. The formation of amide bond prior to the sulfonamide eliminates the protection at amine functional group in substrate **141** and thus reduced the overall number of steps in reaction sequence. The CCK1/CCK2 receptor antagonist **146** was obtained on coupling of intermediate **144** with substituted sulfonyl chloride **145** followed by treatment with lithium hydroxide in THF/H_2O at room temperature. Benzothiazines are widely known due to

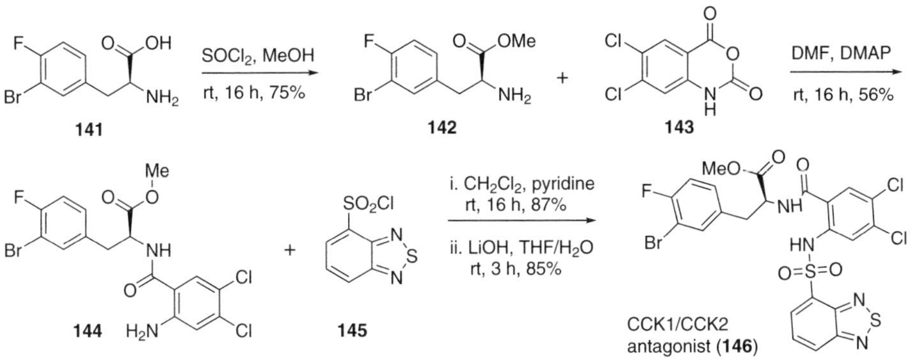

Scheme 5.25 Protection-free synthesis of CCK1/CCK2 inhibitor (**146**).

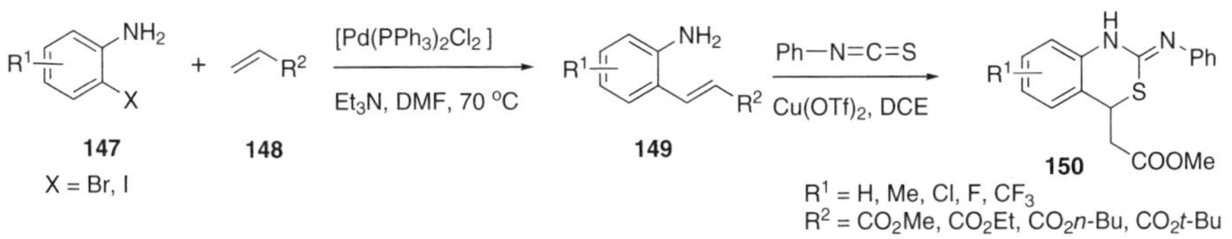

Scheme 5.26 Domino approach for the synthesis of benzothiazine analogs **150**.

their presence in nonracemic building blocks as well as in natural products. A metal-catalyzed PGF domino approach for the synthesis of 2-imino[1,3]benzothiazines has been developed by Verma's group (Scheme 5.26) [52]. The synthesis was initiated with the Heck coupling reaction of acrylates **148** with iodo/bromoaniline **147** in the presence of Pd catalyst in DMF at 70 °C to generate acrylated product **149**. The coupled product **149** on reaction with isothiocyanates under catalytic amount of Cu(OTf)$_2$ gave corresponding benzothiazines **150** via *in situ* thioamidation and concomitant chemoselective intramolecular thia-Michael addition in good yields.

5.10 Protection-Free Synthesis of Macrocyclic Ring Heterocycles

The synthetic macrocycles have gained attention recently, as they were approved as drugs for the treatment of hepatitis C [53]. Particularly, macrocycles have broad relevance in protein–protein interactions (PPI), currently a therapeutic domain mostly covered by antibodies [54]. In the first multicomponent reaction, an unprotected α-amino acid **152** on reaction with benzaldehyde **151** and isocyanide **153** in methanol yielded monoamide intermediate **154** (Scheme 5.27) [55]. Diamine-derived mono-isocyanide was used in order to provide the isocyano-ω carboxylic acid linker **157**, which was macrocyclized with the help of a second MCR, the classical U-4CR, to yield the 21-membered **158**. The overall synthesis is exemplified in not more than five steps and could result in very diverse macrocycles of different sizes and substitution patterns. Importantly, in the U-5C-4CR/U-4CR strategies, the presence of an additive for the final MCR ring closure is crucial.

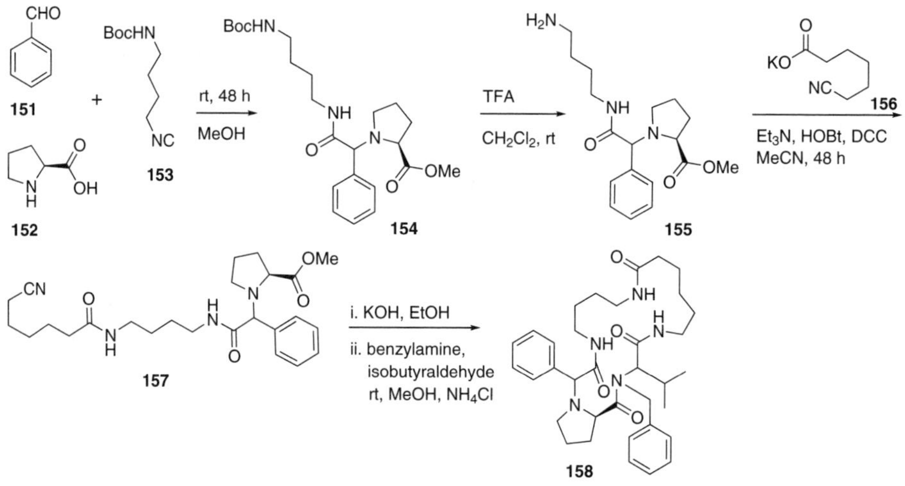

Scheme 5.27 Protection-free synthesis of macrocyclic ring **158**.

5.11 Protection-Free Synthesis of Thiophene Polymer

Protection-free and catalyst-transfer polycondensation using tBu$_4$ZnLi$_2$ for the synthesis of P3HHT (**161**) has been described by Higashihara et al. (Scheme 5.28) [56]. There has been increasing interest in poly(3-alkylthiophene) (PAT) molecules due to their presence in the field of polymer electronic devices [57] as the PAT molecules are best known for their properties useful in balancing high-performance materials. The synthesis was initiated with the halogen–metal exchange of 6-(2-bromo-5-iodothiophen-3-yl) hexan-1-ol **159** with an equivalent molar amount of tBu$_4$ZnLi$_2$ in THF at 0 °C to obtain the zincate complex monomer **160**. Further, the polymerization of **160** was performed through polycondensation using catalytic amount of 1,3-bis(diphenylphosphino) ethanenickel dichloride (Ni(dppe)Cl$_2$) in THF, which afforded the P3HHT (**161**) successfully in 89% yield.

5.12 Protection-Free Synthesis of Azaborine

The 1,2-azaborine synthon **165** has been synthesized by employing PGF strategy (Scheme 5.29) [58]. The potassium allyltrifluoroborate on reaction with allylamine in the presence of TMSCl and acetonitrile gave intermediate **163**. Then, the ring-closing metathesis with Schrock's Mo catalyst furnished the heterocycle **164** in 94% yield with N–H bond intact. Later, the substitution of the allylamino fragment with n-butanol followed by oxidation in the presence of Pd/C afforded the desired synthon **165**. 1,2-Azaborine is a strong inhibitor for the hydroxylation of ethylbenzene by ethylbenzene dehydrogenase (EbDH), thus giving evidence for the concept that azaborine synthon can lead to novel biochemical behavior [59].

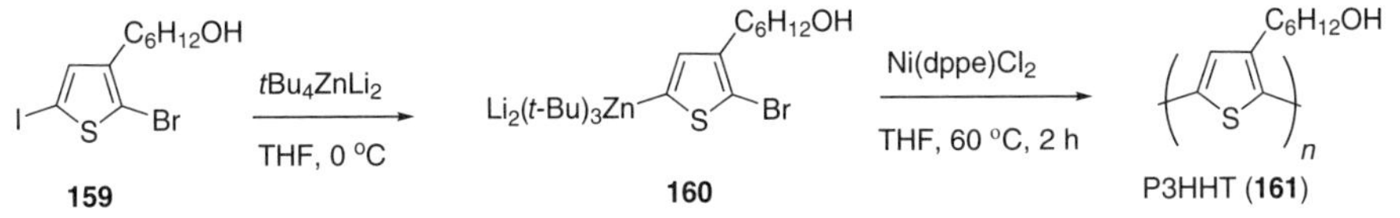

Scheme 5.28 Protection-free synthesis of substituted thiophene polymer **161**.

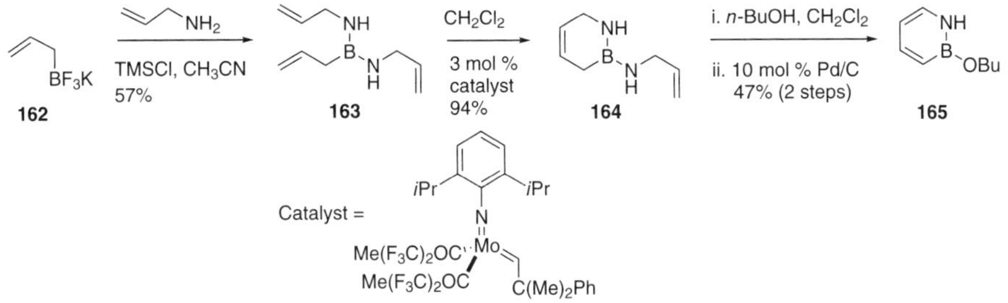

Scheme 5.29 Synthesis of azaborine synthon **165**.

5.13 Conclusion

The introduction of PGF strategy for the synthesis of complex organic molecules has been widely accepted as an important guiding principle of modern synthetic planning, which results in short sequence synthesis. Thus, PGF synthesis will remain a strong stimulus for invention and for the increased efficiency and overall economy of organic synthesis, which will revolutionize and empower chemical synthesis. New reaction inventions and significant optimizations are often necessary to achieve the products through PGF synthesis. Thus, PGF strategy for the total synthesis of natural products and small heterocycles has gained much interest in the past decade among various organic chemists.

References

1 (a) Mittal, A. (2009). *Sci. Pharm.* 77: 497. (b) Joule, J.A. and Mills, K. (2000). *Heterocyclic Chemistry*, 4e, vol. 369. Blackwell Publishing. (c) Sperry, J.B. and Wright, D.L. (2005). *Curr. Opin. Drug. Discov. Devel.* 8: 723. (d) Katritzky, A.R. (1992). *Chem. Heterocycl. Compd.* 28: 241.

2 Dua, R., Shrivastava, S., Sonwane, S.K., and Srivastava, S.K. (2011). *Adv. Biol. Res.* 5: 120 and references cited therein.

3 (a) Chan, C.-Y., Wong, Y.-C., Chan, M.-Y. et al. (2016). *ACS Appl. Mater. Interfaces* 8: 24782. (b) Fan, C. and Yang, C. (2014). *Chem. Soc. Rev.* 43: 6439.

4 (a) Anastas, P.T. and Warner, J.C. (1998). *Green Chemistry: Theory and Practice*. Oxford Univ. Press. (b) Li, C.-J. and Trost, B.M. (2008). *Proc. Natl. Acad. Sci. U. S. A.* 105: 13197.

5 (a) Green, T.W. and Wuts, P.G.M. (1999). *Protective Groups in Organic Synthesis*, 3e. Wiley. (b) Kocienski, P.J. (2005). *Protecting Groups*, 3e. Thieme. (c) Corey, E.J. and Venkateswarlu, A. (1992). *J. Am. Chem. Soc.* 94: 6190.

6 (a) Wender, P.A., Verma, V.A., Paxton, T.J., and Pillow, T.H. (2008). *Acc. Chem. Res.* 41: 40. (b) Trost, B.M. (1991). *Science* 254: 1471.

7 (a) Vallakati, R. and May, J.A. (2012). *J. Am. Chem. Soc.* 134: 6936. (b) Kaur, R., Garg, Y., and Pandey, S.K. (2016). *ChemistrySelect* 1: 4286. (c) Zeldin, R.M. and Toste, F.D. (2011). *Chem. Sci.* 2: 1706.

8 (a) Demaindreville, M., Levy, J., Tillequin, F., and Koch, M. (1983). *J. Nat. Prod.* 46: 310. (b) Yamanaka, E., Shibata, N., and Sakai, S. (1984). *Heterocycles* 22: 371.

9 (a) Fernandez, L.S., Jobling, M.F., Andrews, K.T., and Avery, V.M. (2008). *Phytother. Res.* 22: 1409. (b) Fernandez, L.S., Buchanan, M.S., Carroll, A.R. et al. (2009). *Org. Lett.* 11: 329.

10 (a) Dethe, D.H., Erande, R.D., and Ranjan, A. (2011). *J. Am. Chem. Soc.* 133: 2864. (b) Dethe, D.H., Erande, R.D., and Ranjan, A. (2013). *J. Org. Chem.* 78: 10106.

11 Hesse, R., Krahl, M.P., Jäger, A. et al. (2014). *Eur. J. Org. Chem.* 4014.

12 Furukawa, C., Ito, C., Yogo, M., and Wu, T.-S. (1986). *Chem. Pharm. Bull.* 34: 2672.

13 (a) Louie, J., Driver, M.S., Hamann, B.C., and Hartwig, J.F. (1997). *J. Org. Chem.* 62: 1268. (b) Wolfe, J. and Buchwald, S. (1996). *J. Org. Chem.* 61: 1133.

14 Wu, T.-S., Wang, M.-L., and Wu, P.-L. (1995). *Tetrahedron Lett.* 36: 5385.

15 Dethe, D.H., Das, S., Dherange, B.D., and Mahapatra, S. (2015). *Chem. Eur. J.* 21: 8347.

16 Tan, S.H., Banwell, M.G., and Willis, A.C. (2016). *J. Org. Chem.* 81: 8022.

17 Chen, J., Chen, J.-J., Yao, X., and Gao, K. (2011). *Org. Biomol. Chem.* 9: 5334.

18 Naveen, B., Mudiraj, A., Khamushavalli, G. et al. (2016). *Eur. J. Med. Chem.* 113: 167.

19 Jaeger, R.J.R., Lamshöft, M., Gottfried, S. et al. (2013). *J. Nat. Prod.* 76: 127.

20 (a) Verma, A.K., Danodia, A.K., Saunthwal, R.K. et al. (2015). *Org. Lett.* 17: 3658. (b) Saunthwal, R.K., Patel, M., Kumar, S. et al. (2015). *Chem. Eur. J.* 21: 18601. (c) Saunthwal, R.K., Saini, K.M., Patel, M., and Verma, A.K. (2017). *Tetrahedron* 73: 2415.

21 (a) Ohashi, M., Takeda, I., Ikawa, M., and Ogoshi, S. (2011). *J. Am. Chem. Soc.* 133: 18018. (b) Ohashi, M., Ikawa, M., and Ogoshi, S. (2011). *Organometallics* 30: 2765.

22 Gao, Y.-C., Jia, Y.-H., Li, T.-T. et al. (2015). *ARKIVOC* (v): 153.

23 Davison, E.K. and Sperry, J. (2015). *Org. Biomol. Chem.* 13: 7911.

24 (a) Enomoto, A., Rho, M.C., Komiyama, K., and Hayashi, M. (2004). *J. Nat. Prod.* 67: 2070. (b) Qin, T.J., Zhao, X.H., Yun, J. et al. (2008). *World J. Gastroenterol.* 14: 5210. (c) Ko, W.S., Park, T.Y., Park, C. et al. (2005). *Oncol. Rep.* 14: 475. (d) Zhang, P., Cui, Z., Liu, Y. et al. (2005). *Chem. Pharm. Bull.* 53: 1582.

25 Dangerfield, E.M., Gulab, S.A., Plunkett, C.H. et al. (2010). *Carbohydr. Res.* 345: 1360.

26 Clive, D.L.J. and Cheng, P. (2013). *Tetrahedron* 69: 5067.

27 Stodulski, M., Kohlhepp, S.V., Raabe, G., and Gulder, T. (2016). *Eur. J. Org. Chem.* 2170.

28 Byrne, L.T., Guevara, B.Q., Patalinghug, W.C. et al. (1992). *Aust. J. Chem.* 45: 1903.

29 Kalaitzakis, D., Noutsias, D., and Vassilikogiannakis, G. (2015). *Org. Lett.* 17: 3596.

30 Garcia-Barrantes, P.M., Harp, J.R., and Lindsley, C.W. (2016). *Tetrahedron Lett.* 57: 2194.

31 Selig, P. and Raven, W. (2014). *Org. Lett.* 16: 5192.

32 (a) Saunthwal, R.K., Patel, M., and Verma, A.K. (2016). *Org. Lett.* 18: 2200. (b) Saunthwal, R.K., Patel, M., and Verma, A.K. (2016). *J. Org. Chem.* 81: 6563.

33 Kommi, D.N., Kumar, D., Seth, K., and Chakraborti, A.K. (2013). *Org. Lett.* 15: 1158.

34 (a) Bock, M.G., Dipardo, R.M., Pitzenberger, S.M. et al. (1987). *J. Org. Chem.* 52: 1644. (b) Goetz, M.A., Monaghan, R.L., Chang, R.S.L. et al. (1988). *J. Antibiot.* 39: 875. (c) Liesch, J.M., Hensens, O.D., Zink, D.L., and Goetz, M.A. (1988). *J. Antibiot.* 41: 878.

35 (a) Dunlop, J. (1998). *Gen. Pharmacol.* 31: 519. (b) Herranz, R. (2003). *Med. Res. Rev.* 23: 559.

36 Huang, P.-Q., Wang, Y., Luo, S.-P. et al. (2015). *Tetrahedron Lett.* 56: 1255.

37 Saunthwal, R.K., Patel, M., Tiwari, R.K. et al. (2015). *Green Chem.* 17: 1434.

38 (a) Tatsuta, K., Takahashi, H., Amemiya, Y., and Kinoshita, M. (1983). *J. Am. Chem. Soc.* 105: 4096. For reviews, see also: (b) Tatsuta, K. and Hosokawa, S. (2005). *Chem. Rev.* 105: 4707. (c) Tatsuta, K. and Hosokawa, S. (2006). *Sci. Technol. Adv. Mater.* 7: 397.

39 Nemoto, H., Tanimoto, K., Kanao, Y. et al. (2012). *Tetrahedron* 68: 7295.

40 Mahajan, P.S., Gonnade, R.G., and Mhaske, S.B. (2014). *Eur. J. Org. Chem.* 8049.

41 Samwel, S., Mdachi, S.J.M., Nkunya, M.H.H. et al. (2007). *Nat. Prod. Commun.* 2: 737.

42 Charette, A.B., Molinaro, C., and Brochu, C. (2001). *J. Am. Chem. Soc.* 123: 12168.

43 Takenaka, K., Tanigaki, Y., Patil, M.L. et al. (2010). *Tetrahedron: Asymmetry* 21: 767.

44 (a) Manners, G.D. and Jurd, L. (1977). *J. Chem. Soc. Perkin Trans. 1.* 405. (b) Targett, N.M. and Keeran, W.S. (1984). *J. Nat. Prod.* 47: 556.

45 Tarman, K., Palm, G.J., Porzel, A. et al. (2012). *Phytochem. Lett.* 5: 83.

46 (a) Krishna, P.R., Mallula, V.S., and Kumar, P.V.A. (2012). *Tetrahedron Lett.* 53: 4997. (b) Haydl, A.M., Berthold, D., Spreider, P.A., and Breit, B. (2016). *Angew. Chem. Int. Ed.* 55: 5765.

47 Reddy, C.R., Rao, N.N., Sujitha, P., and Kumar, C.G. (2012). *Synthesis* 1663.

48 Bothwell, I.R. and Luo, M. (2014). *Org. Lett.* 16: 3056.

49 Aa Perrone, P., Daverio, F., Valente, R. et al. (2007). *J. Med. Chem.* 50: 5463.

50 (a) Herranz, R. (2003). *Med. Res. Rev.* 23: 559. (b) Tonini, M., De Giorgio, R., and De Ponti, F. (2004). *Drugs* 64: 347. (c) Varnavas, A. and Lassiani, L. (2006). *Expert Opin. Ther. Pat.* 16: 1193. (d) Berna, M.J., Tapia, J.A., Sancho, V., and Jensen, R.T. (2007). *Curr. Opin. Pharmacol.* 7: 583.

51 Liu, J., Deng, X., Fitzgerald, A.E. et al. (2011). *Org. Biomol. Chem.* 9: 2654.

52 Saunthwal, R.K., Patel, M., Kumar, S., and Verma, A.K. (2015). *Tetrahedron Lett.* 56: 677.

53 Raboisson, P., De Kock, H., Rosenquist, A. et al. (2008). *Bioorg. Med. Chem. Lett.* 18: 4853.

54 Khoury, K., Holak, T.A., and Dömling, A. (2013). *Protein–Protein Interactions in Drug Discovery. Methods and Principles in Medicinal Chemistry*, vol. 56, 129. Wiley-VCH.

55 Liao, G.P., Abdelraheem, E.M.M., Neochoritis, C.G. et al. (2015). *Org. Lett.* 17: 4980.

56 Higashihara, T., Goto, E., and Ueda, M. (2012). *ACS Macro Lett.* 1: 167.

57 (a) Kim, J.Y., Lee, K., Coates, N.E. et al. (2007). *Science* 317: 222. (b) Thompson, B.C. and Fréchet, J.M.J. (2008). *Angew. Chem. Int. Ed.* 47: 58. (c) Wei, Q., Nishizawa, T., Tajima, K., and Hashimoto, K. (2008). *Adv. Mater.* 20: 2211. (d) Chen, J. and Cao, Y. (2009). *Acc. Chem. Res.* 42: 1709. (e) Cheng, Y.-J., Yang, S.-H., and Hsu, C.-S. (2009). *Chem. Rev.* 109: 5868. (f) Green, M.A., Emery, K., Hishikawa, Y., and Warta, W. (2010). *Prog. Photovolt. Res. Appl.* 18: 144.

58 Abbey, E.R., Lamm, A.N., Baggett, A.W. et al. (2013). *J. Am. Chem. Soc.* 135: 12908.

59 Knack, D.H., Marshall, J.L., Harlow, G.P. et al. (2013). *Angew. Chem. Int. Ed.* 52: 2599.

6

Protecting-Group-Free Synthesis of Drugs and Pharmaceuticals

Remya Ramesh¹, Swapnil Sonawane², D. Srinivasa Reddy¹, and Rakeshwar Bandichhor²

¹ *Division of Organic Chemistry, CSIR-National Chemical Laboratory, Pune, India*
² *API Research & Development, Integrated Product Development, Dr. Reddy's Laboratories Ltd., Hyderabad, India*

6.1   Introduction

Environmental factor (E-factor) or its more meaningful version, process mass intensity (PMI), is the biggest stigma attached to industrial operations. By definition, it is how much waste gets generated in a process of making a unit quantity of desired material, roughly +100 units. Isn't it awful? Doesn't it pose a challenge to chemists and engineers? We think it does, but how many of us really take it as a challenge to design green synthetic routes that might eventually help sustain the processes when these become part of industrial operations? In chemical industry, route selection is one of the most important components of successful execution. Protection–deprotection has a reputation of contributing to the E-factor, and intelligent synthetic design can help to avoid this, eventually leading to the development of much greener, cost-effective, and operationally simple processes. In particular, a commercial route has to be cost effective, and a protection–deprotection sequence adds at least two steps, which makes it more expensive. These two additional steps lead to poor process efficiency and yield. In order to avoid these losses, one has to come up with the protection- and deprotection-free synthesis by adopting route selection through scientific guidelines.

In the recent scenario, optimal use of protecting groups toward efficient syntheses of active pharmaceuticals with high atom and step economy is a formidable challenge [1–4]. To complicate matters further, the material that corresponds to the protecting group (and reagents used for its introduction/removal) must be separated from the desired compound and discarded. This increases the overall waste production, leading to violation of ideal synthesis rules and the 12 green chemistry principles [5] that encompass many factors besides reaction yield when evaluating the efficiency of a route. Despite the obvious limitations, majority of modern-day total syntheses use protecting groups, and comprehensive books dedicated to their use are commonly found in synthetic laboratories [1]. To our knowledge, Robinson's

Protecting-Group-Free Organic Synthesis: Improving Economy and Efficiency, First Edition.
Edited by Rodney A. Fernandes.
© 2018 John Wiley & Sons Ltd. Published 2018 by John Wiley & Sons Ltd.

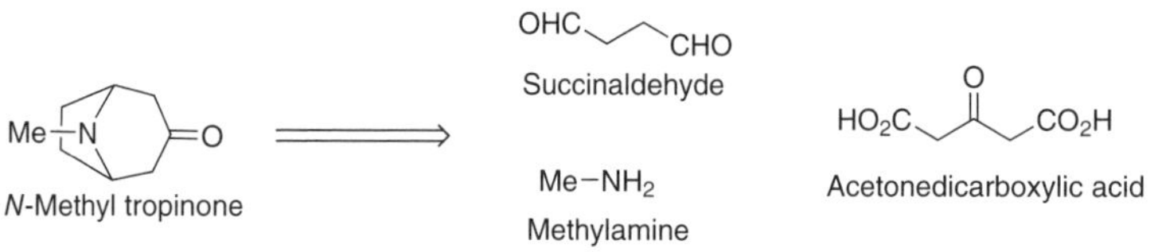

Figure 6.1 Synthesis of N-methyl tropinone.

synthesis of N-methyl tropinone (precursor to atropine, a CNS drug) is the first protection-free synthesis accomplished in 1917 [6]. The one-step synthesis of N-methyl tropinone involves a reaction of succinaldehyde, methylamine, and acetonedicarboxylic acid in water, which initiates a cascade sequence invoking imine and Mannich chemistry, followed by a final double decarboxylation in acidic media to furnish N-methyl tropinone (Figure 6.1).

While other green chemistry features of drugs and pharmaceuticals synthesis are touched upon, we highlight the significance of syntheses that have been managed by avoiding protection–deprotection strategies through case studies. In this chapter we have rationalized and chosen to include the syntheses of ten pharmaceutically relevant molecules of interest that feature protection–deprotection-free (or minimal) approaches. The first is raltegravir potassium (MK-0518), essentially a member of a new class of compounds that is found to be extremely effective in managing the life-threatening disease AIDS. This works as an inhibitor of HIV integrase, an enzyme that catalyzes the integration of viral DNA into the host genomic DNA, thus preventing further virus replication. Moreover, the synthesis of this molecule involves minimal protection and deprotection chemistry.

The second molecule that has been included is levetiracetam, an antiepileptic drug. Epilepsy is estimated to have an occurrence of around 1% of the general population and requires extended treatment. In this context, considering the status of this disease, the effectiveness in terms of affordability of the medicine becomes stringent. Moreover, there are many synthetic approaches precedented in the literature, but most of them suffer from either high cost or environmental burden due to protection–deprotection strategies employed. Protecting-group-free synthesis of levetiracetam is discussed in this section.

The third molecule, phosphate salt of sitagliptin (Januvia), is discussed in detail. This was discovered by Merck and approved by the US FDA for the management of type 2 diabetes mellitus. Sitagliptin inhibits the dipeptidyl peptidase-4 enzyme, thereby increasing insulin secretion. It is being manufactured by using one of the most fascinating process, and it is one of the most prestigious molecules in the pharmaceutical world due to the fact that the synthetic efforts toward this has earned two US Presidential Green Chemistry awards. The biocatalytic route for the synthesis of sitagliptin (asymmetric reduction part) is truly protection-free.

The fourth one is paroxetine, an antidepressant drug developed by GlaxoSmithKline. It is a selective serotonin reuptake inhibitor (SSRI) that increases the extracellular level of neurotransmitter serotonin. There is no truly competing medicine to this available in the market. The greener approach toward this molecule features enzymatic desymmetrization, which qualifies to be an almost protection–deprotection-free approach.

Apitolisib is the fifth case, which is a molecule in advanced stages of approval. Clinical results are extremely encouraging, and if approved, it might become a very important medicine to treat cancers. Apitolisib (GDC-0980) inhibits the PI3K/AKT/ mTOR signaling pathway that has been found to be overactive in cancer cells. Moreover, the first-generation route for the synthesis of apitolisib involved protection–deprotection steps and also suffered from low yield in the final step. Accommodating the synthesis of this molecule in this chapter would allow scientists to design greener route in the future.

The sixth molecule, azepinomycin, is a natural product that exhibited potent guanine deaminase inhibitory activity. Various pathogens require guanine deaminase for the synthesis of nucleotides. A one-step protecting-group-free synthesis of azepinomycin was reported by Coggins' group. Although four syntheses were known for the said natural product, all the reported strategies were lengthy, consisting of more than six steps. Inclusion of this molecule adds elegance to this chapter due to the fact that there is only one step involved in its synthesis.

The synthesis of sulfanyl-histidine developed by Daunay et al. that features a short protecting-group-free synthesis of a rare naturally occurring amino acid, 5-sulfanyl-histidine, starting from histidine is described as the seventh case. This is a privileged molecule as it has been used in the study of adrenochrome. This synthesis provides access to gram-scale quantities of thioester in two steps and good yields without the use of any chromatographic purification. The synthesis for this satisfies the criteria of a protection–deprotection-free approach.

Syn-Ake® is a tripeptide that reduces the appearance of wrinkles. It is a muscular nicotinic acetylcholine membrane receptor (mnAChR) antagonist and prevents the intake of sodium ions, keeping the muscles relaxed. Inclusion of this peptide adds to the element of diversity in this chapter. Conventional synthesis of peptides involves protecting groups and is lengthy. A short protecting-group-free synthesis of Syn-Ake was reported by Balaev and coworkers. The evolving trend toward developing a protection- and deprotection-free synthesis is quite impressive and is discussed as the eighth case.

Rhodanines and thiazolidinediones are scaffolds present in several medicinally active compounds. Thiazolidinediones (glitazones) are peroxisome proliferator-activated receptor-γ (PPARγ) agonists useful in managing type 2 diabetes mellitus. Drugs such as ciglitazone, pioglitazone, rosiglitazone, darglitazone, and rivoglitazone belong to this class. Recent advances considering synthetic development, particularly protection and deprotection approaches, on these important molecules having high clinical potential would be interesting; therefore it has been selected as the ninth case.

S-Adenosyl-L-methionine (SAM) acts as a methyl donor in transmethylation reactions catalyzed by methyltransferase enzymes. This reaction is involved in various disease pathways and protein regulation. Hence, analogs of SAM have been designed to inhibit methyltransferase and find application as chemical biology tools. Se-based SAM analogs have also gained interest and are expected to be more reactive because of the weaker Se—C bond. In this context, Bothwell's group developed a protecting-group-free synthesis of Se-alkyl Se-adenosyl-L-selenomethionine (SeAM) analogs. This part of the chapter as the tenth case would feature Se-alkyl chemistry highlighting protection–deprotection approaches.

6.2 Raltegravir

Raltegravir potassium (MK-0518) belongs to a new class of compounds discovered for development as inhibitors of HIV integrase, an enzyme catalyzing the integration of viral DNA into the host genomic DNA, thus preventing further virus replication [7–9]. It was approved on 12 October 2007 and is marketed as ISENTRESS, as a complementary agent to existing antiretroviral therapies (Figure 6.2).

The first-generation route to **1** commenced with the Strecker reaction in which the commercially available acetone cyanohydrin **2** was reacted with ammonia to afford **3** (Scheme 6.1). Subsequent Cbz protection gave the *N*-Cbz derivative **4** in 85% yield, which was treated with hydroxylamine to effect its addition across —CN and provide the amidoxime **5** in 88% yield. The coupling reaction between **5** and dimethyl acetylenedicarboxylate (DMAD) and cyclization at elevated temperature yielded the advance intermediate **7** in 52% yield. The *N*-methylation step to produce **8** in 70% yield was critical and brilliantly executed by using $Mg(OMe)_2$ and CH_3I in DMSO, followed by chromatographic purification to conveniently remove the undesired 6-*O*-methylated side product **9**. Amidation of **8** to obtain **10** in 90% yield followed by Cbz deprotection to get **11** in 96% yield, and coupling with oxadiazole **12** gave raltegravir (**1**) in 91% yield. In this synthesis the *N,O*-diacylated derivative **13** was formed as the major side product. The synthesis of compound **12** as shown in Scheme 6.2 was achieved in 90% yield. A critical analysis of the route reveals that the synthesis involved (i) many chromatographic purification steps, (ii) excess of reagents and intermediates, and (iii) poor selectivity during methylation. The synthesis involved 10 linear steps with approximately 24% overall yield [10].

In order to circumvent the poor selectivity observed during the methylation step, Zhong et al. used altogether different reagents (Scheme 6.3) [11]. For example, *N*-methylation of **7** was carried out with NMP and 2 equiv of $Mg(OH)_2$, while 2 equiv. of trimethylsulfoxonium iodide was the methylating agent. This combination restricted the formation of unwanted product **9**. The second important aspect of this route was the protection of free –OH group with pivalate, which enabled them to reduce the quantity of the oxadiazole **12** from 2.2 equivalents (Scheme 6.1) to stoichiometric amount. The overall yield was 33%. However the major shortcomings of this route were its inability to reduce the number of steps and protection–deprotection sequences.

Although both Schemes 6.1 and 6.3 are logically correct, they are associated with protection–deprotection issues. Raltegravir (**1**) is indeed a critical new tool to address the worldwide HIV/AIDS epidemic. In order to overcome these issues, Gurjar et al. (Scheme 6.4) came up with a new synthetic strategy, which removed the necessity of protecting groups, and the *N*-methylation at the end of the sequence was an interesting event with good selectivity [12].

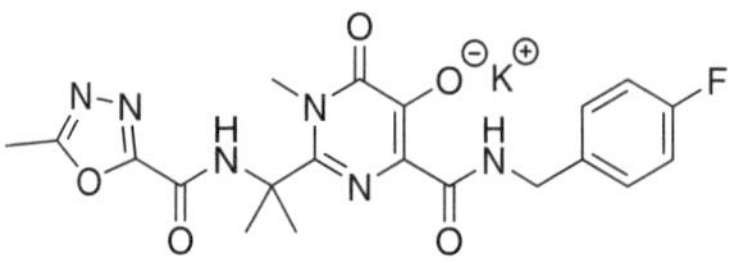

Raltegravir potassium (**1**)

Figure 6.2 Structure of raltegravir potassium (**1**).

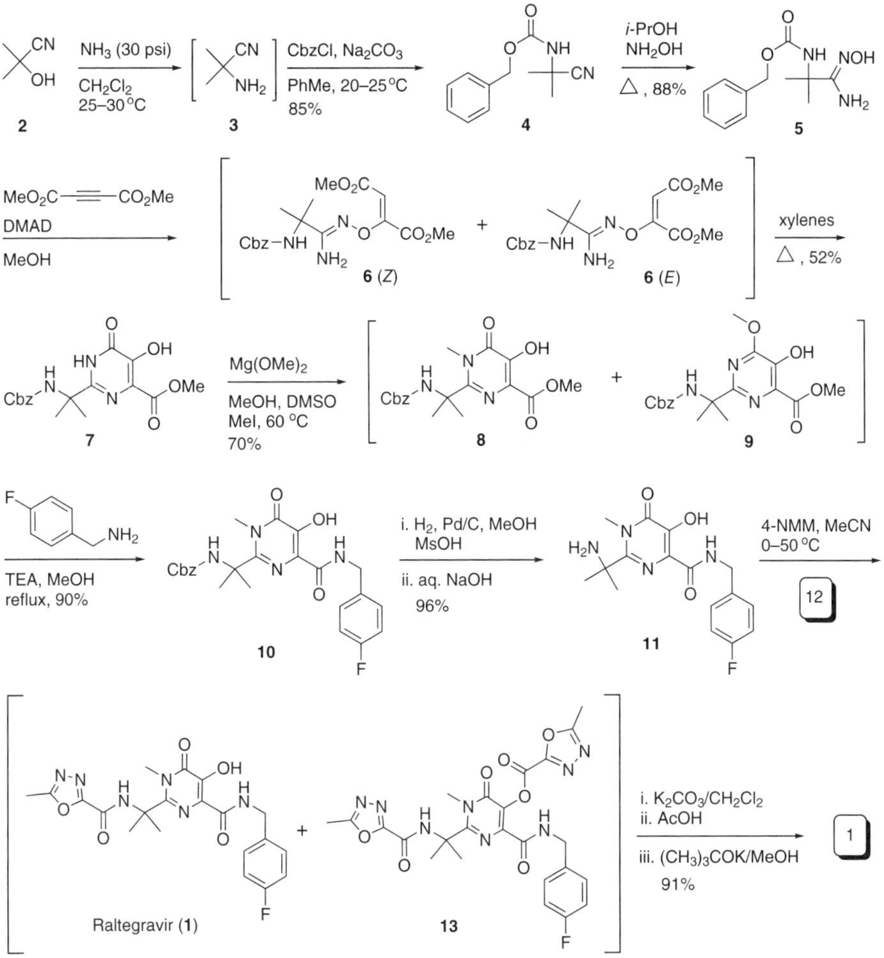

Scheme 6.1 First-generation synthesis of raltegravir potassium (**1**).

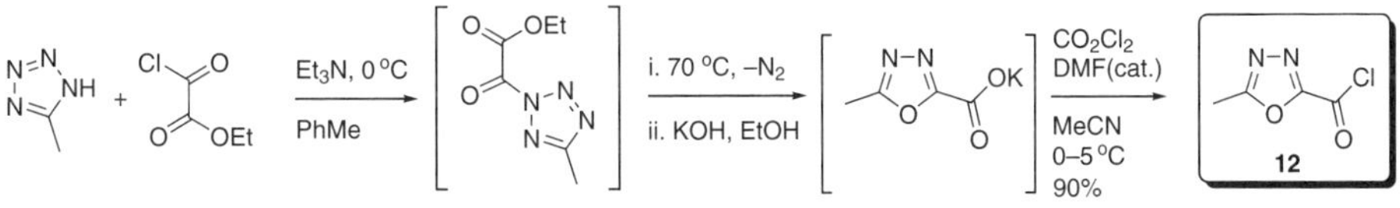

Scheme 6.2 Synthesis of **12**.

Thus, aminonitrile **3** was acylated with oxadiazole chloride **12** in the presence of *N*-methylmorpholine and a catalytic amount of DMAP in acetonitrile to obtain oxadiazole intermediate **16** in 70% yield. Thereafter, intermediate **16** was subjected to hydroxylamine to offer oxime intermediate **17** in 94% yield, which is well equipped to undergo cycloaddition reaction with DMAD to afford advanced heterocyclic intermediate **18** in 66% yield. Intermediate **18** was allowed to react with parafluorobenzylamine to give amide intermediate **19** in 90% yield. Methylation of **19** completed the synthesis of

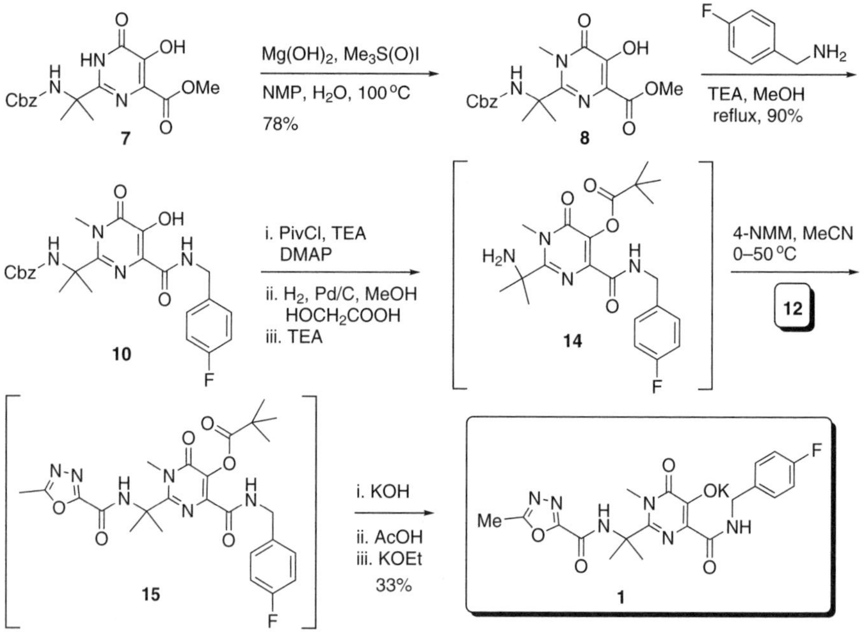

Scheme 6.3 Second-generation synthesis of raltegravir potassium (**1**).

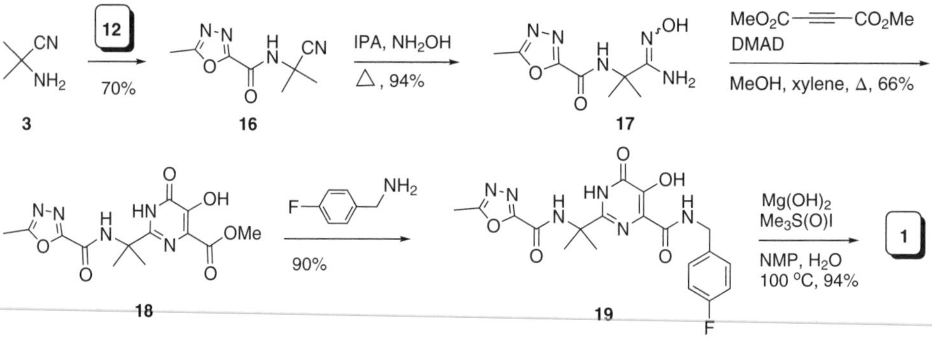

Scheme 6.4 Gurjar's synthesis of raltegravir (**1**).

raltegravir (**1**) with 94% yield. The lack of *O*-methylation with substrate **19** was surprising, but the reason could be attributed to the steric and stereoelectronic factors that the oxadiazole group might have exerted. The synthesis was found to be the most effective among the reported synthetic strategies.

6.3 Levetiracetam

Levetiracetam (**20**; Figure 6.3) is used to manage epileptic disorders since few decades. Epilepsy is a widely known medical disorder with an occurrence of around 1% of the general population that requires extended or sometimes lifelong treatment [13]. In this

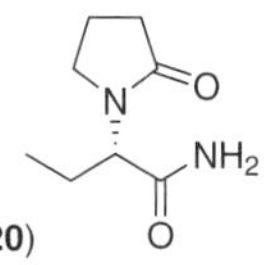

Levetiracetam (**20**)

Figure 6.3 Structure of levetiracetam (**20**).

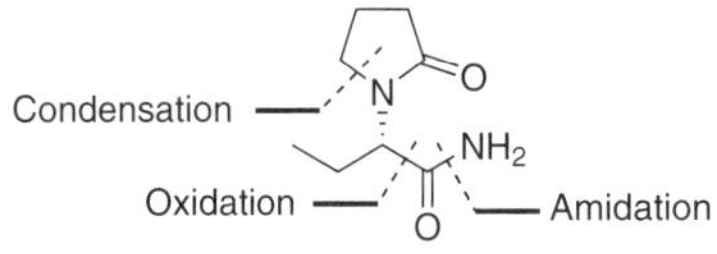

Condensation

Oxidation ——— Amidation

20

Figure 6.4 Novel approach to the synthesis of levetiracetam (**20**).

context, the effectiveness in terms of affordability of the medicine becomes stringent. Moreover, there are many synthetic approaches precedented in the literature, but most of them suffer from either high cost or environmental burden due to protection–deprotection strategies employed [14].

A protection–deprotection-based approach starts with benzoyl protection followed by oxidation of (*S*)-aminobutyric acid to obtain the corresponding *N*-benzoyl-protected (*S*)-aminobutyric acid [15]. After *N*-benzoyl deprotection and subsequent amidation, (*S*)-aminobutyramide was obtained. A chemoselective butyrolactam ring formation using 4-chlorobutyryl chloride and amide was then performed to obtain levetiracetam (**20**). In this step, 2 mol of corrosive hydrochloric acid was generated, which adds unattractive features to the process. Alternatively, biocatalysis has gained reputation as a means for making industrial processes for APIs more cost effective and eco-friendly [16]. Recently, a biocatalytic process was developed that involves resolution of a *rac*-2-pyrrolidinonyl nitrile catalyzed by nitrile hydratases to obtain **20**. The nitrile intermediate was prepared by an alkylation of 2-pyrrolidinone with 2-chlorobutanenitrile. A mutant nitrile hydratase enzyme was used to render the resolution, which offered excellent output, good resolution yield (43%), and high stereoselectivity (94% ee). The atom efficiency of this process was found to be excellent due to the fact that the undesired (*R*)-enantiomer was recycled [16]. A drawback of this process is that most of the synthetic steps were carried out in aqueous media, making the isolation of intermediates or end products tedious, since it involves extraction using excessive amounts of solvent or distillation of water that requires high energy. In order to develop a noninfringing route, it was imperative to develop a scalable process that involved solvent-free condensation, metal-catalyzed oxidation, and amidation. The authors attempted to develop a cost-competitive, atom-economical, and protection-free synthesis of levetiracetam (**20**; Figure 6.4). The synthesis started with the solvent-free condensation of γ-butyrolactone **21** and (*S*)-aminobutanol **22** under heating or microwave irradiation conditions to afford intermediate **23** (Scheme 6.5). These two units (**21** and **22**) were bridged through an amide linkage, which may be sensitive to racemization at higher temperature or under microwave along with heating conditions. The RuO_2/NaOCl-catalyzed oxidation of alcohol **23** was performed to obtain the penultimate acid intermediate **24**. Amidation of **24** was achieved through a mixed anhydride that was formed *in situ*. Eventually, treatment of the mixed anhydride with various ammonium salts afforded levetiracetam (**20**).

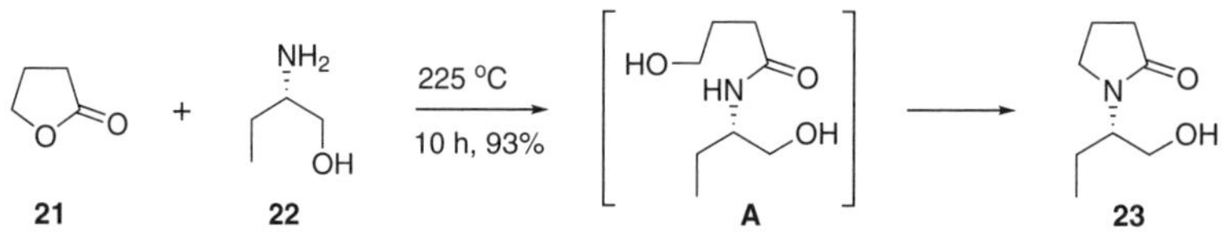

Scheme 6.5 Condensation reaction to obtain intermediate **23**.

The condensation of γ-butyrolactone with amines is precedented in the literature [17]. Considering the nature of the reaction that requires higher temperature, the condensation employing high boiling solvents such as toluene can help distill out water azeotropically. However, the best results were obtained under solvent-free conditions. The maximum yield (93%) for condensation to afford intermediate **23** was achieved at 225 °C in 10 h. There were some experiments conducted thermally in the presence of catalytic amounts of 4 Å MS, phosphoric, *p*-toluenesulfonic, methanesulfonic, or boric acids, all of which offered no advantage in terms of yield or purity.

Condensation under microwave irradiation conditions also provided exciting results. When performed thermally, longer reaction times (several hours) were required. However, in the case of microwave conditions, an accelerated reaction rate was observed, and the transformation was very clean and high yielding. Thus, the best yield (81.7%) was obtained at 200 °C under 100 W microwave irradiation in 1 h time. Notably, the condensation between lactone **21** and amino alcohol **22** passes through intermediate **A** in the synthesis of **23**, as shown in Scheme 6.5. However, there was no need to isolate the intermediate **A** as it was found to be more than 82–93% pure. Interestingly, there was no signature of any racemization during condensation of **21** and **22** to obtain **23** as downstream chemistry afforded enantiomerically pure acid derivative **24** and thereafter title compound **20**.

The oxidation of an alcohol to acid is a well-known reaction [18] that can be applied to this system. In the next step, oxidation of the alcohol was performed by employing a catalytic amount of RuO_2 in the presence of sodium hypochlorite (NaOCl). The method was moderately effective but more environmentally benign in comparison with the method where $KMnO_4$ has been used. In the $KMnO_4$-mediated oxidation [19], the isolation of the product was very difficult due to the acid **24** being trapped in the MnO_2 sludge. Advantageously, in the catalytic method, the isolation of the hygroscopic acid product **24** was possible with reasonably better yield (64.8%) and purity (91.9%) as shown in Scheme 6.6.

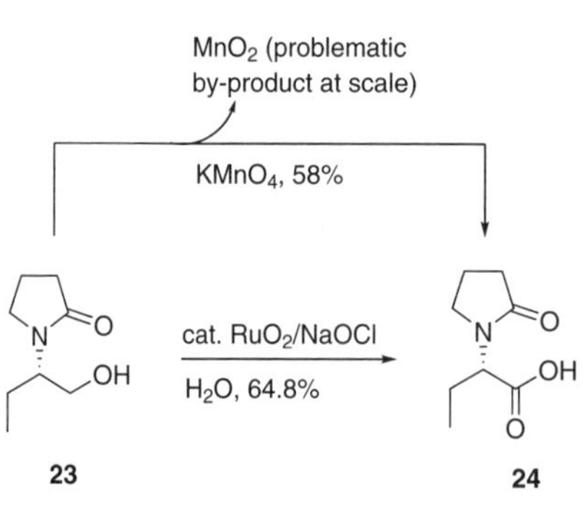

Scheme 6.6 RuO_2-catalyzed oxidation of alcohol **23** to acid **24**.

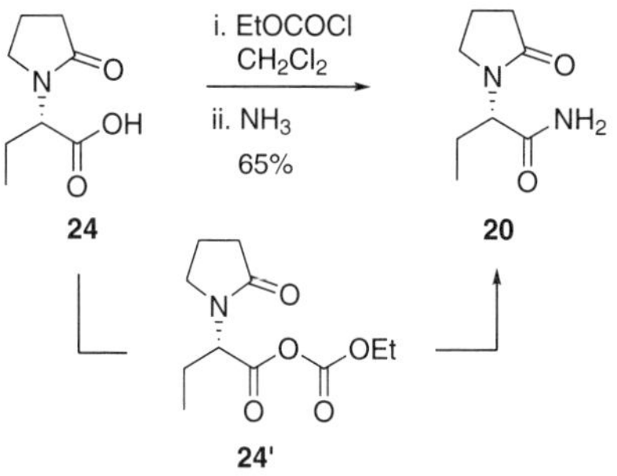

Scheme 6.7 Amidation to obtain levetiracetam (**20**).

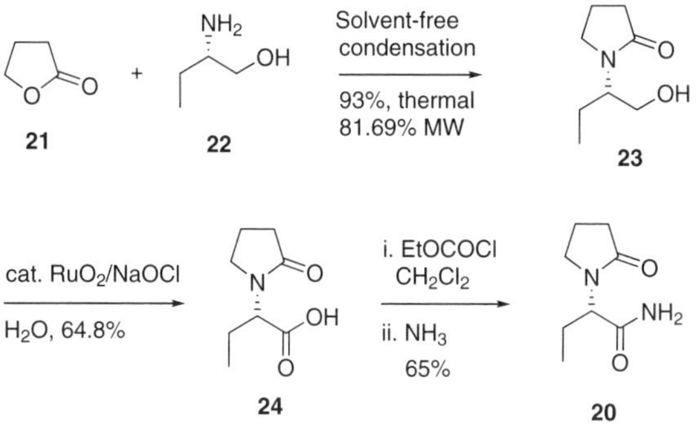

Scheme 6.8 Green approach for the synthesis of levetiracetam (**20**).

In the final step, an amidation was performed first by preparing the mixed anhydride of **24** with ethyl chloroformate and a subsequent nucleophilic substitution with ammonia to afford desired product **20** in good yields as shown in Scheme 6.7. The authors employed a number of ammonia sources including ammonia gas, ammonium acetate, and ammonium chloride. In most cases, the yield and purity varied by not more than 2–3%. The reaction employing ammonia gas afforded the product in the best yield and purity at the expense of affording ethyl formic acid, essentially producing no by-product from the ammonia source.

After optimizing all these three steps, a comparatively greener and novel synthesis was developed as shown in Scheme 6.8.

In conclusion, the second-generation synthesis of levetiracetam (**20**) presented in Scheme 6.8 features solvent-free condensation and no protection–deprotection steps, and the generation of salt waste is completely avoided. Moreover, the metal-catalyzed oxidation of alcohol **23** to advanced intermediate **24** makes the synthesis attractive and eco-friendly.

6.4 Sitagliptin

The phosphate salt of sitagliptin, discovered by Merck, has been approved by the US FDA for the management of type 2 diabetes mellitus. Sitagliptin inhibits (Figure 6.5) the dipeptidyl peptidase-4 enzyme that breaks down the incretins, which play a key role in

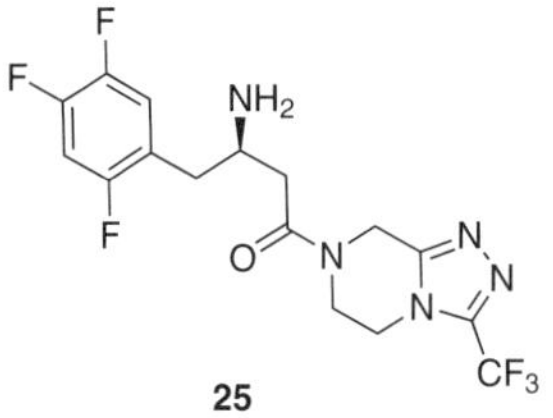

Figure 6.5 Structure of sitagliptin (**25**).

glucoregulation. This is widely known as Januvia and was approved by the US FDA for the treatment of type 2 diabetes mellitus in late 2006 [20].

6.4.1 Medicinal Chemistry Route

An early medicinal chemistry route for sitagliptin (**25**) was achieved by coupling of β-amino acid **33** with fused heterocycle **38** [21]. The required β-amino acid **33** was synthesized starting from α-amino acid **30** (Scheme 6.9) [22, 23]. Alkylation of **26** with the bromide derivative **27** afforded intermediate **28** in 75% yield, which was treated with hydrochloric acid followed by protection of the resultant amine with di-*tert*-butyl dicarbonate to give ester **29** in 62% yield. Hydrolysis afforded the α-amino acid **30** in 99% yield, which was converted into diazo ketone **31** in 91% yield. Thereafter, rearrangement of diazo ketone **31** to ester **32** followed by hydrolysis offered β-amino acid **33** in 59% yield (from **31**). Direct conversion of **31** also yielded product **33** in >95% yield [24].

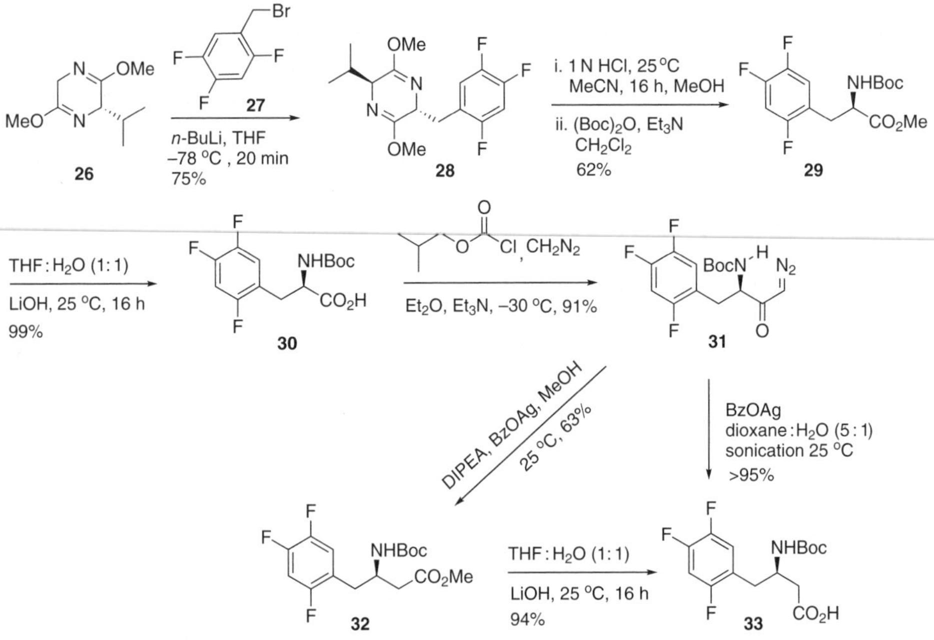

Scheme 6.9 Synthesis of β-amino acid **33**.

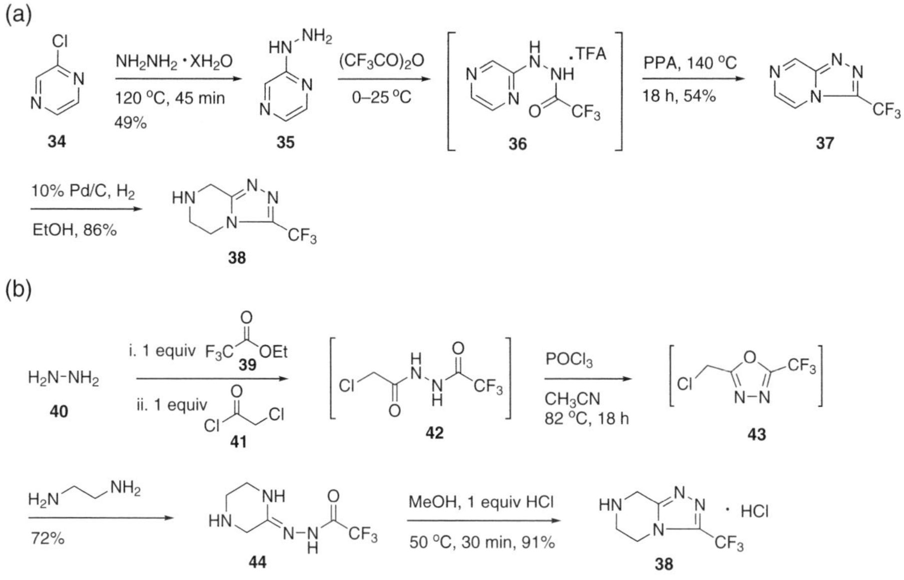

(a)

(b)

Scheme 6.10 Synthesis of piperazine derivative **38**.

The synthesis of piperazine derivative **38** is demonstrated as shown in Scheme 6.10a. Hydrazinopyrazine **35** was prepared from chloropyrazine **34** [25]. Intermediate **35** was acylated to afford **36**, which was then treated with polyphosphoric acid (PPA) to obtain triazolopyrazine **37** followed by Pd/C-catalyzed hydrogenation, affording desired heterocycle **38**. Alternatively, reaction of ethyl trifluoroacetate **39** with hydrazine **40** afforded the corresponding hydrazide (Scheme 6.10b). The latter was condensed with chloroacetyl chloride **41** to yield the bishydrazide **42**. Phosphorus oxychloride (POCl$_3$)-mediated cyclization of compound **42** in acetonitrile afforded the oxadiazole **43** that was treated with ethylenediamine to provide **44** in 72% yield. Eventually, cyclization of hydrazide **44** yielded the HCl salt of **38** in 91% yield.

Coupling of the β-amino acid **33** with triazolopiperazine **38** followed by Boc deprotection provided the HCl salt of desired compound **25** (Scheme 6.11) in 76% yield [26].

The medicinal chemistry route for **25** was found to be inefficient due to the usage of a sacrificial expensive chiral auxiliary and protection–deprotection strategy. The chemistry turned out to be environmentally unfriendly and less cost attractive.

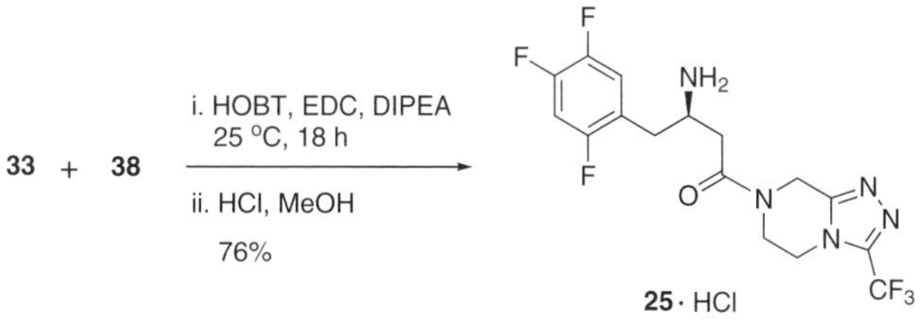

Scheme 6.11 Synthesis of HCl salt of sitagliptin (**25**).

6.4.2 Process Chemistry Route

Scheme 6.12 features the process chemistry route for synthesis of sitagliptin. In this strategy, the synthesis started with the condensation of 2-(2,4,5-trifluorophenyl)acetic acid **45** with Meldrum's acid **46** that gave rise to the adduct **47**, which was then treated with refluxing methanol to yield methyl ester **48** in 94% yield. Enantioselective reduction of ester **48** gave rise to 3-(*S*)-hydroxy-4-(2,4,5-trifluorophenyl)butyric acid methyl ester **49**, which was further hydrolyzed with LiOH to provide the corresponding free carboxylic acid **50** in 83% yield. Reaction of acid **50** with benzyloxyamine afforded the benzyl derivative **51**, which was cyclized to yield azetidin-2-one derivative **52** in 81% yield. Cleavage of the azetidine ring afforded acid **53**, which was further condensed with **38** to yield the adduct **54**. Finally, this compound was debenzylated to obtain **25** in 78% yield [27].

This process chemistry route for **25** did not employ the use of expensive chiral auxiliary and was found to gradually evolve from the medicinal chemistry route. In order to introduce chirality, asymmetric metal-catalyzed reduction protocol was employed. This route also involves protection and deprotection strategy, which does not make it very attractive. The chemistry was significantly better than that used in the medicinal chemistry route; however there still remained a few challenges that needed to be addressed.

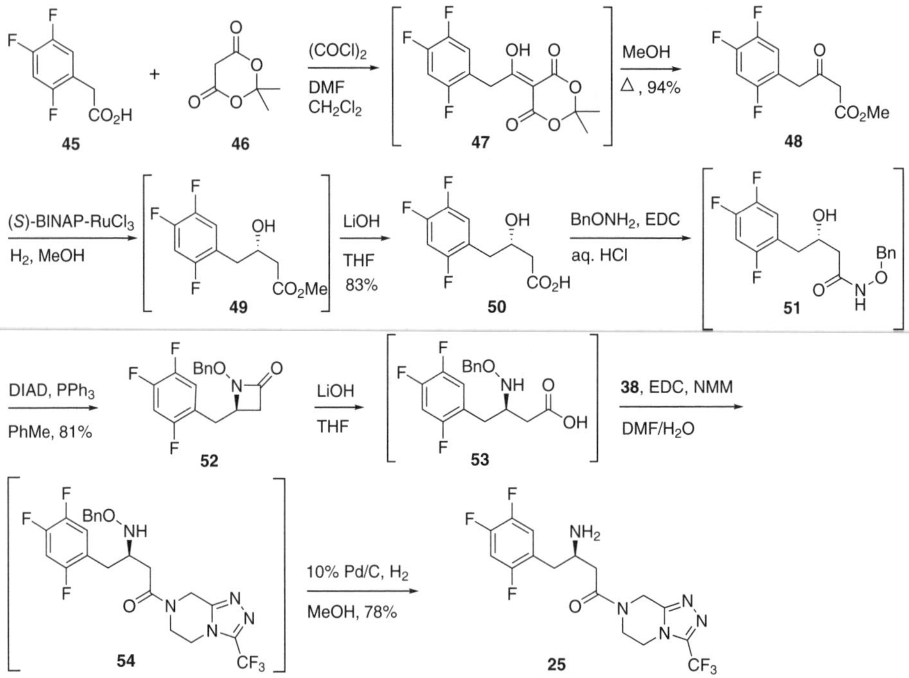

Scheme 6.12 Process chemistry route for the synthesis of sitagliptin (**25**).

6.4.3 Greener Approach

The synthesis of enamine **55** includes the reaction of Meldrum's acid **46** and 2,4,5-trifluorophenyl acetic acid **45** to give rise to adduct **47** (Scheme 6.13). This process features the activation of **45** by formation of a mixed anhydride with pivaloyl chloride and reaction with **46** to form Meldrum's adduct **47**. Adduct **47** could be obtained as Hünig's salt and as well as in free acid form after treatment with trifluoroacetic acid. Treatment of **47** with cyclic amine **38** resulted in the formation of keto amide **56** that passes through ketene intermediate **57**. Although the keto amide **56** could be isolated at this point as a crystalline solid, it was more desirable to prepare the enamine amide **55** directly. Ammonium acetate (NH$_4$OAc) along with methanol (MeOH) was mixed with the crude reaction mixture of **56**, and upon heating to 45 °C, the desired product **55** crystallized from the reaction mixture as the reaction proceeded. Upon cooling the reaction mixture to 0–5 °C, **55** was directly isolated as a white crystalline solid through simple filtration, thereby eliminating the need for aqueous work-up and minimizing waste generation. Noticeably, only the Z-isomer was formed during the reaction. Thus, **55**, which contain the entire structure of sitagliptin (**25**), can be prepared in one-pot process in 82% yield.

Considering all the process components, the (S,S)-t-Bu-JOSIPHOS **58** as ligand in combination with [Rh(COD)Cl]$_2$ was selected for further development. Hydrogenation of **55** in the presence of NH$_4$Cl (0.15 mol%), [Rh(COD)Cl]$_2$ (0.15 mol%), and (S,S)-t-Bu-JOSIPHOS **58** (0.155 mol%) in MeOH under 250 psi of hydrogen at 50 °C for 16–18 h proved to be extremely robust and afforded **25** in 98% yield and 95% ee reproducibly (Scheme 6.14) [28, 29].

In this green chemistry route, an asymmetric catalytic hydrogenation of unprotected enamine **55** employed rhodium salts of a ferrocenyl-based ligand as the catalyst, resulting in sitagliptin (**25**) in high optical purity and yield. Almost 95% of the precious rhodium metal was claimed to be recovered and recycled. This route also avoids protection–deprotection steps. The synthesis as such has only three steps, and the overall yield was increased by 50%. This route, on a manufacturing scale, allowed reduction of

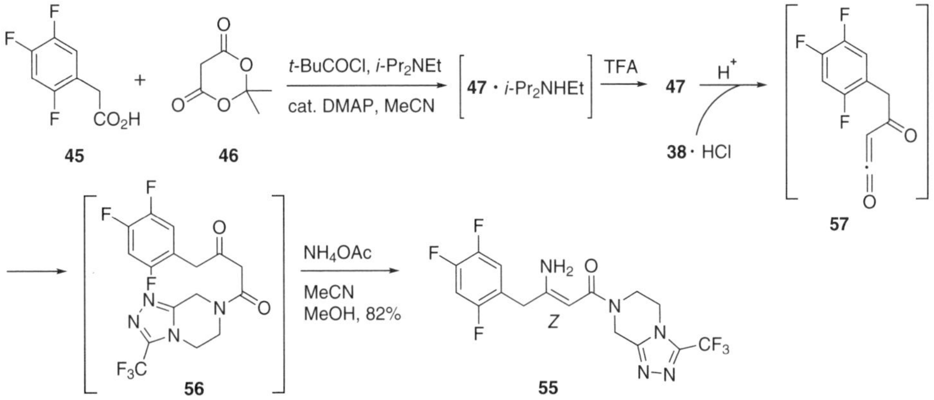

Scheme 6.13 Synthesis of advanced precursor enamine **55**.

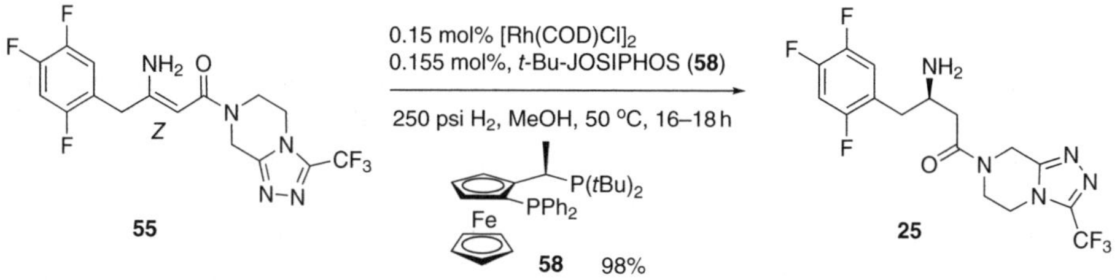

Scheme 6.14 Greener synthesis of **25**.

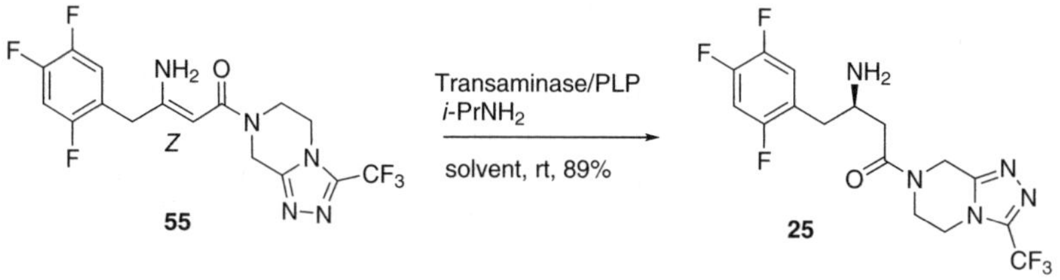

Scheme 6.15 Enzymatic synthesis of **25**.

the amount of waste by over 80% and completely eliminated the aqueous work-up. This process turned out to be scalable, greener, and cost effective.

Another approach directed toward a synthesis that could avoid the use of expensive metal and ligands that eventually led to an elegant biocatalytic route to sitagliptin utilizing an engineered transaminase enzyme to convert enamine **55** to **25** in 89% yield was developed and commercialized (Scheme 6.15), which is an ideal protection–deprotection-free and the greenest synthesis so far reported for this molecule [30a]. Recently, phenylethylamine chiral auxiliary-based synthesis of sitagliptin has also been reported by Bandichhor et al. [30b].

6.5 Paroxetine

Paroxetine (**59**; Figure 6.6), under trade names Seroxat and Paxil, is a GlaxoSmithKline-developed drug. It is used to treat major depressive and premenstrual dysphoric disorder. In general, it is found to be effective in first-line therapy for generalized anxiety disorder [31].

6.5.1 Medicinal Chemistry Route

Methyl ester **60** was treated with 4-fluorophenylmagnesium bromide **61** to obtain a mixture of 1-methyl-4-(4-fluorophenyl)piperidine-3-carboxylic acid methyl esters **62a** and **62b** as shown in Scheme 6.16. The hydrolytic equilibration of this mixture of **62a/62b** by means of base gave the corresponding *trans* carboxylic acid derivative **63**, which was activated with thionyl chloride to obtain acid chloride derivative **64**.

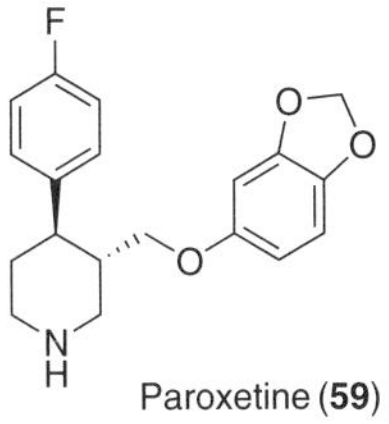

Figure 6.6 Structure of paroxetine (**59**).

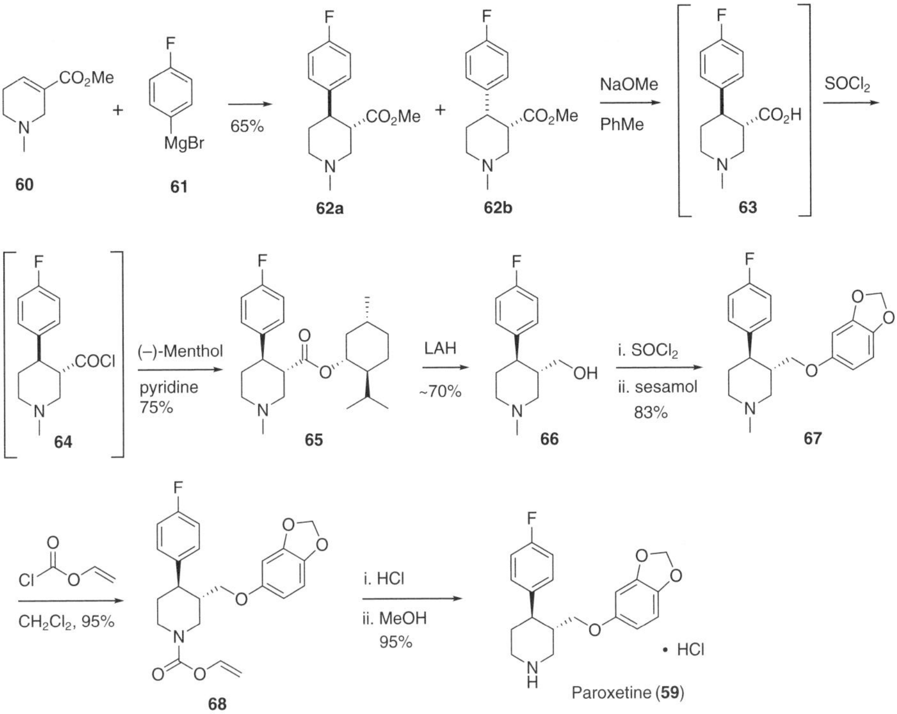

Scheme 6.16 Medicinal chemistry route for **59**.

(−)-Menthol was allowed to react with **64** to afford enantiopure menthol ester **65** in 75% yield (based on available desired *trans* isomer). The carbinol **66** was obtained by reduction in the presence of lithium aluminum hydride in about 70% yield. The reaction of **66** with thionyl chloride and treatment with 3,4-methylenedioxyphenoxide (sesamol) afforded **67** in 83% yield. Demethylation of **67** was achieved by treatment with vinyl chloroformate to obtain vinylurethane derivative **68** in 95% yield. Intermediate **68** was further treated with HCl gas to yield the chlorourethane derivative that was subsequently hydrolyzed in refluxing methanol to afford HCl salt of paroxetine (**59**) in 95% yield [32].

The medicinal chemistry route for **59** involves protection–deprotection (*N*-methyl and vinylformate), resolution strategies, and a lot of work-up. Moreover, after resolution the recycling of undesired diastereomer appears to be impossible; therefore this route is considered to be inefficient.

6.5.2 Process Chemistry Route

In this route, bis-ester **70**, which was prepared in 75% yield by reaction of *p*-fluoroben-zaldehyde **69** with ethyl acetoacetate followed by esterification, was considered as starting point (Scheme 6.17). Screening of the enzymes for hydrolysis of **70** resulted in the application of pig liver esterase for desymmetrization that afforded acid ester **71** in excellent yield (86%) and purity. In order to invert the stereochemistry at C3 center and achieve the reduction of the acid functionality of **71** simultaneously, the reactions were performed by using firstly lithium hydride-mediated deprotonation and secondly reduction with lithium borohydride that afforded alcohol **72** in 94% yield. Alcohol **72** was mesylated and treated with benzylamine to provide lactam **73** in 82% yield. Homologation of **73** afforded **74** in 88% yield that was subsequently reduced to obtain amino alcohol **75** in 92% yield. Reaction of **75** with sesamol afforded penultimate precursor **76** in 80% yield, which was subjected to hydrogenolysis to yield paroxetine **59** in 93% yield [33].

The process chemistry route for **59** involves biocatalytic desymmetrization and avoids resolution but involves significant amount of aqueous work-up. However, this route is comparatively greener than the medicinal chemistry route. It was found to be amenable to kilogram-scale synthesis, but needed improvement to make it eco-friendly.

6.5.3 Greener Approach

A more efficient enzymatic desymmetrization strategy was recently reported as shown in Scheme 6.18 [16]. The key step involved a protease (isolated from *Bacillus subtilis*)

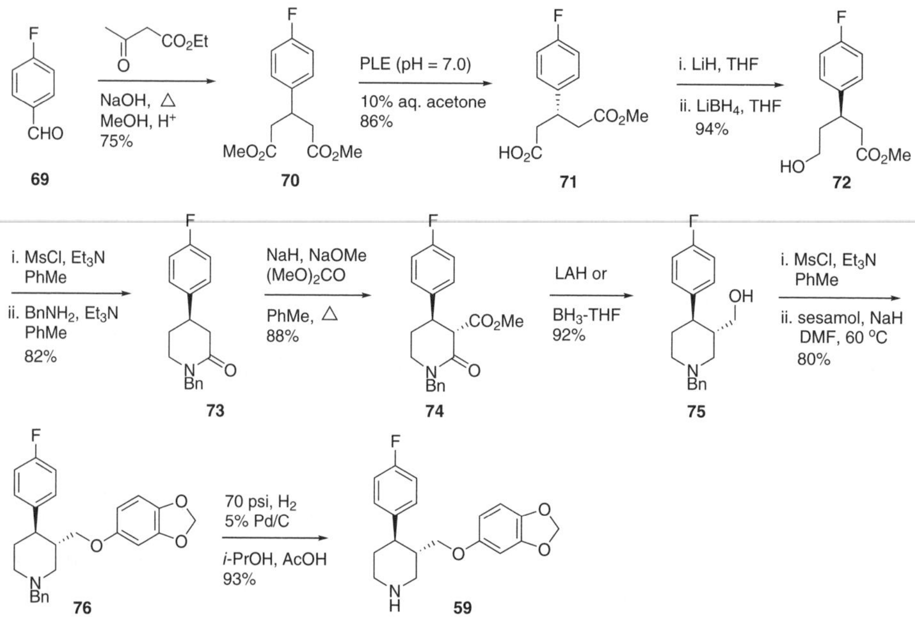

Scheme 6.17 Process chemistry route for **59**.

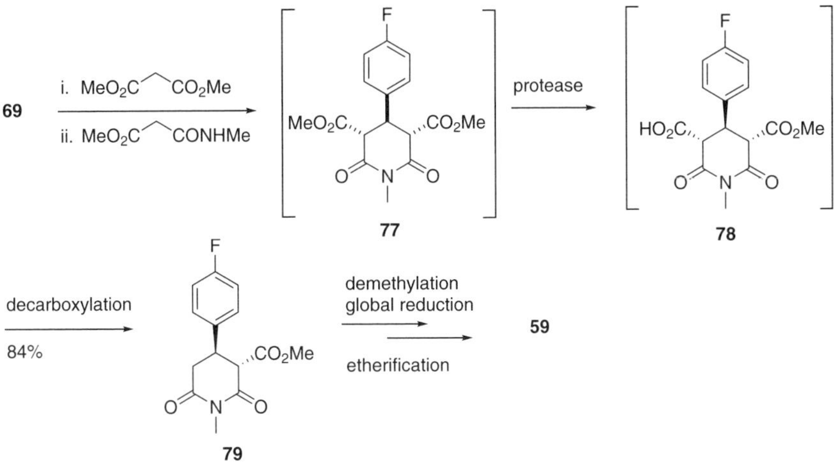

Scheme 6.18 Greener route for **59**.

catalyzed desymmetrization of meso-diester **77**. The latter was synthesized from **69** via cascades of reactions viz. Knoevenagel condensation, Michael addition, and intramolecular cyclizations. The protease-catalyzed desymmetrization of **77** led to the (3*R*,4*S*)-ester **79** as a single enantiomer after decarboxylation of intermediate **78**. Further deprotection and global reduction followed by etherification with sesamol afforded **59**.

This best possible green chemistry route for **59** based on the enzymatic desymmetrization concept had the process efficiency that was much better than the medicinal and process chemistry routes. Interestingly, by practicing this route, the yield of the overall transformation almost doubled in comparison with the process chemistry route, resulting in a greener, shorter, and cost-efficient synthesis of paroxetine (**59**).

6.6 Synthesis of PI3K/mTOR Inhibitor Apitolisib

Cancers are caused by abnormal cell growth and spread to other body parts. Nearly 15% of the overall human deaths are caused by cancer. Inhibition of PI3K/AKT/mTOR signaling pathway is found to be useful in cancer therapy. This pathway regulates the cell cycle and if overactive can result in cancer [34]. Apitolisib (GDC-0980) is a PI3K/mTOR inhibitor that went to human clinical trials for the treatment of cancer. The first-generation route for the synthesis of apitolisib (**86**) involved protection–deprotection steps and also suffered from low yield in the final step (Scheme 6.19) [35].

Formylation of compound **80** to **81** in 87% yield followed by reductive amination gave the *in situ* Boc-protected amine **83** (Scheme 6.19). Suzuki coupling with protected boronate **84** led to the coupled product **85** in 90% yield. The protecting group facilitated isolation of the product by improving its solubility. Boc deprotection followed by coupling with (L)-lactic acid gave the final product apitolisib (**86**) in 61% yield. When this final reaction performed on a large scale, complete conversion of starting material was not observed, and additional reagents have to be added to drive the reaction. Further two impurities were also formed in the final steps, diamine **87** and dilactate **88**

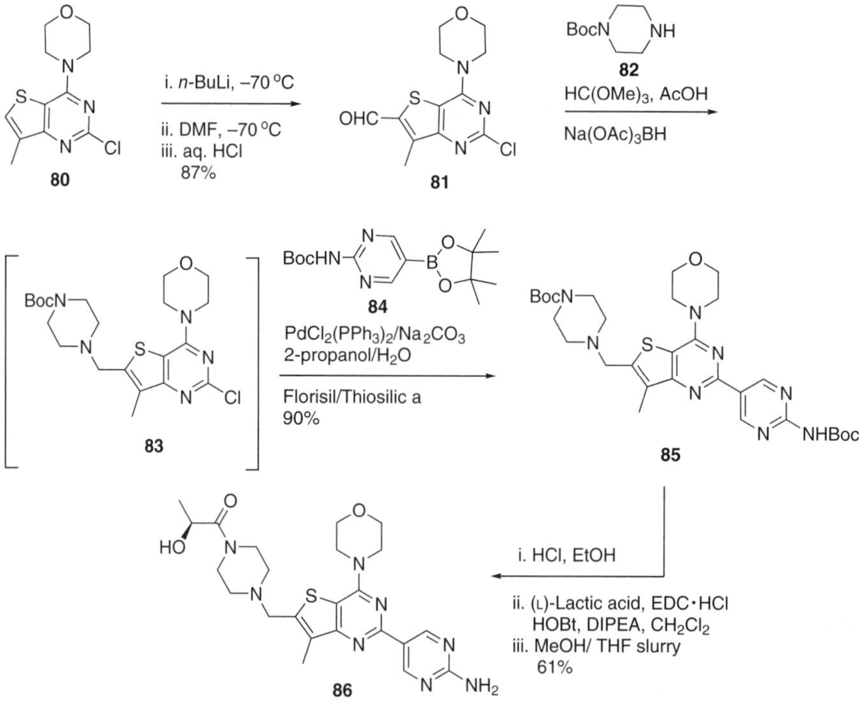

Scheme 6.19 First-generation synthesis of apitolisib (**86**).

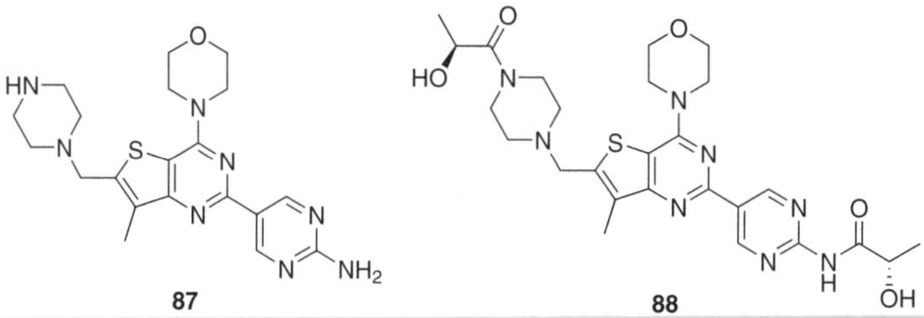

Figure 6.7 Process impurities in the final step of apitolisib synthesis.

(Figure 6.7). This route was used to produce the API for GLP tox batch (material generation in GLP for toxicological tests), but was not ideal for large-scale synthesis.

Gosselin and coworkers developed a kilogram-scale protecting-group-free synthesis of apitolisib (**86**) in overall four steps [36]. The formyl group was introduced on thienopyrimidine heterocycle **80** using DMF as the carbon source to give aldehyde **81** in 96% yield (Scheme 6.20). The reaction was carried out at −8 °C, whereas the previous synthesis was done under cryogenic conditions. This aldehyde was then subjected to a reductive amination with the oxalate **89** to produce compound **90** in 82% yield. The oxalate salt was employed for the synthetic sequence because it was less hygroscopic and performed well. Trimethyl orthoformate acts as a dehydrating agent, facilitating the formation of iminium ion, and 2-picoline-BH$_3$ was

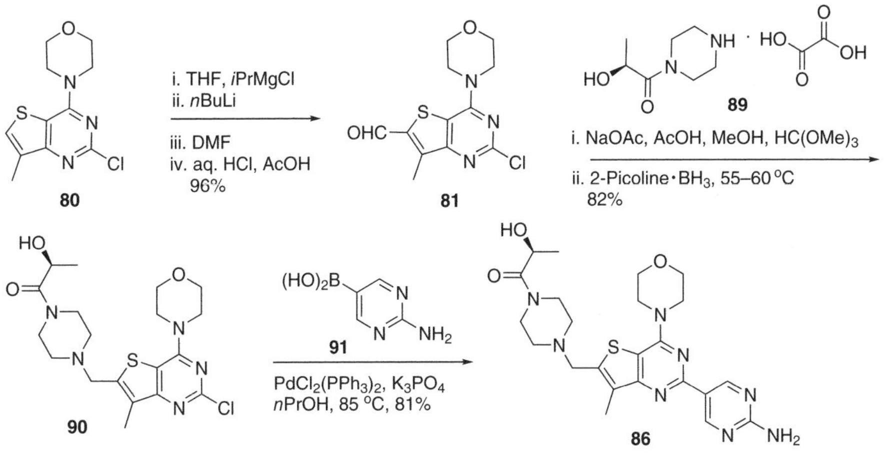

Scheme 6.20 Manufacturing-scale synthesis of apitolisib (**86**).

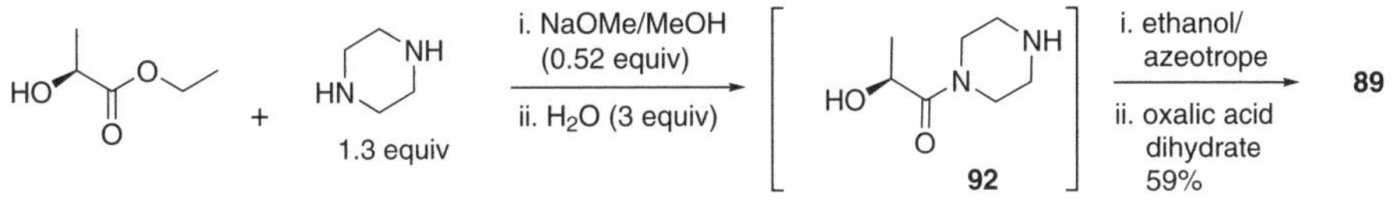

Scheme 6.21 Synthesis of piperazine lactamide oxalate **89**.

used as the reducing agent. Although Na(OAc)$_3$BH performed well in the reductive amination, it had to be added as a solid, which was inconvenient. After screening various conditions, the present condition using 2-picoline-BH$_3$ was chosen. Under the optimized Suzuki–Miyaura conditions, compound **90** and boronic acid **91** gave the required product **86** with a yield of 81% after crystallization from propanol/water. The expensive boronate **84** (Scheme 6.19) was replaced with boronic acid **91**, which was prepared from 2-amino-5-bromopyrimidine in one step. Also, the catalyst loading was reduced to 0.15 mol%, which was possible due to the increased rate of the reaction at higher temperature. The crude product was further purified by API crystallization to obtain the product with >99% purity.

The efficient synthesis of piperazine lactamide oxalate **89** is also interesting (Scheme 6.21). The previous synthesis consisted of multiple steps involving protection–deprotection. Here, the amidation reaction of piperazine with (*S*)-ethyl lactate afforded the product along with the bis-lactamide. The ratio of both the products could be controlled by varying the equivalents of starting materials, with 1.3 equiv of (*S*)-ethyl lactate being optimal. Addition of water improved the yield of the required product presumably by hydrolysis of the bis-lactamide to compound **92**. By the addition of 0.25 equivalents of oxalic acid, the disodium oxalate salt precipitated in ethanol, which was then filtered off, and more oxalic acid was added while adjusting the pH to obtain the oxalate **89** in 59% yield. By employing this new route to piperazine lactamide, a 55% reduction in PMI and 41% increase in atom economy could be achieved. Use of toxic reagents as well as the volume of waste was reduced drastically.

6.7 Azepinomycin

Azepinomycin is a natural product isolated from *Streptomyces* sp., which showed potent guanine deaminase inhibitory activity [37]. Guanine deaminase catalyzes the conversion of guanine to xanthine. Various pathogens require this enzyme for the synthesis of nucleotides.

The previously reported synthesis [37] of azepinomycin (**96**) starts with *N*-substituted triacetylribosyl imidazole derivative **93** (Scheme 6.22). This was subjected to reaction with ethyliodo acetate **94** in the presence of Ag$_2$O and DMF, resulting in the intermediate **95** in 25% yield. DIBALH reduction of **95** led to an advanced intermediate, and treatment of this with ammonium hydroxide offered deacylated ribosyl azepinomycin, which was treated with phosphoric acid to afford azepinomycin (**96**) in 60% yield.

Coggins et al. reported a protecting-group-free synthesis of natural product azepinomycin (**96**) in one step (Scheme 6.23) [38]. Although four syntheses were known for the said natural product, all the reported strategies were lengthy, consisting of more than six steps [37, 39]. The required starting material **97** for this synthesis could be prepared from hydrogen cyanide tetramer by a photochemical reaction. Reaction of **97** and hydroxylacetaldehyde in the presence of phosphate buffer gave the desired product **96** in a yield of 83% at pH 4. The product was purified by solid-phase extraction using Dowex® 50 W ion-exchange resin. The reaction proceeds through the formation of an imine that undergoes an Amadori rearrangement to form the product. This report provides a high-yielding protecting-group-free synthesis of azepinomycin in water.

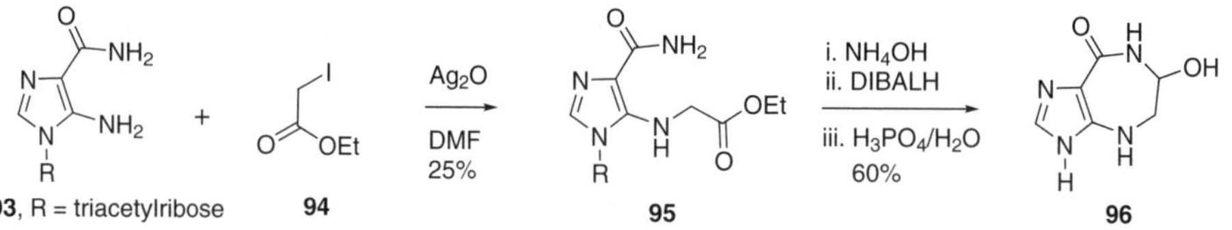

Scheme 6.22 Synthesis of azepinomycin (**96**).

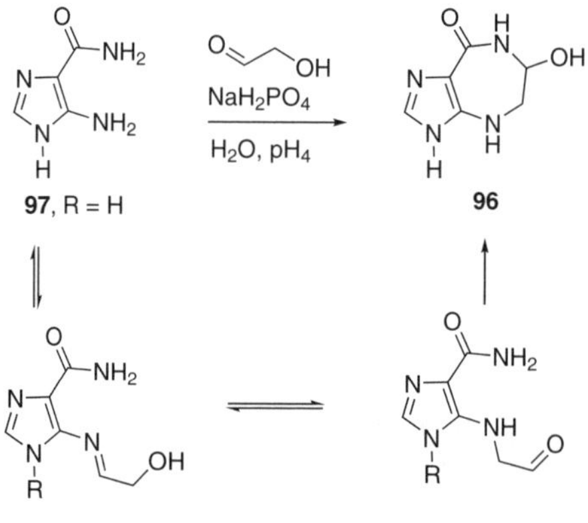

Scheme 6.23 Improved synthesis of azepinomycin (**96**).

6.8 One-Pot Synthesis of Sulfanyl-histidine

Ovothiol A [(2*S*)-2-amino-3-(1-methyl-4-sulfanyl-1*H*-imidazol-5-yl)propanoic acid], found in the eggs of certain marine invertebrates, is known to act as a reducing antioxidant. It is found to scavenge hydrogen peroxide released during fertilization. It is also found in pathogens, e.g. trypanosomes and *Leishmania*. Its antioxidant properties have attracted pharmaceutical interest to create prospective antitrypanosomal chemotherapeutic agents [40a, b]. Thioester **100**, a surrogate of 4-thiohistidine **99**, is considered a new amino acid (Scheme 6.24). This amino acid has also found significance in the biogenesis of adrenochrome, an unusual Fe(III)-chelating peptide from *O. vulgaris* and has the potential to gain importance in mercaptohistidine-based pharmaceuticals [40a–d]. Daunay et al. developed a short protecting-group-free synthesis of this rare naturally occurring amino acid, 4-sulfanyl-histidine, starting from histidine (**98**; Scheme 6.24) [40e]. Thioacetic acid reacted with the bromolactone intermediate formed from histidine (**98**) at the C5 position to form the thioester **100**. This thioester, on hydrolysis in presence of reducing agent, gave the amino acid **99**. The reducing agent 3-mercaptopropionic acid was employed to prevent oxidation of sulfur. This synthesis provides access to gram-scale quantities of **99** in two steps (overall yield = 59%) without the use of any chromatographic purification. After the formation of thioester (60 min), 3-mercaptopropionic acid was added and heated in the same pot to obtain **100** in 74% yield.

6.9 Synthesis of an Antiwrinkle Venom Analog

Syn-Ake is a tripeptide that reduces the appearance of wrinkles. This compound is a mnAChR antagonist that prevents the intake of sodium ions, keeping the muscles relaxed. The mechanism of action is the same as that of Waglerin-1, a compound present in the venom of the temple viper [41]. Conventional synthesis of peptides involves protecting groups and is lengthy. A short protecting-group-free synthesis of Syn-Ake (**107**) was reported by Balaev and coworkers (Scheme 6.25) [42].

Initial attempts for the acylation of proline (**101**) gave the desired product **102** along with the dehydrohalogenated side product (*N*-acryloyl proline). After changing various

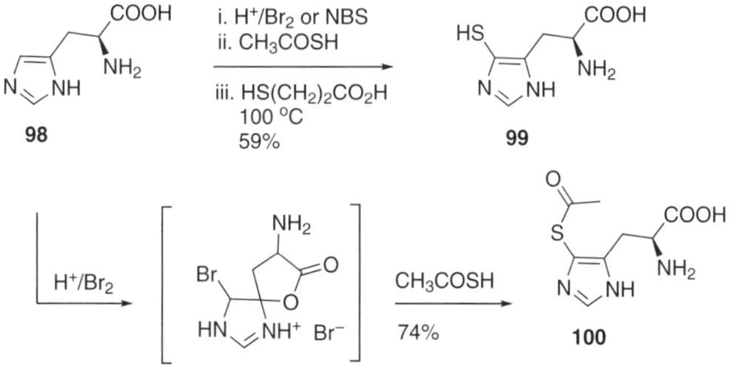

Scheme 6.24 One-pot synthesis of 5-sulfanyl-histidine **100**.

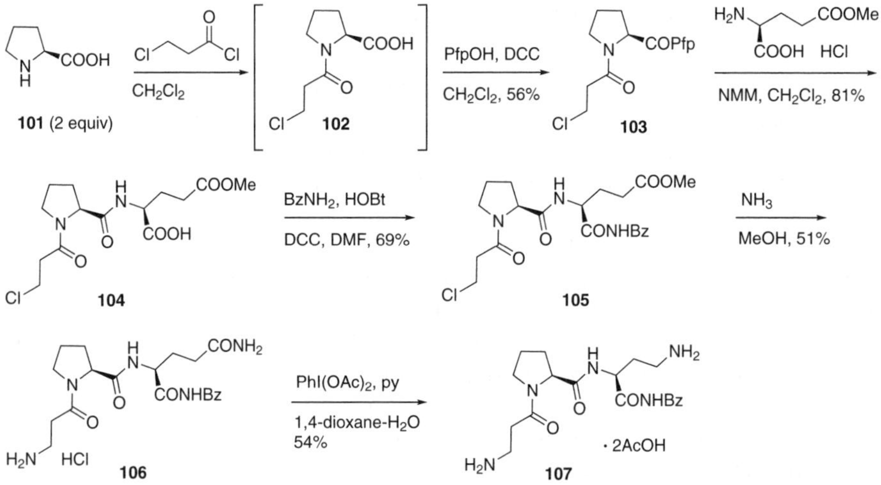

Scheme 6.25 Synthesis of antiwrinkling agent Syn-Ake (**107**).

bases, it was found that the reaction went smoothly by the addition of one more equivalent of proline (Scheme 6.25). The carboxylic acid was then esterified with pentafluorophenol using DCC as coupling reagent to obtain **103** in 56% yield. The activated ester **103**, on reaction with glutamic ester, gave dipeptide **104** in 81% yield. Benzylation followed by ammonolysis gave the tripeptide **106** in 51% yield. The key Hofmann-type rearrangement was achieved using (diacetoxyiodo)benzene and pyridine to give the final product **107** with a yield of 54%.

6.10 Preparation of 5-Arylidene Rhodanine and 2,4-Thiazolidinediones

Rhodanines and thiazolidinediones are pharmacologically important scaffolds present in medicinally active compounds [43]. Thiazolidinediones (glitazones) are PPARγ agonists used for treating type 2 diabetes mellitus. Clinically relevant agents such as ciglitazone, pioglitazone, rosiglitazone, darglitazone, and rivoglitazone belong to this class [44]. Dhruva Kumar et al. reported an environmentally benign protocol to access 5-arylidene rhodanines and 2,4-thiazolidinediones from nitrones using polyethylene glycol (PEG) as solvent (Figure 6.8) [45]. PEG is water soluble and is considered to be safe.

Nitrones are 1,3-dipoles useful in cycloaddition reactions. A mixture of nitrone **108** and rhodanine/thiazolidinedione **109** in PEG, on heating at 80 °C, gave the required

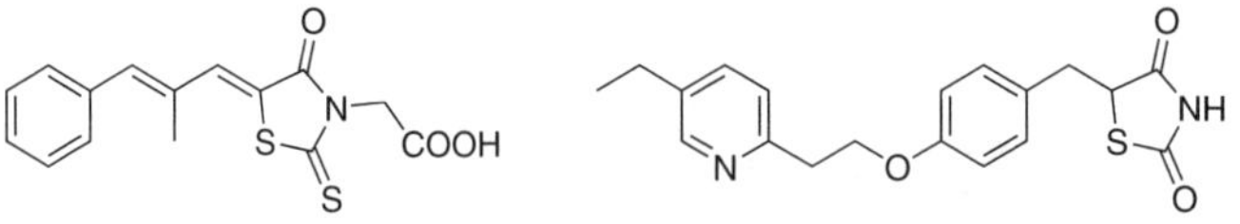

Epalrestat

Pioglitazone

Figure 6.8 Clinically relevant molecules of 5-arylidene rhodanine and 2,4-thiazolidenedione class.

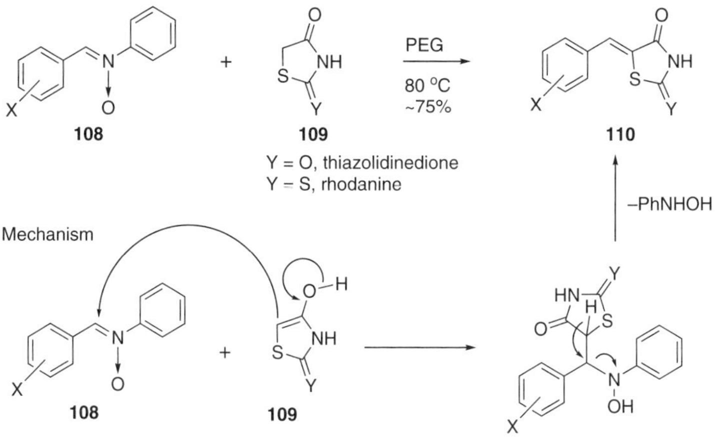

Scheme 6.26 Synthesis of rhodanines and thiazolidinediones.

product **110**, which was purified by precipitation followed by crystallization (Scheme 6.26) to obtain it in around 75% yield. Different solvents were screened for the reaction and PEG gave the best results. PEG recovered after filtration of product could be successfully recycled for five runs. The reaction proceeds through a nucleophilic addition followed by elimination (Scheme 6.26). The developed procedure is efficient, short, environmentally friendly, and devoid of any protecting groups.

6.11 *Se*-Adenosyl-L-Selenomethionine and Analogs

SAM is a methyl donor required for the reaction catalyzed by the methyltransferase enzymes. These transmethylation reactions are involved in various disease pathways and protein regulation. Various SAM analogs have been designed to inhibit methyltransferase and also find application as chemical biology tools [46]. Se-based SAM analogs have also gained interest and are expected to be more reactive because of the weaker Se—C bond [47]. In this context, Bothwell et al. developed a protecting-group-free synthesis of SeAM analogs [48].

In the body, SAM is synthesized from L-methionine and ATP. Inspired by the biosynthetic pathway, it was proposed that SeAM and analogs could be synthesized from L-methionine and adenosine. L-Methionine (**111**) on reaction with methyl iodide gave the *N*-methyl derivative, which underwent an intramolecular displacement to form homoserine lactone. This lactone on hydrolysis gave the hydroxy amino acid **112** in 74% yield (Scheme 6.27). The alcohol was converted to bromide **113** in 62% yield and then treated with Na_2Se_2 formed *in situ* to produce selenohomocystine **114** in 84% yield. $NaBH_4$ reduction of **114** gave the anion **115**, which then displaced the iodo group from iodo derivative **116** to afford product **117** in 58% yield. The latter was immediately subjected to alkylation to generate a library of SeAM analogs **118a–g** in around 50% yields. The compound **117** is prone to intramolecular lactonization and hence should be handled with care. SeAM and six new analogs were synthesized by this efficient, short protecting-group-free route starting from inexpensive starting materials.

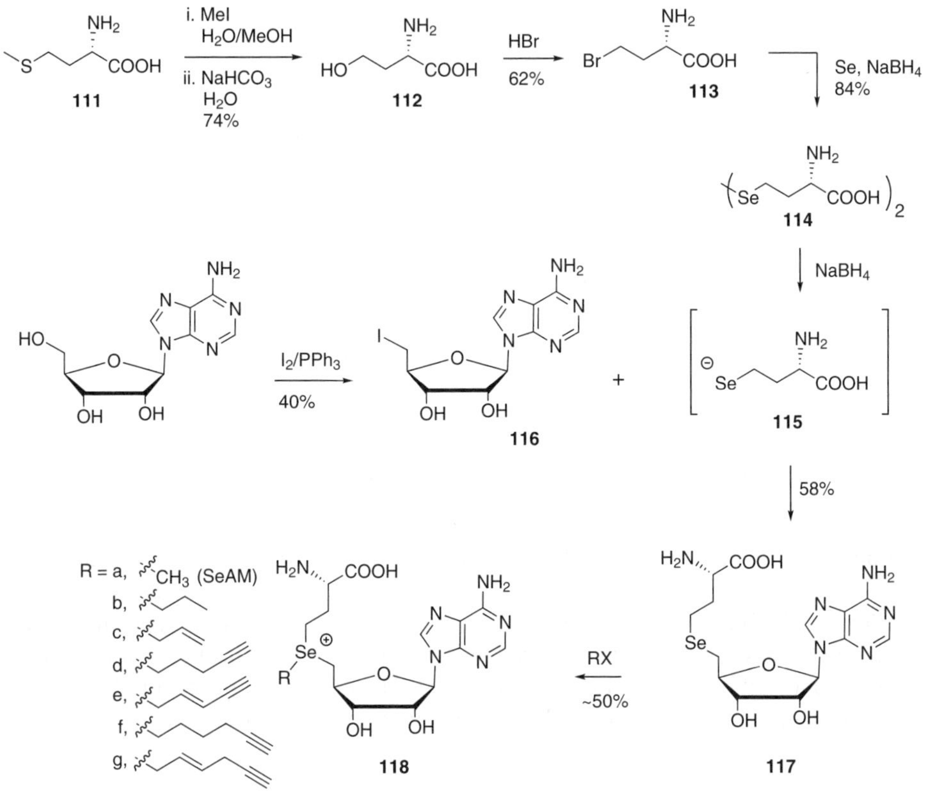

Scheme 6.27 Synthesis of SeAM and analogs.

6.12 Conclusions

A review on the synthetic strategies for a number of drug molecules that have been developed and evolved from protection–deprotection approaches to greener synthetic protocols is presented. Protection–deprotection-free approaches have several advantages, e.g. less energy intensive that requires less number of manpower, E-factor that gets minimized, and overall process efficiency that improves. In this particular chapter, we have discussed several approaches for the synthesis of various medicines, i.e. raltegravir, levetiracetam, sitagliptin, paroxetine, apitolisib, azepinomycin, and sulfanyl-histidine, reflecting that there is a possibility of changing any route in order to make it protection–deprotection-free. It is evident from this compilation that early technical intervention during synthetic design of the molecules would help to develop protection–deprotection-free synthesis contributing to sustainable manufacturing processes. Right route selection aiming to avoid protection–deprotection, to target to minimize PMI and genotoxic impurities (GTIs), and to increase atom economy by employing innovative and nonstoichiometric methods (catalysis) at scale would be the way forward for sustainable manufacturing of not only the medicines but various materials in general.

Acknowledgment

The authors are thankful to the management of the Dr. Reddy's Laboratories Ltd. and CSIR-NCL, Pune, for supporting this work.

References

1 Young, I.S. and Baran, P.S. (2009). *Nat. Chem.* 1: 193–205.
2 Trost, B.M. (1991). *Science* 254: 1471–1477.
3 Wender, P.A., Verma, V.A., Paxton, T.J., and Pillow, T.H. (2008). *Acc. Chem. Res.* 41: 40–49.
4 Grubbs, R.H. (2007). *Chem. Sci.* 4: C69.
5 Wender, P.A. and Miller, B.L. (1993). *Organic Synthesis: Theory and Applications* (ed. T. Hudlicky), 27–66. JAI Press.
6 Robinson, R. (1917). *J. Chem. Soc. Trans.* 111: 762–768.
7 Patil, G.D., Kshirsagar, S.W., Shinde, S.B. et al. (2012). *Org. Process. Res. Dev.* 16: 1422–1429.
8 Pace, P., Spieser, S.A.H., and Summa, V. (2008). *Bioorg. Med. Chem. Lett.* 18: 3865–3869.
9 Croxtall, J.D., Lyseng-Williamson, K.A., and Perry, C.M. (2008). *Drugs* 68: 131–138.
10 Angelaud, R., Belyk, K.M., Cooper, V.B. et al. (2004). Potassium salt of an hiv integrase inhibitor. WO 2006060730 A3, filed 2 December 2005 and published 17 August 2006.
11 Humphrey, G.R., Pye, P.J., Zhong, Y.-L. et al. (2011). *Org. Process. Res. Dev.* 15: 73–83.
12 Gurjar, M.K., Sonawane, S.P., Maikap, G.S. et al. (2011). Synthesis of raltegravir. US patent, US9403809 B2, filed 21 December and published 2 August 2016.
13 Boschi, F., Camps, P., Comes-Franchini, M. et al. (2005). *Tetrahedron: Asymmetry* 16: 3739–3745.
14 (a) Sarma, K.D., Zhang, J., Huang, Y., and Davidson, J.G. (2006). *Eur. J. Org. Chem.* 3730–3737. (b) Kotkar, S.P. and Sudalai, A. (2006). *Tetrahedron Lett.* 47: 6813–6815. (c) Forcato, M., Michieletto, I., Maragni, P. et al. (2006). Procédé de préparation de lévétiracétam. WO2008/012268 A1, filed 20 July 2007 and published 31 January 2008. (d) Mandal, A.K., Mahajan, S.W., Ganguly, P. et al. (2005). Process for preparing levetiracetam and racemization of (r)- and (s)-2-amino butynamide and the corresponding acid derivatives. WO2006/103696 A2, filed 20 January 2006 and published 5 October 2006. (e) Rizzulti, G. and Gianolli, E. (2005). Synthesis of (S)-α-ethyl-2-oxo-1-pyrrolidineacetamide. WO 2006127300 A1, filed 12 May 2006 and issued 30 November 2006.
15 (a) Marian, E. and Cegla, M. (1981). *Pol. J. Chem.* 55: 2205–2210. (b) Mandal, A.K., Mahajan, S.W., Ganguly, P. et al. PCT US2006/1034400. (c) Gobert, J., Greets, J. P., and Bodson, G. Eur. Pat. Appl. E0162036; Chem. Abstr.105, 018467. (d) Acharyulu, P.V.R., Raju, C.M.H. PCT US2005/0818.
16 Tao, J. and Xu, J.-H. (2009). *Curr. Opin. Chem. Biol.* 13: 43–50.
17 (a) Hensman, J.R. and Khambati, R.E. (2004). Process for the preparation of n-akyl-pyrrolidones. WO2005/121083 A1, filed 1 June 2005 and published 22 December 2005. (b) Yoon, Y.-S., Shin, H.K., and Kwak, B.-S. (2002). *Catal. Commun.* 3: 349–355. (c) Kulig, K., Holzgrabe, U., and Malawska, B. (2001). *Tetrahedron: Asymmetry* 12: 2533–2536. (d) Rao, Y.V.S., Kulkarni, S.J., Subrahmanyam, M., and Rao, A.V.R. (1994). *J. Org. Chem.* 59: 3998–4000. (e) Zienty, F.B. and Steahly, G.W. (1947). *J. Am. Chem. Soc.* 69: 715–716.

18 (a) Overman, L.E., Ricca, D.J., and Tran, V.D. (1997). *J. Am. Chem. Soc.* 119: 12031–12040. (b) Clinch, K., Vasella, A., and Schauer, R. (1987). *Tetrahedron Lett.* 28: 6425–6428. (c) Lee, J.C., Lee, K., and Cha, J.K. (2000). *J. Org. Chem.* 65: 4773–4775.

19 Mylavarapu, R., Anand, R.V., Kondaiah, G.C.M. et al. (2010). *Green Chem. Lett. Rev.* 3: 225–230.

20 Drucker, D., Easley, C., and Kirkpatrick, P. (2007). *Nat. Rev. Drug Discov.* 6: 109–110.

21 (a) Holst, J.J. (1994). *Gastroenterology* 107: 1048–1055. (b) Drucker, D.J. (1998). *Diabetes* 47: 159–169. (c) Deacon, C.F., Holst, J.J., and Carr, R.D. (1999). *Drugs of Today* 35: 159–170. (d) Livingston, J.N. and Schoen, W.R. (1999). *Annu. Rep. Med. Chem.* 34: 189–198.

22 Evans, D.A., Britten, T.C., Ellman, J.A., and Dorow, R.L. (1990). *J. Am. Chem. Soc.* 112: 4011–4030.

23 Schoellkopf, U., Groth, U., and Deng, C. (1981). *Angew. Chem. Int. Ed.* 20: 798–799.

24 Müller, A., Vogt, C., and Sewald, N. (1988). *Synthesis* 837–841.

25 Huynh-Dinh, T., Sarfati, R.S., Gouyette, C., Igolen, J. et al. (1979). *J. Org. Chem.* 44: 1028–1035.

26 Kim, D., Wang, L., Beconi, M. et al. (2005). *J. Med. Chem.* 48: 141–151.

27 Hansen, K.B., Balsells, J., Dreher, S. et al. (2005). *Org. Process. Res. Dev.* 9: 634–639.

28 Lee, N.E. and Buchwald, S.L. (1994). *J. Am. Chem. Soc.* 116: 5985–5986.

29 Hansen, K.B., Hsiao, Y., Xu, F. et al. (2009). *J. Am. Chem. Soc.* 131: 8798–8804.

30 (a) Savile, C.K., Janey, J.M., Mundorff, E.C. et al. (2010). *Science* 329: 5989. (b) Gutierrez, O., Metil, D., Dwivedi, N. et al. (2015). *Org. Lett.* 17: 1742–1745.

31 Johnson, T.A., Curtis, M.D., and Beak, P. (2001). *J. Am. Chem. Soc.* 123: 1004–1005.

32 (a) Crowe, D., Jones, D.A., and Ward, N. (1999). Process for the preparation of 1-methyl-3-carbomethoxy-4-(4'-fluorophenyl)-piperidine. WO Patent WO 2001017966 A1, filed 8 September 2000 and published 15 March 2001. (b) Murthy, K.S.K. and Rey, A.W. (1997). Preparation stereoselective de composes piperidine a 4-aryle 3-substitue. WO 1999007680 A1, filed 31 July 1998 and published 18 February 1999.

33 Yu, M.S., Lantos, I., Peng, Z.-Q. et al. (2000). *Tetrahedron Lett.* 41: 5647–5651.

34 (a) Sabbah, D.A., Brattain, M.G., and Zhong, H. (2011). *Curr. Med. Chem.* 18: 5528–5544. (b) Engelman, J.A. (2009). *Nat. Rev. Cancer* 9: 550–562.

35 Sutherlin, D.P., Bao, L., Berry, M. et al. (2011). *J. Med. Chem.* 54: 7579–7587.

36 Tian, Q., Hoffmann, U., Humphries, T. et al. (2015). *Org. Process. Res. Dev.* 19: 416–426.

37 Isshiki, K., Takahashi, Y., Iinuma, H. et al. (1987). *J. Antibiot.* 40: 1461–1463.

38 Coggins, A.J., Tocher, D.A., and Powner, M.W. (2015). *Org. Biomol. Chem.* 13: 3378–3381.

39 (a) Fuji, T., Saito, T., and Fujisawa, T. (1988). *Heterocycles* 27: 1163–1166. (b) Chakraborty, S., Shah, N.H., Fishbein, J.C., and Hosmane, R.S. (2012). *Bioorg. Med. Chem. Lett.* 22: 7214–7218.

40 (a) Schmidt, A. and Krauth-Siege, R.L. (2002). *Curr. Top. Med. Chem.* 2: 1239–1259. (b) Steenkamp, D.J. (2002). *Life* 53: 243–248. (c) Mirzahosseini, A., Hosztafi, S., Tóth, G., and Noszál, B. (2014). *ARKIVOK* vi: 1–9. (d) Rossi, F., Nardi, G., Palumbo, A., and Prota, G. (1985). *Comp. Biochem. Physiol. B: Biochem. Mol. Biol.* 80B: 843–845. (e) Daunay, S., Lebel, R., Farescour, L. et al. (2016). *Org. Biomol. Chem.* 14: 10473–10480.

41 (a) Namjoshi, S., Caccetta, R., and Benson, H.A.E. (2008). *J. Pharm. Sci.* 97: 2524–2542. (b) Chhipa, N.M.R. and Chaudhari, B.G.J. (2012). *Curr. Pharm. Res.* 9: 11–18.

42 Balaev, A.N., Okhmanovich, K.A., and Osipov, V.N. (2014). *Tetrahedron Lett.* 55: 5745–5747.

43 (a) Tomasic, T. and Masic, L.P. (2012). *Expert Opin. Drug Discovery* 7: 549–560. (b) Cantello, B.C.C., Cawthorne, M.A., Cottam, G.P. et al. (1994). *J. Med. Chem.* 37: 3977–3985.

44 Krische, D. (2000). *West J. Med.* 173: 54–57.

45 Kumar, D., Narwal, S., and Sandhu, J.S. (2013). *Int. J. Med. Chem.* 2013: 4, Article ID 273534.

46 (a) Zhang, J. and Zheng, Y.G. (2016). *ACS Chem. Biol.* 11: 583–597. (b) Islam, K., Zheng, W., Yu, H. et al. (2011). *ACS Chem. Biol.* 6: 679–684.

47 (a) Singh, S., Zhang, J., Huber, T.D. et al. (2014). *Angew. Chem. Int. Ed.* 53: 3965–3969. (b) Bothwell, I.R., Islam, K., Chen, Y. et al. (2012). *J. Am. Chem. Soc.* 134: 14905–14912.

48 Bothwell, I.R. and Luo, M. (2014). *Org. Lett.* 16: 3056–3059.

7

Protecting-Group-Free Synthesis in Carbohydrate Chemistry

Alejandro Cordero-Vargas[1] and Fernando Sartillo-Piscil[2]

[1]*Instituto de Química, Universidad Nacional Autónoma de México, Mexico*
[2]*Centro de Investigación de la Facultad de Ciencias Químicas and Centro de Química de la Benemérita, Universidad Autónoma de Puebla, Mexico*

7.1 Introduction

Carbohydrates and their derivatives are probably the preferred starting materials for chemists to synthesize and to determine the absolute configuration of biologically important natural compounds [1]. Due to their well-established stereochemistry and because they prefer the cyclic conformation to the acyclic form, these naturally occurring compounds are extraordinary templates and building blocks for rationally designed synthesis [2]. However, in their intrinsic and beautiful oxygenated nature is found their genuine malice: usually some of their hydroxyl and carbonyl groups are not required, and they have to be either eliminated or protected, turning the synthesis into a tedious and complicated endeavor. By following the statement of Professor Grubbs [3] alluding to the ideal synthesis, that "The major challenges in chemistry are the construction of molecules without using protecting groups...," the carbohydrates, in their natural form, are considered as the most challenging substrates in organic synthesis, especially for total synthesis. In fact, this is the reason why there exist only few reports describing protecting-group-free total synthesis (PGF-TS) employing carbohydrates. Almost all the total syntheses with the greatest protecting-group-free (PGF) spirit are those where simple carbohydrate derivatives, such as methyl glycosides, carbohydrate-derived lactones, or *O*-isopropylidene furanoses/pyranoses, were employed as starting materials. Moreover, since the main premise for the construction of molecules without using protecting groups must be the lowering of economic cost and reduction of environmental impact, a chemist must not vacillate on this endeavor, even for moral reasons; otherwise he/she will regret it forever. A remarkable example that attempts to put forward the advantages of using a PGF synthesis with carbohydrates as starting materials is the total synthesis of the naturally occurring cephalosporolides E and F (two secondary metabolites that contain an unprecedented central 1,6-dioxaspiro[4,4]nonane core). Both natural compounds have been frequently synthesized from carbohydrate derivatives (D-glucose derivatives);

Protecting-Group-Free Organic Synthesis: Improving Economy and Efficiency, First Edition.
Edited by Rodney A. Fernandes.

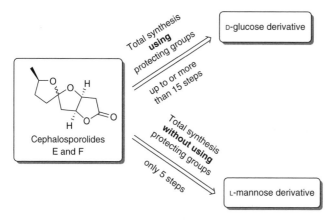

Figure 7.1 The total synthesis of cephalosporolides E and F with or without using protecting groups.

up to or more than 15 steps have been required under a classical non-PGF synthesis [4]. However, by using the PGF approach from an L-mannose derivative, chemists have been able to synthesize them in only five steps (see Figure 7.1 and *vide infra* for details) [5]. Obviously, the marked improvement in the synthesis of cephalosporolides E and F under the latter approach should motivate all synthetic organic chemists to look to carbohydrates and derivatives for the synthesis of even more complex biologically active products under such economic and environmentally friendly approaches.

7.2 Protecting-Group-Free Total Synthesis (PGF-TS)

It is somewhat logical that in the seminal review article of Young and Baran [6] regarding PGF-TS, only one invoked the use of a carbohydrate as starting material. However, from a more critical point of view, the only PGF-TS discussed in the review, which displayed the synthesis of polyhydroxylated pyrrolidines from either D-ribose or D-xylose [7], invoked the use of Fischer *O*-glycosylation as the first step. Furthermore, the preparation of a glycosylated derivative might be viewed as the protection of one hydroxyl group and one carbonyl group of the respective carbohydrate precursors. In this regard, the group of Stocker converted D-ribose (**1**) into carbohydrate-protected iodofuranose **2** in two steps via the eventual formation of an acetal center in **2** (Scheme 7.1) [7]. The successful application of the Vasella reductive amination [8] of **2** to the corresponding amino alcohol **3** represented one of the key steps for the synthesis of 1,4-dideoxy-1,4-aminol-L-lyxitol **4**. From compound **3**, iodocyclization with I_2/NaHCO$_3$ followed by basic hydrolysis led to pyrrolidine **4** (Scheme 7.1).

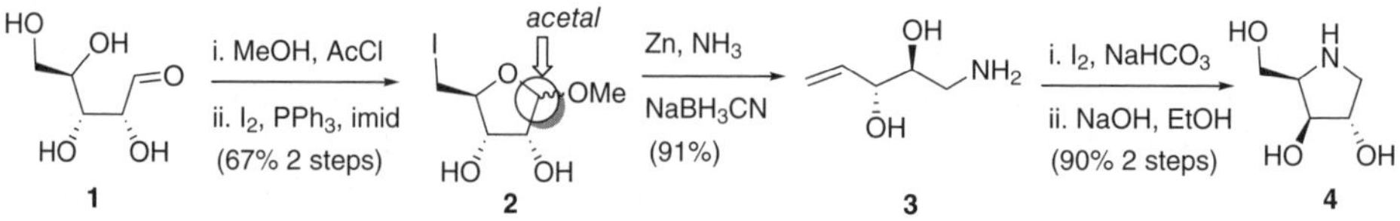

Scheme 7.1 PGF-TS of hydroxypyrrolidine **4** from D-ribose (**1**).

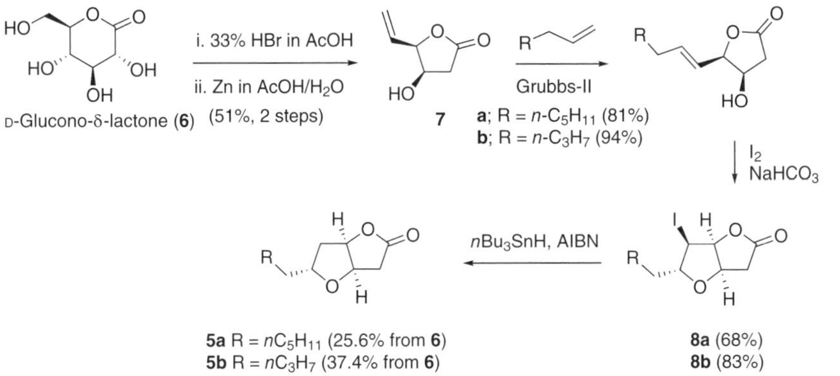

Scheme 7.2 Concise PGF-TS of Hagen's gland lactones from D-glucono-δ-lactone **6**.

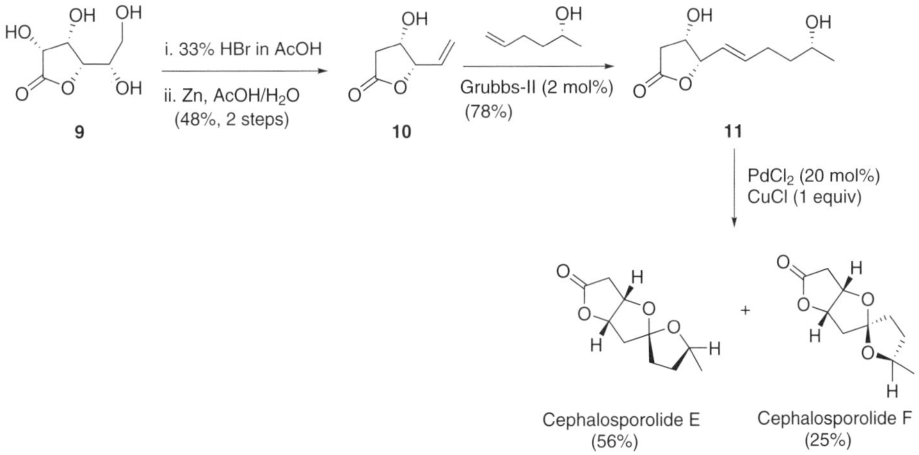

Scheme 7.3 Concise PFG-TS of cephalosporolides E and F from unnatural carbohydrate derived γ-lactone **10**.

In 2012, the group of Fernandes reported a practical and concise PGF-TS of Hagen's gland lactones (**5a** and **5b**) from D-glucose (in its δ-lactone form, **6**) [9]. Crucial for this PGF-TS was the "one-pot two-step" conversion of D-glucono-δ-lactone (**6**) into vinyl-γ-lactone **7** in acceptable overall yield (Scheme 7.2). Application of a cross-metathesis with a Grubbs second-generation catalyst followed by iodocyclization under standard conditions afforded iodobicyclic lactones **8a** and **8b** in good yields. These were deiodinated under tin radical conditions to natural compounds **5a** and **5b**. With an overall yield of 26% for **5a** and 37% for **5b**, this five-step PGF-TS remains the most concise for Hagen's gland lactones.

Few years later, the same group reported another notable PGF-TS of naturally occurring compounds. Staring from the nonnatural L-mannose (in its γ-lactone form, **9**), diastereomeric cephalosporolides E and F were obtained in four linear steps in 22.8 and 10.5% overall yields, respectively (Scheme 7.3) [5]. Using a "two-step one-pot" protocol, γ-lactone **9** was transformed into γ-lactone **10** in acceptable yield. The latter was subjected to a cross-coupling metathesis to afford the advanced intermediate **11** in good yield. Finally, the spiroketal framework of cephalosporolides E and F was forged via a Pd-catalyzed Wacker-type oxidative spiroketalization. This concise and elegant PGF-TS of cephalosporolides E and F highlighted the compatibility of substrates bearing hydroxyl and carbonyl groups with transition metal catalysts, which were employed in Wacker and alkene metathesis reactions [5].

In 2015, Du and coworkers not only reported an efficient PGF-TS of cephalosporolides H and I, but they also revised the absolute configuration at the spiroketal center of both naturally occurring products [10]. Cross-coupling metathesis between γ-lactone **12** (which was prepared in two steps from D-glucono-δ-lactone **6**) and chiral olefin **13** mediated by Grubbs II catalyst provided key diol **14** (Scheme 7.4). Pd-catalyzed spiroketalization of **14** gave surprisingly a single diastereoisomer **15** in good yield, which was transformed into iodobicyclic lactone **16** by simple iodination with NaI. Iodolactone **16** was then converted into cephalosporolide H by means of CuCl-catalyzed cross-coupling with Grignard reagent **17** and to cephalosporolide I by radical Michael addition with benzyl acrylate **18** in the presence of $NaBH_3CN$ and AIBN followed by debenzylation with Pd/C, H_2 (Scheme 7.4).

In a similar fashion, the group of van Delft developed a PGF formal synthesis of sphydrofuran (Scheme 7.5) [11]. Although the selective Wittig olefination of D-xylose **19** followed by an oxycyclization of the olefin **20** to the corresponding D-threose-C-glycoside derivative **21** was conducted under PGF conditions; further transformations of C-glycoside **21** required the use of protecting groups to obtain an advanced intermediate **22** of the natural product sphydrofuran [12]. Consequently, this formal PGF-TS of sphydrofuran is in fact a PGF synthesis of 3,4-dihydroxytetrahydrofurans from D-furanoses.

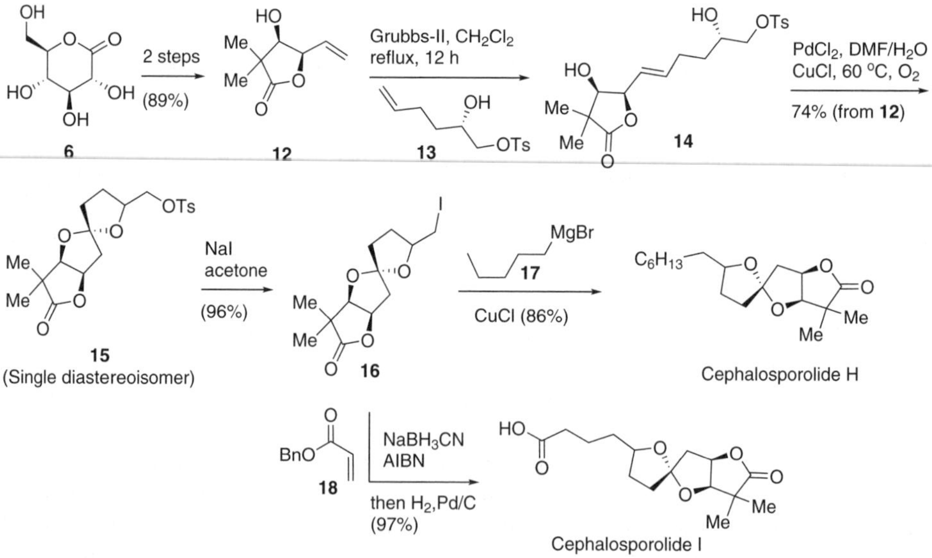

Scheme 7.4 PGF-TS of cephalosporolides H and I from D-glucono-δ-lactone **6** by Du and coworkers.

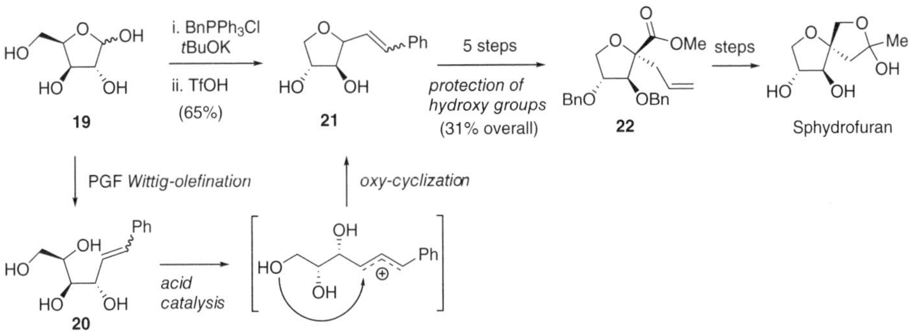

Scheme 7.5 PGF synthesis of 3,4-dihydroxytetrahydrofuran 22 from D-xylose (19) and its application to an advanced intermediate of sphydrofuran.

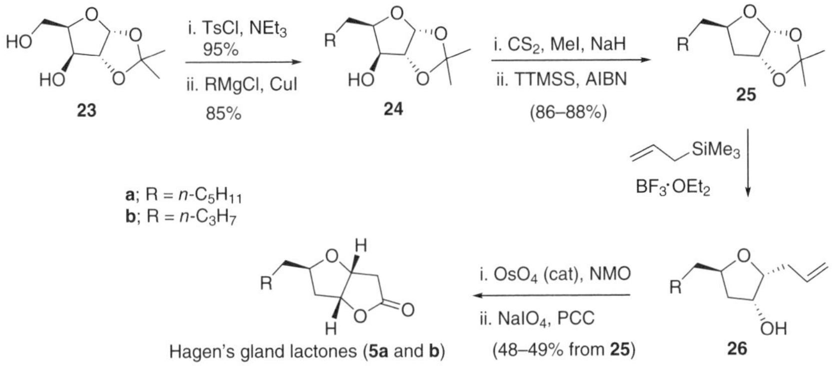

Scheme 7.6 Total synthesis of Hagen's gland lactones (5a and 5b) from D-xylofuranose derivative 23.

In some cases, chemists have taken advantage of the *locked* cyclic furanose form of carbohydrates and have transformed furanose carbohydrate derivatives into tetrahydrofurans, which are present in many naturally occurring compounds [13]. In this sense, the total synthesis of Hagen's gland lactones 5a and 5b was completed from 1,2-*O*-isopropylidene-α-D-xylofuranose 23 (Scheme 7.6) [14]. Featuring a selective carbon chain elongation at C5 position of 23 by means of a substitution reaction on the primary tosyl group with the respective organocuprate nucleophile, 5-deoxy-D-xylofuranose 24 was obtained without the need of protection of the secondary hydroxyl group [15]. A Barton–McCombie deoxygenation reaction (first xanthate formation with CS$_2$, MeI, and NaH and then deoxygenation with TTMSS and AIBN at toluene reflux) on the secondary hydroxyl group of 24 gave 25 in good yields. Deoxygenated compounds 25 were subjected to a stereoselective nucleophilic substitution at the anomeric position with allyltrimethylsilane in the presence of BF$_3$·OEt$_2$ [16] to obtain 26. The latter, without chromatographic purification, was subjected to olefin cleavage (dihydroxylation and NaIO$_4$ treatment) and then oxidized with PCC to give the lactone compounds, Hagen's gland lactones (5a and 5b) [14].

Moreover, this work was not disclosed as a PGF-TS, probably because the authors considered that the use of 1,2-*O*-isopropylidene-D-xylofuranose invoked the protection of carbonyl group and two hydroxyl groups from the natural xylose carbohydrate. Indeed, this situation is quite similar to that in the PGF synthesis of hydroxypyrrolidine from D-ribose (*vide supra*) [7] wherein the Fischer *O*-glycosylation plays the same role as the 1,2-*O*-isopropylidene moiety. Accordingly, it should be stated here that if the *apparent protection* of the carbonyl and hydroxyl groups of any carbohydrate leads to the formation of locked furanose carbohydrate derivatives and this structural form remains in the final target compound, then this chemical transformation should not be considered as a functional group protection. Under this premise, we believe that the total synthesis of the Hagen's gland lactones from 1,2-*O*-isopropylidene-D-xylofuranose **23** should be classified as a PGF-TS.

Apparently, the only way to achieve a genuine PGF-TS starting from carbohydrates in the natural form is that the carbohydrates be transformed into compounds with fewer oxygenated groups, for instance, conversion of biomass-derived sugars to mixed liquid alkenes with eight carbons or more (Scheme 7.7) [17]. In this regard, the groups of Lange et al. [18] and Bond et al. [19] have reported the effective conversion of hexoses to levulinic acid (**27**), which is eventually converted into γ-valerolactone (**28**). A catalytic CO_2 elimination process of **28** produced a mixture of isomeric butenes, which can be catalytically dimerized into a mixture of alkenes with eight carbons or more (Scheme 7.7) [17].

Since one of the prime premises of total synthesis is to make bigger molecules from smaller ones, and not the other way round, the conversion of carbohydrates even if under PGF fashion into hydrocarbons represents a significant breakthrough to the petroleum chemistry, but not much for the "state-of-the-art total synthesis."

Furthermore, there have been a couple of PGF syntheses of nonnaturally occurring compounds with promising biological activity. An elegant, short PGF synthesis of thiosugars 1-deoxythionojirimycin (**29**), 1-deoxythiomannojirimycin (**30**), and 1-deoxy-thioallonojirimycin (**31**) was achieved by Chandrasekaran's group in 2010 [20]. The crucial step for the synthesis of thio-carbohydrates is introduction of the sulfur atom, which is traditionally performed by nucleophilic displacement with Na_2S [21], KSAc [22], NaSH [23], $(NH_2)_2CS$ [24], or AcSH [25]. This transformation usually requires multiple protection–deprotection steps, making the preparation of these compounds very tedious. The unified strategy for the synthesis of these thio-carbohydrate derivatives (**29–31**) consisted in the formation of bicyclic lactones **32**, **33**, and **34** from bromolactones **35**, **36**, and **37**, respectively, by treating them with benzyltriethylammonium tetrathiomolybdate, $[BnEt_3N]_2MoS_4$, a mild sulfur transfer reagent (Scheme 7.8). Final

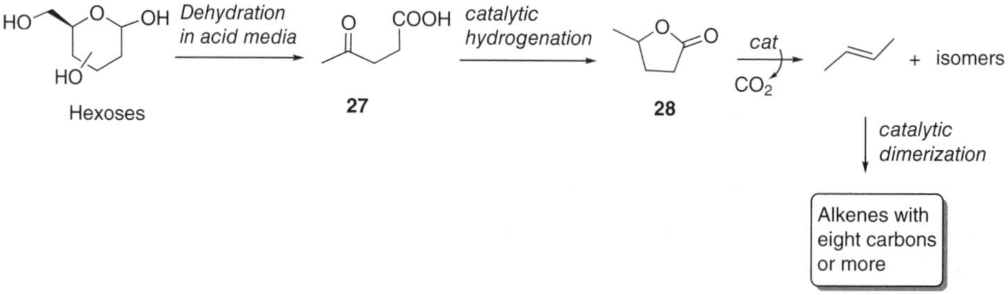

Scheme 7.7 PGF-TS of mixture of alkenes with eight carbons or more.

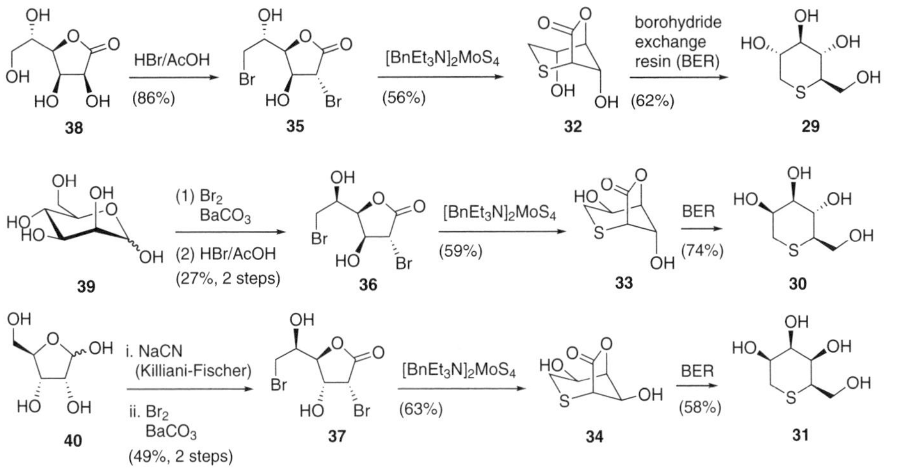

Scheme 7.8 PGF synthesis of thio-sugars 1-deoxythionojirimycin (**29**), 1-deoxythiomannojirimycin, (**30**) and 1-deoxythioallonojirimycin (**31**).

reduction of the lactones **32–34** with borohydride exchange resin (BER) afforded the desired compounds **29–31** in good overall yields from simple carbohydrates (**38**, **39**, and **40**).

7.3 Selective PGF Functionalization at the Anomeric Position (*O*-, *N*-, and *C*-Glycosylation)

In general, carbohydrates contain three types of hydroxyl groups (primary, secondary, and anomeric). Due to their structural similarity, it is quite difficult to achieve either selective functionalization or even protection [26]. Therefore, it is a common practice to make long protection–deprotection routes in order to obtain a specific free or functionalized hydroxyl group, even for glycosylation reactions, where the reactivity of the anomeric center is unique. The first PGF glycosylation reaction is attributed to Emil Fischer [27], who reported the synthesis of methyl α-D-glucopyranoside (**41**) from D-glucose (**42**) by means of an acid-catalyzed reaction (Scheme 7.9).

Despite the effectiveness of Fischer's glycosylation reaction, the need for strong protic acidic conditions sometimes limits this method. This is particularly problematic when oligosaccharides are employed as starting materials, because protonation of the intracyclic oxygen could cause acetal exchanges cleaving the existing glycosidic bonds [28]. Besides, the predominance of the thermodynamically favored pyranoside products has inspired the synthetic community to search for milder and more selective glycosylation

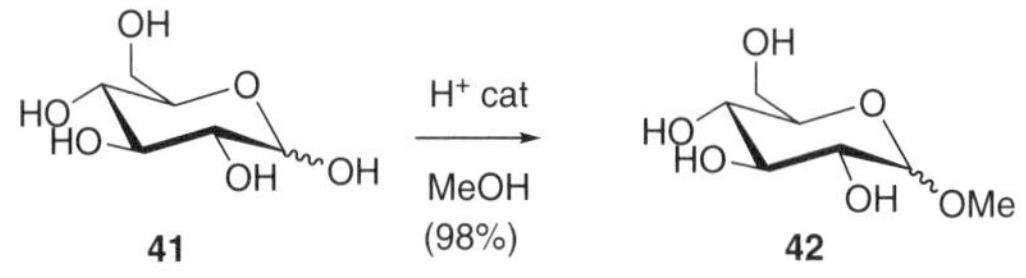

Scheme 7.9 Fischer *O*-glycosylation.

methods. Nevertheless, real improvements on the original Fischer's method took many years to come [29]. By using Lewis acids, such as $FeCl_3$ [29] or $BF_3 \cdot Et_2O$ [30], not only is the Fischer *O*-glycosylation conducted under PGF conditions, but also the thermodynamic nature (e.g. **41** to **43**) of the reaction is shifted toward a kinetic control (e.g. **41** to **44**, Scheme 7.10) [29].

In general, the PGF Fischer *O*-glycosylation provides mixtures of α/β-anomers; however, Linhardt and coworkers [31] demonstrated that glycosylation of both D-glucose **41** and D-mannose **45** can be stereoselectively conducted to α-anomers **46** and **47**, respectively, at room temperature in ionic liquid and in the presence of either Amberlite IR-120 (H⁺) resin or *p*TSA (Scheme 7.11).

Like the PFG *O*-glycosylations, PGF *N*-glycosylations are difficult to achieve, especially in a stereoselective fashion [32]. An interesting way for introducing nitrogen atom at the anomeric position in a single step and in a stereoselective way was reported by the group of Demailly [33]. Accordingly, when unprotected D-glucose **41** is treated with Ph_3P, *N*-chlorosuccinimide, and LiN_3, glycosyl azide **48** is obtained in good yield and high 1,2-*trans* stereoselectivity (Scheme 7.12).

The mentioned difficulties for the preparation of oligosaccharides through the traditional Fischer glycosidation have forced chemists to conduct selective modification of simple oligosaccharides by enzymatic methods. For example, Csuk and coworkers [34] reported a PGF synthesis of maradolipid (**49**), an unsymmetric 6,6′-trehalose

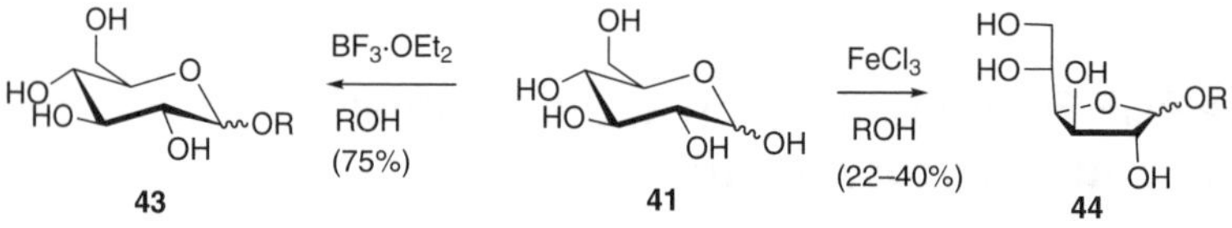

Scheme 7.10 Modified PGF Fischer *O*-glycosylation.

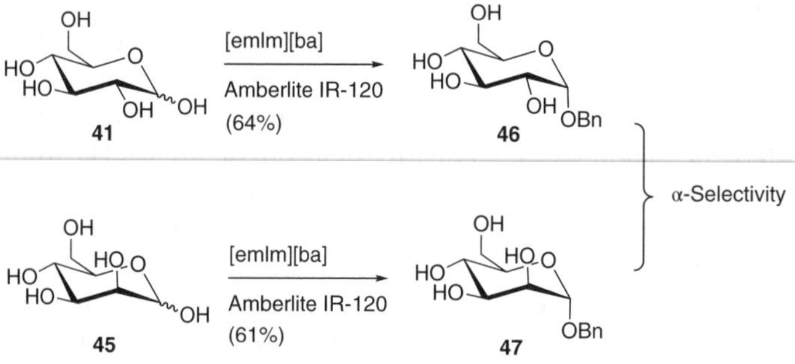

Scheme 7.11 Stereoselective PGF Fischer *O*-glycosylation.

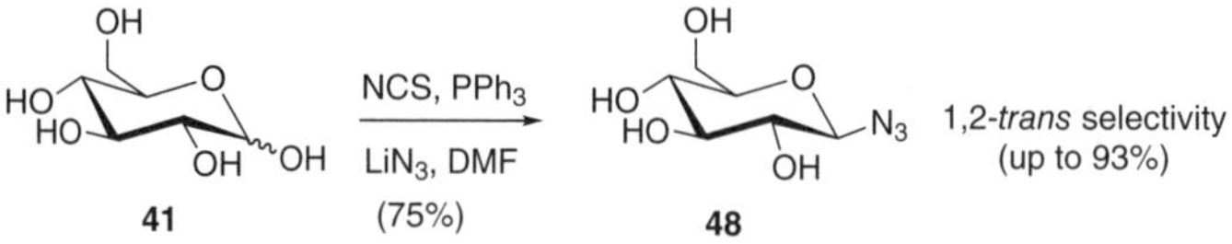

Scheme 7.12 Stereoselective PGF *N*-glycosylation.

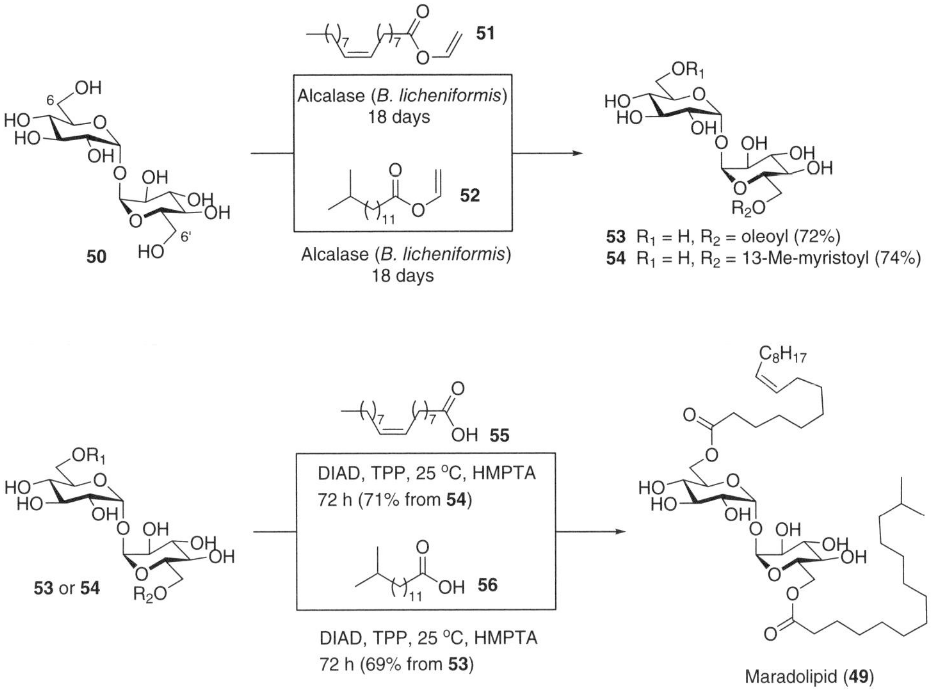

Scheme 7.13 Enzymatic and chemical synthesis of maradolipid (**49**).

diester that is an important component of the outer membrane of mycobacteria. Starting from trehalose dihydrate **50**, differentiation of the two primary hydroxyl groups was achieved by treating **50** with a 1 : 1 mixture of vinyl esters **51** and **52** in the presence of Alcalase from *B. licheniformis* to give a separable mixture of monoacylated products **53** and **54** in good yields (Scheme 7.13). A final esterification of either **53** or **54** with acids **55** or **56** by chemical means under PGF conditions (Mitsunobu reaction) afforded maradolipid (**49**) in 69–71% yields. Even though this route does not employ protecting groups, the key selective monoesterification could not be achieved by chemical methods.

As expected, the total synthesis of maradolipid (**49**) without the use of enzymes was longer and required the use of protecting groups. However, Sarpe and Kulkarni [35] developed a clever strategy involving the use of *minimal protecting groups*. This concept proposes the use of a single protecting group, preferentially a labile one that could be easily removed at the end of the route under mild conditions. In Kulkarni's total synthesis of maradolipid (**49**), per-*O*-trimethylsilyl trehalose **57** was easily prepared from **50**, and the crude compound was subjected to controlled alkaline hydrolysis using a catalytic amount of potassium carbonate, allowing the selective deprotection of the primary alcohols (**58**) (Scheme 7.14). Sequential selective esterification of the resulting diol first with oleic acid (to **59**) and then with 13-methylmyristic acid under DCC/DMAP conditions gave **60**, which was finally treated with Dowex H$^+$ exchange resin to render the desired maradolipid (**49**) in excellent overall yield. This *minimal protecting-group* strategy has also been successfully applied to the synthesis of other oligosaccharides [36].

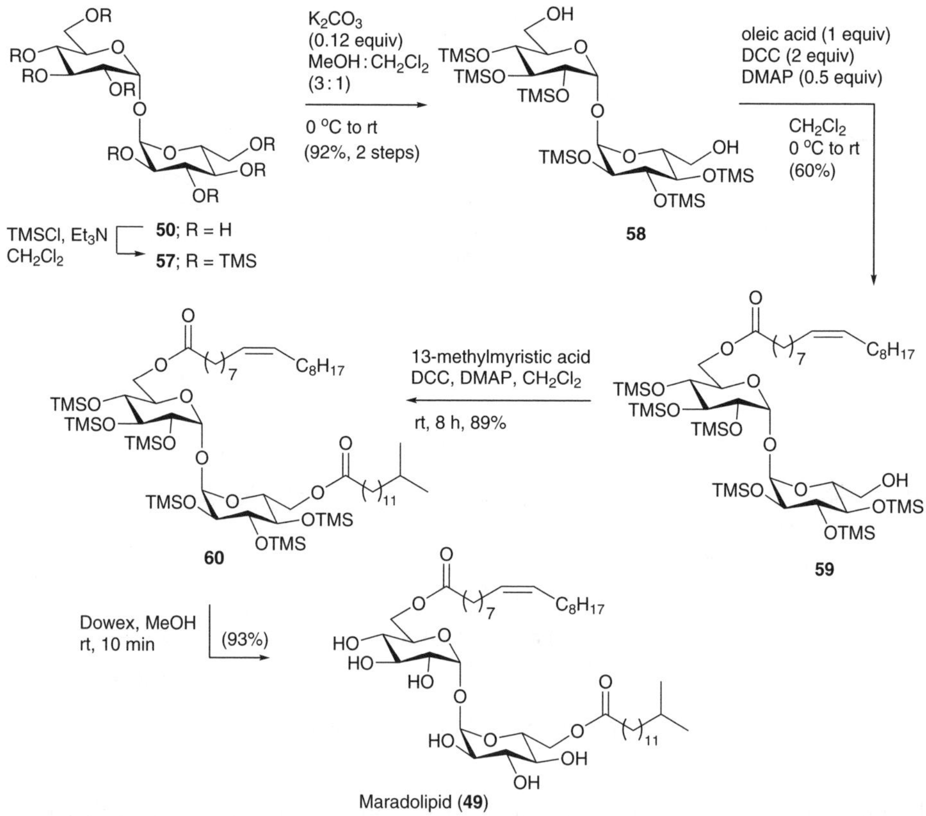

Scheme 7.14 Total synthesis of maradolipid (**49**) with *minimal protecting groups*.

In a similar way, PGF *N*-glycosylations in the form of glycoconjugates have found applications in medicinal chemistry [37]. Although it is not a peptide bond, the urea glycosyl linkage has emerged as an uncommon and important structural motif present in some amino sugar antibiotics [38]. In this context, a PGF synthesis of urea-linked glycoconjugates **61** was reported by Ichikawa (Scheme 7.15) [39]. The high stereoselectivity for the β-anomer suggests that the process is not thermodynamically controlled and the anomeric effect has a minor contribution. For purification and characterization purposes, a complete acetylation of the product was performed, albeit the approach was conducted under complete PGF conditions, isolating the nonacylated *N*-methylurea glycoside after treating the crude reaction mixture with methanol and ether.

In a similar context, Peluso and Imperiali [40] synthesized aspartyl-derived glycopeptide **62**, an inhibitor of oligosaccharide transferase, from *N*-acetylglucosamide **63** to peptide **64** by simply stirring a mixture of the starting materials in acetate buffer with DMSO (Scheme 7.16).

Fairbanks and coworkers reported another remarkable example of the conversion of sugars into glycopeptides in 2014 [41]. Instead of using a traditional nitrogen atom source (such as NaN$_3$), they prepared 2-azido-1,3-dimethylimidazolinium hexafluorophosphate (ADMP) **65** [42], which was used both as the anomeric activating group in **41** and also as

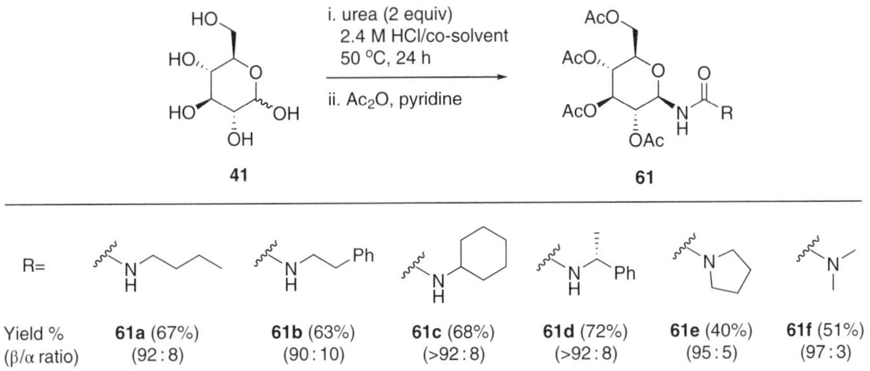

Scheme 7.15 Stereoselective PGF synthesis of urea-linked glycoconjugates.

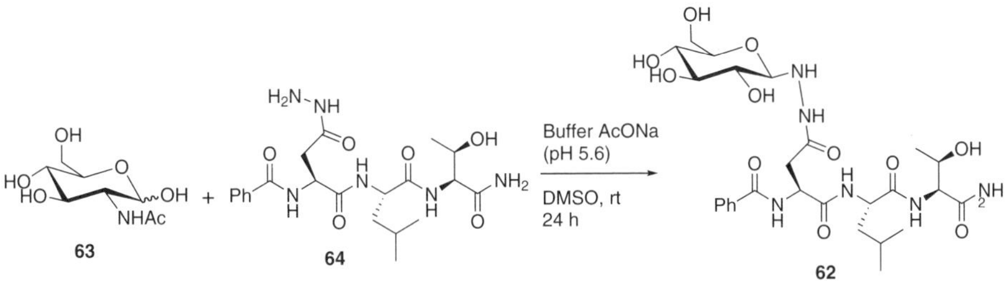

Scheme 7.16 Stereoselective PGF synthesis of aspartyl-derived glycopeptide (62).

a source for azide group to thus obtain azide **48**. The latter was directly transformed into triazole acid **66** by a *click* chemistry sequence with alkyne **67** (Scheme 7.17). A wide number of substituted alkynes and carbohydrates reacted under these reaction conditions, giving a variety of new compounds in good yields. In a spectacular example of this PGF triazole-linked glycoside synthesis, the reaction between alkyne-bearing peptide **68** and tetrasaccharide **69** afforded a triazole-linked glycopeptide **70**. Simpler and more complex saccharides and even peptides were evaluated using this method, always giving high yields and complete stereoselectivity for the β-anomer. It is important to note that both the peptide chain and the sugar moiety are completely unprotected (Scheme 7.17).

PGF C-glycosylation reactions can be performed in moderate yields and stereoselectivity. Taking advantage of the electrophilic character of the carbohydrate in the equilibrated open form, PGF Wittig olefination is a common transformation for carbon chain elongation in carbohydrates. For instance, higher sugars (carbohydrates containing seven or more consecutive carbon atoms) are compounds with biological significance and are important synthetic intermediates [43]. Kochetkov and Dmitrev [44] performed the first PFG Wittig olefination in 1965. They carried out Wittig olefination of various free aldoses with phosphonium salt PPh$_3$=CHCO$_2$Et (**71**). Although the olefination proceeded in acceptable yield (**41** to **72**), an unexpected intramolecular oxy-Michael cyclization produced a considerable amount of the C-glycosides **73** and **74**. Besides, the subsequent dihydroxylation with OsO$_4$ resulted in poor selectivity for the higher sugars **75** and **76** (Scheme 7.18).

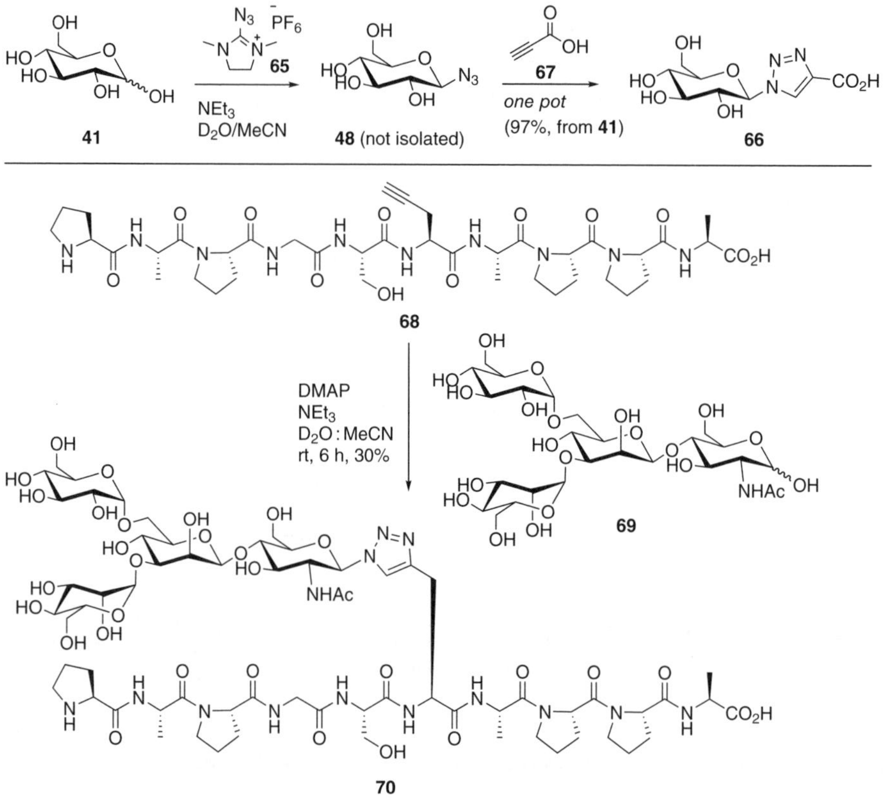

Scheme 7.17 PGF synthesis of triazole-linked glycosides **66** and **70**.

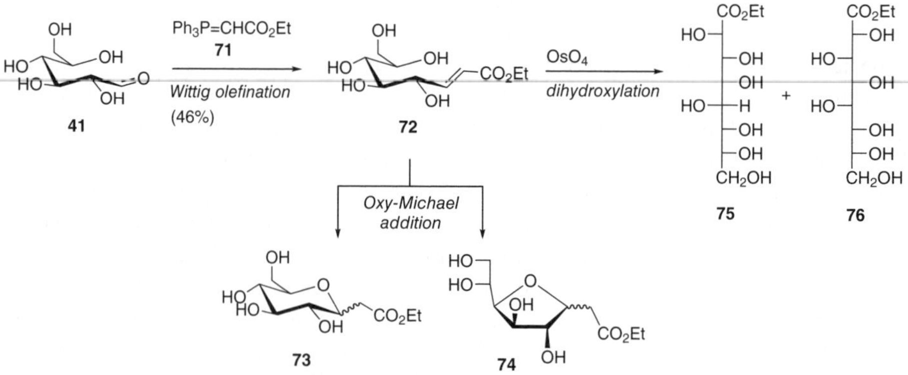

Scheme 7.18 PGF Wittig olefination.

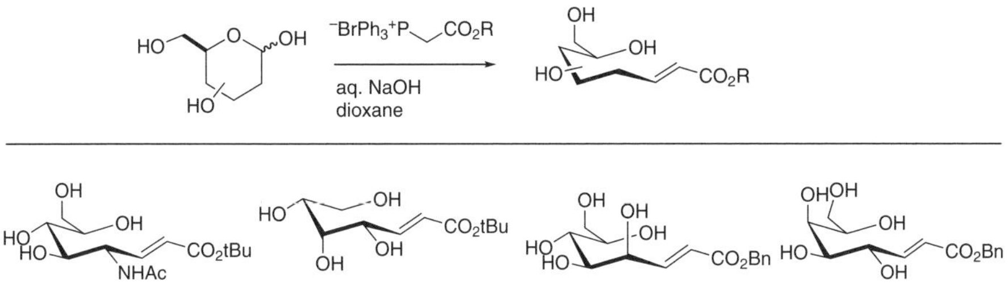

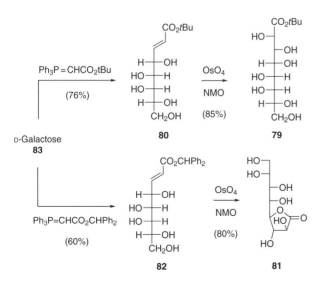

Scheme 7.19 Selective PGF Wittig olefination.

Later, Railton and Clive [45] found that the use of bulky *O*-alkyl groups in the Wittig reagent diminished the amount of Michael cyclization products. Thus, phosphorus ylides bearing *tert*-butyl and diphenylmethyl ester functionality provided the corresponding alkenes in good yields. Compounds **75–78** were respectively prepared with high (*E*)-selectivity and good yields under PGF conditions from their corresponding carbohydrates precursors (Scheme 7.19).

The group of Madsen [46] later reported a simple modification of Clive's procedure for preparing higher carbohydrates. Owing to the bulky ester groups, the OsO_4-catalyzed dihydroxylation under Upjohn conditions proceeded in high stereoselectivity giving rise exclusively to octanoic acid ester **79** from olefin **80**. In contrast, when more labile diphenylmethyl ester was used, octonolactone **81** was isolated directly by crystallization in high yield from **82** (Scheme 7.20). In addition to D-galactose (**83**), other carbohydrates like D-glucose, D-mannose, D-ribose, D-xylose, D-lyxose, D-glycero-D-guloheptose, or D-glycero-D-galactoheptose were homologated to their corresponding higher sugars.

Scheme 7.20 Selective PGF Wittig olefination for carbohydrate homologation.

7.4 Selective PGF Functionalization at the Anomeric and Nonanomeric Positions (Oxidations)

The oxidation of hydroxyl groups represents a basic and fundamental chemical transformation in organic synthesis. However, the carbohydrates in their natural form represent the most challenging substrates for selective oxidations. In this regard, selective oxidation at the anomeric center is a well-documented transformation [1b], which was originally disclosed by Killiani [47]. When L-arabinose (**84**) was treated with bromine in water, the corresponding calcium arabinate **85** was isolated (Scheme 7.21). After this seminal report, some applications [48] and modifications have been reported, most of them including the use of a buffer [49] and the complete evaporation of the solvent, leading to the corresponding γ-lactone. This is a very common and scalable procedure [50] for the preparation of carbohydrate-derived 5-membered lactones, such as L-lyxono-γ-lactone **86** from L-lyxose (**87**) [51].

Uronic acids, carbohydrates containing a hemiacetal moiety and a carboxylic acid group, are important building blocks in oligosaccharide and glycoprotein synthesis [52]. These carbohydrate derivatives were obtained by oxidation of hexoses, which typically utilize chromium-based reagents. However, the oxidation is not selective, and the secondary hydroxyl groups can also be transformed into their corresponding ketones. Thus, oxidation of D-glucose to its respective uronic acid typically requires at least five steps (selective protection of the primary alcohol, protection of the remaining secondary hydroxyl groups, selective deprotection of the primary alcohol, oxidation, and complete deprotection of the carbohydrate). Fortunately, Flitsch described a pivotal selective oxidation of the primary hydroxyl group [53]. Thus, using NaOCl and a catalytic amount of TEMPO [54], methyl ester **88** is obtained in modest yield from methyl-D-glucopyranoside (**42**) after esterification (Scheme 7.22). Although the methyl-D-glucopyranoside (**42**) can be seen as a protected carbohydrate, its preparation can be achieved in a PGF fashion; therefore, the overall transformation of D-glucose to uronic methyl ester **88** can be considered as a PGF synthesis.

Whereas the oxidation of the anomeric and primary hydroxyl groups is somewhat well described, the oxidation of secondary alcohols in sugars is considerably more difficult and barely known. One of the occasional reports for the preceding process was described by de Vries and Minnaard [55], who used a cationic palladium complex ([[(neocuproine)PdOAc]$_2$OTf$_2$) as the catalyst for the selective oxidation at C3 position of various *O*- or *S*-glycosides **89–93** (Scheme 7.23). The reaction is highly

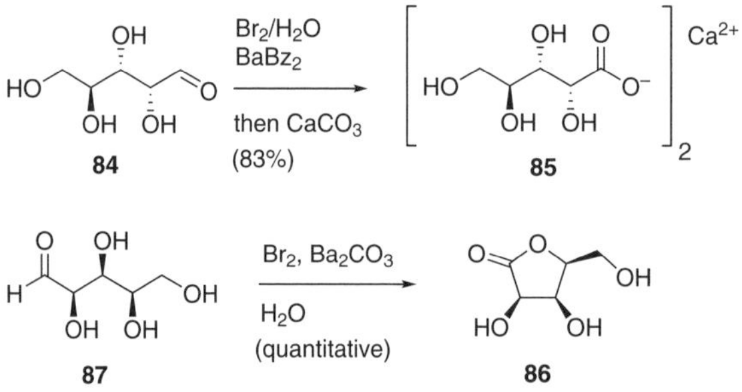

Scheme 7.21 Selective PGF oxidation at the anomeric position.

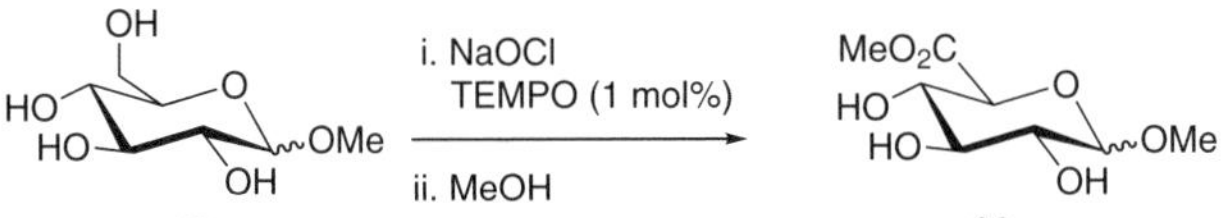

Scheme 7.22 Selective PGF oxidation at primary hydroxyl group of D-glucose derivative to **88**.

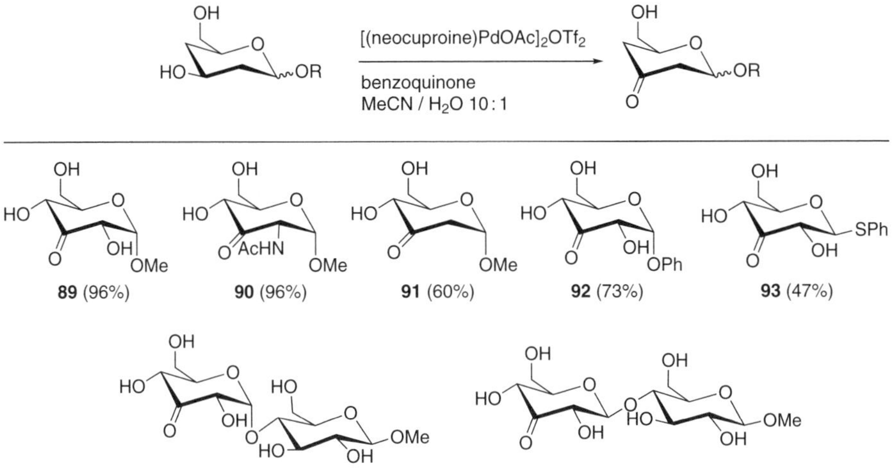

Scheme 7.23 Selective PGF oxidation at C-3 position of *O*- or *S*-glycosides.

regioselective, albeit in variable yields. However, these variations were not attributed to the lack of selectivity, but to purification issues. It is worth noting that the oxidation is also highly regioselective on disaccharides methyl maltoside and methyl cellobioside, providing ketones **94** and **95** in practical yields.

Recently, the same author applied this oxidizing system to the selective oxidation of oligosaccharides and glycosyl azides [56] and to oxidation–reduction sequences for obtaining different sugars. For example, the synthesis of D-allose and allitol was performed in two steps from unprotected glucose [57]. Regioselective oxidation at C3 of D-glucose (**41**) provided 3-ketoglucose **96**, which depending on the reagent and the reaction conditions employed provided either D-allose **97** (LS-selectride) or allitol **98** (NaBH₄) (Scheme 7.24).

7.5 Conclusion

Although the price for using carbohydrates as starting materials in synthesis is quite low, especially if we take into consideration that chirality and optical purity are included in the budget, the intrinsic nature of carbohydrates, which is defined by the presence of reactive hydroxyl and carbonyl functional groups, tends to increase both the economic and environmental cost by the need of modulating their reactivity by using protecting-group strategy. Despite the remarkable efforts for making relevant organic compounds

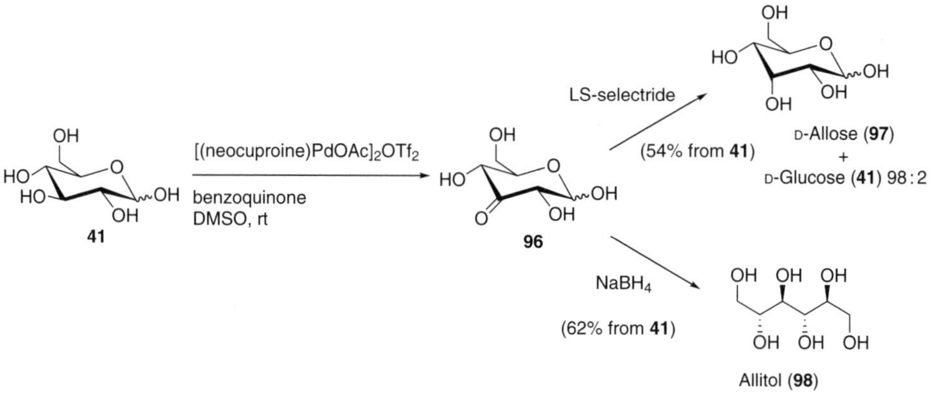

Scheme 7.24 Regioselective oxidation–reduction sequence for the synthesis of D-allose (**97**) and allitol (**98**) from D-glucose (**41**).

starting from carbohydrates without the use of any protecting group, the state of the art of this organic synthesis approach should be considered as "in progress." Therefore, much further efforts in this area by designing more efficient strategies for functional group differentiation without protection remain to be accomplished.

References

1 (a) Hanessian, S., Giroux, S., and Merner, B.L. (2013). *Design and Strategy in Organic Synthesis: From the Chiron Approach to Catalysis*. Weinheim: Wiley-VCH. (b) Collins, P. and Ferrier, R. (1995). *Monosaccharides: Their Chemistry and Their Roles in Natural Products*. Chichester: Wiley.

2 (a) Hanessian, S. (2012). *J. Org. Chem.* 77: 6657. (b) Hanessian, S. (1984). *Total Synthesis of Natural Products: The "Chiron Approach"*. Oxford: Pergamon. (c) Hanessian, S. (1989). *Aldrichim. Acta* 22: 3.

3 Grubbs, R.H. (2007). *Chem. Sci.* 4: C69.

4 Fernandes, R.A. and Kattanguru, P. (2013). *Asian J. Org. Chem.* 2: 74.

5 Chaudhari, D.A., Kattanguru, P., and Fernandes, R.A. (2015). *RSC Adv.* 5: 42131.

6 Young, I.S. and Baran, P.S. (2009). *Nat. Chem.* 1: 193.

7 Dangerfield, E.M., Timmer, M.S.M., and Stocker, B.L. (2009). *Org. Lett.* 11: 535.

8 Bernet, B. and Vasella, A. (1979). *Helv. Chim. Acta* 62: 1990.

9 Fernandes, R.A. and Kattanguru, P. (2012). *J. Org. Chem.* 77: 9357.

10 Li, J., Zhao, C., Liu, J., and Du, Y. (2015). *Tetrahedron* 71: 3885.

11 van Kalkeren, H.A., van Rootselaar, S., Haasjes, F.S. et al. (2012). *Carbohydr. Res.* 362: 30.

12 Di Florio, R. and Rizzacasa, M.A. (1998). *J. Org. Chem.* 63: 8595.

13 Lorente, A., Lamriano-Merketegi, J., Albericio, F., and Álvarez, M. (2013). *Chem. Rev.* 113: 4567.

14 Paz-Morales, E., Melendres, R., and Sartillo-Piscil, F. (2009). *Carbohydr. Res.* 344: 1123.

15 Toshima, H., Sato, H., and Ichihara, A. (1999). *Tetrahedron* 55: 2581.

16 (a) Sánchez-Eluterio, A., Quintero, L., and Sartillo-Piscil, F. (2011). *J. Org. Chem.* 76: 5466. (b) García-Tellado, F., de Armas, P., and Marrero-Tellado, J.J. (2000). *Angew. Chem. Int. Ed.* 39: 2727.

17 Bozell, J.J. (2010). *Science* 329: 522.

18 Lange, J.-P., Price, R., Ayoub, P.M. et al. (2010). *Angew. Chem. Int. Ed.* 49: 4479.

19 Bond, J.Q., Alonso, D.M., Wang, D. et al. (2010). *Science* 327: 1110.

20 Gunasundari, T. and Chandrasekaran, S. (2010). *J. Org. Chem.* 75: 6685.

21 (a) Halila, S., Benazza, M., and Demailly, G. (2001). *Tetrahedron Lett.* 42: 3307. (b) Benazza, M., Halila, S., Viot, C. et al. (2004). *Tetrahedron* 60: 2889. (c) Merrer, Y.L., Fuzier, M., Dosbaa, I. et al. (1997). *Tetrahedron* 53: 16731.

22 (a) Cubero, I.I., López-Espinosa, M.T.P., Richardson, A.C., and Suárez-Ortega, M.D. (1993). *Carbohydr. Res.* 242: 109. (b) Chou, W.-C., Chen, L., Fang, J.-M., and Wong, C.-H. (1994). *J. Am. Chem. Soc.* 116: 6191.

23 Crombez-Robert, C., Benazza, M., Fréchou, C., and Demailly, G. (1997). *Carbohydr. Res.* 303: 359.

24 Ermert, P. and Vasella, A. (1993). *Helv. Chim. Acta* 76: 2687.

25 Hashimoto, H., Kawanishi, M., and Yuasa, H. (1996). *Carbohydr. Res.* 282: 207.

26 Bols, M. (1995). *Carbohydrate Building Blocks*. New York: Wiley.

27 Fischer, E. (1893). *Chemische Berichte* 26: 2400.

28 Hanessian, S. and Lou, B. (2000). *Chem. Rev.* 100: 4443.

29 Lubineau, A. and Fischer, J.-C. (1991). *Synth. Commun.* 21: 815.

30 Bertho, J.-N., Ferrières, V., and Plusquellec, D. (1995). *J. Chem. Soc. Chem. Commun.* 1391.

31 Park, T.-J., Weïwer, M., Yuan, X. et al. (2007). *Carbohydr. Res.* 342: 614.

32 Likhosherstov, L.M., Novikova, O.S., Derevitskaja, V.A., and Kochetkov, N.K. (1986). *Carbohydr. Res.* 146: C1–C5.

33 Larabi, M.-L., Fréchou, C., and Demailly, G. (1994). *Tetrahedron Lett.* 35: 2175.

34 Csuk, R., Schultheiß, A., Sommerwerk, S., and Kluge, R. (2013). *Tetrahedron Lett.* 54: 2274.

35 Sarpe, V.A. and Kulkarni, S.S. (2011). *J. Org. Chem.* 76: 6866.

36 Sarpe, V.A. and Kulkarni, S.S. (2014). *Org. Lett.* 16: 5732.

37 Kwase, Y.A., Cochran, M., and Nitz, M. (2014). *Modern Synthetic Methods in Carbohydrate Chemistry: From Monosaccharides to Complex Glycoconjugates* (ed. D. Werz and S. Vidal), 67–96. Weinheim: Wiley-VCH.

38 (a) Ellestad, G.A., Cosulich, D.B., Broschard, R.W. et al. (1978). *J. Am. Chem. Soc.* 100: 2515. (b) Chen, R.H., Whittern, D.N., Buko, A.M., and McAlpine, J.B. (1989). *J. Antibiot.* 42: 533. (c) Dobashi, K., Nagaoka, K., Watanabe, Y. et al. (1985). *J. Antibiot.* 38: 1166. (d) Greenstein, M., Speth, J.L., and Maiese, W.M. (1981). *Antimicrob. Agents Chemother.* 20: 425. (e) Osburne, M.S., Maiese, W.M., and Greenstein, M. (1990). *Antimicrob. Agents Chemother.* 34: 1450.

39 Ichikawa, Y., Minami, T., Kusaba, S. et al. (2014). *Org. Biomol. Chem.* 12: 3924.

40 Peluso, S. and Imperiali, B. (2001). *Tetrahedron Lett.* 42: 2085.

41 Lim, D., Brimble, M.A., Kowalczyk, R. et al. (2014). *Angew. Chem. Int. Ed.* 53: 11907.

42 Kitamura, M., Tashiro, N., Miyagama, S., and Okauchi, T. (2011). *Synthesis* 1037.

43 Danishefsky, S.J. and DeNinno, M.P. (1987). *Angew. Chem. Int. Ed.* 26: 15.

44 Kochetkov, N.K. and Dmitriev, B.A. (1965). *Tetrahedron* 21: 803.

45 Railton, C.J. and Clive, D.L.J. (1996). *Carbohydr. Res.* 281: 69.

46 Jørgensen, M., Iversen, E.H., and Madsen, R. (2001). *J. Org. Chem.* 66: 4625.

47 Kiliani, H. (1886). *Chemische Berichte* 19: 3029.

48 (a) Fischer, E. and Meyer, J. (1889). *Chemische Berichte* 22: 1941. (b) Ruff, O. (1899). *Chemische Berichte* 32: 550. (c) Ruff, O. (1901). *Chemische Berichte* 34: 1362. (d) Wohl, A. and List, E. (1897). *Chemische Berichte* 30: 3101.

49 (a) Hudson, C.S. and Isbell, H.S. (1929). *J. Am. Chem. Soc.* 51: 2225. (b) Nelson, W.L. and Cretcher, L.H. (1930). *J. Am. Chem. Soc.* 52: 403. (c) Chaveriat, L., Stasik, I., Demailly, G., and Beaupère, D. (2004). *Tetrahedron* 60: 2079.

50 Chaudhuri, N.C., Moussa, A., Stewart, A. et al. (2005). *Org. Process. Res. Dev.* 9: 457.

51 Taylor, C.M., Barker, W.D., Weir, C.A., and Park, J.H. (2002). *J. Org. Chem.* 67: 4466.

52 Robyt, J.F. (1998). *Essentials of Carbohydrate Chemistry*. New York: Springer.

53 Davis, N.J. and Flitsch, S.L. (1993). *Tetrahedron Lett.* 34: 1181.

54 (a) Siedlecka, R., Skarzewski, J., and Mlochowski, J. (1990). *Tetrahedron Lett.* 31: 2177. (b) Anelli, P.L., Banfi, S., Montanari, F., and Quici, S. (1989). *J. Org. Chem.* 54: 2970.

55 Jäger, M., Hartmann, M., de Vries, J.G., and Minnaard, A.J. (2013). *Angew. Chem. Int. Ed.* 52: 7809.

56 Einsik, N.N.H.M., Lohse, J., Witte, M.D., and Minnaard, A.J. (2016). *Org. Biomol. Chem.* 14: 4859.

57 Jumde, V.R., Einsik, N.N.H.M., Witte, M.D., and Minnaard, A.J. (2016). *J. Org. Chem.* 81: 11439.

8

Protecting-Group-Free Synthesis of Glycosyl Derivatives, Glycopolymers, and Glycoconjugates

Tomonari Tanaka

Department of Biobased Materials Science, Graduate School of Science and Technology, Kyoto Institute of Technology, Japan

8.1   Introduction

Carbohydrates are an important class of natural resource. Polysaccharides such as cellulose and starch find wide use in materials science and food chemistry, and fully or partially modified protected polysaccharides are useful as materials such as acetylated celluloses. Although low molecular weight saccharides including mono- and disaccharides are relatively easily modified, the modification of oligosaccharides, which have higher molecular weight, requires advanced synthetic techniques executed under carefully controlled conditions. This has hampered the development of glycomaterials using oligosaccharides, and consequently, glycotechnology remains an academic pursuit.

Direct conversion of the hydroxy group at the anomeric position (1-position) of free saccharides is important for obtaining glycosyl derivatives and is quite easily achieved for low molecular weight saccharides using classical methods. For example, the Fischer glycosylation reaction is the classic method for modifying the hydroxy group of a free saccharide at the anomeric position. In addition, many other direct methods for obtaining glycosyl derivatives from free saccharides have been reported. However, these methods are difficult to apply to higher molecular weight oligosaccharides due to cleavage of the glycosyl linkage in severe reaction environments such as acidic conditions and high temperature. Recently, several novel and efficient methodologies for direct conversion at the anomeric position of free saccharides have been developed, but most have only been applied to low molecular weight mono- and disaccharides such as glucose, galactose, maltose, and lactose in a polar organic solvent because these saccharides contain only hydroxy groups and are soluble in polar organic solvents. Although water is the best solvent for solubilizing simple saccharides, it is difficult to perform chemical reactions in water. Furthermore, high molecular weight oligosaccharides are not soluble in polar organic solvents. There are many different saccharide moieties in nature, for example, acidic saccharides with carboxy, sulfate, and phosphate groups and

Protecting-Group-Free Organic Synthesis: Improving Economy and Efficiency, First Edition.
Edited by Rodney A. Fernandes.

amino saccharides with amino groups. In these cases, specific derivatization of the hydroxy group is more laborious due to the necessity to protect the other functional groups prior to derivatization of the hydroxy group. This makes it difficult to synthesize glycosyl derivatives directly from free saccharides. Therefore, to obtain glycosyl derivatives from various free saccharides, a simpler and efficient synthetic method applicable to not only mono- and disaccharides but also to oligosaccharides with higher molecular weight and containing various functional groups is clearly desired. In addition, the resulting glycosyl derivatives should be amenable to the synthesis of glycomaterials. In this chapter, we describe the protecting-group-free synthesis of glycosyl derivatives, glycopolymers, and glycoconjugates. We first introduce approaches developed by other research groups for the direct conversion of an anomeric hydroxy group to synthesize glycosyl derivatives from free saccharides. Then, we review novel direct anomeric activations using dehydrative condensing agents in water. This is followed by an introduction to our previous and novel methods for synthesizing glycopolymers from free saccharides. Finally, we introduce recent reports describing the preparation of other glycoconjugates, i.e. glycosylated peptides and polyamino acids, glycodendrimers, and glycoproteins.

8.2 Protecting-Group-Free Synthesis of Glycosyl Derivatives from Free Saccharides

In general, in order to replace the anomeric hydroxy group by a substitute group, a multistep process involving protection, activation, substitution, and deprotection is necessary (Figure 8.1 b→c→d→e). This process is easily applied to the synthesis of saccharide derivatives having a functional group at the anomeric position. However, it is difficult to apply this synthetic strategy to higher molecular weight oligosaccharides

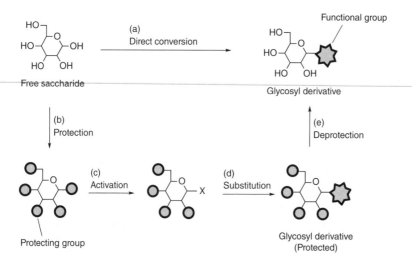

Figure 8.1 Synthetic routes to glycosyl derivatives from free saccharide. (a) Direct synthesis of glycosyl derivative. (b) Protection of hydroxy groups. (c) Activation of anomeric position. (d) Substitution by a functional group at anomeric position. (e) Deprotection.

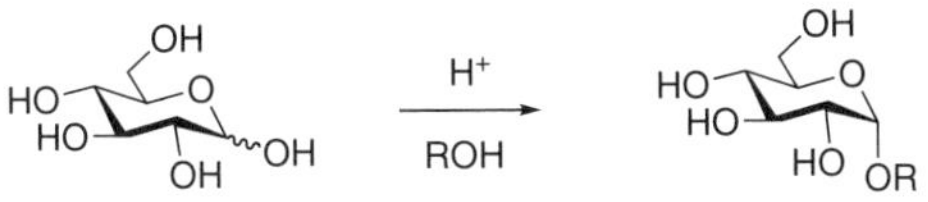

Scheme 8.1 Fischer glycosylation.

(a)

(b)

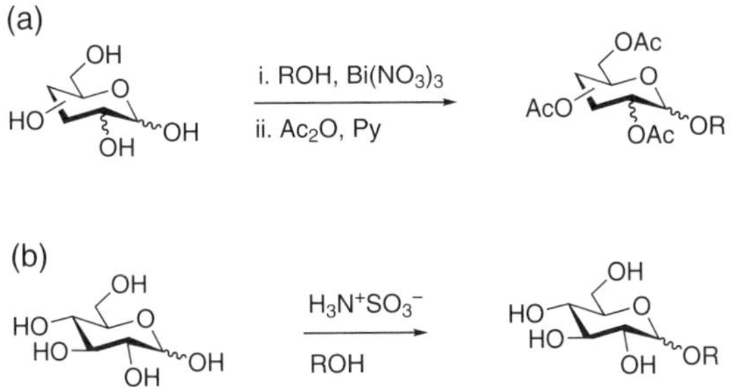

Scheme 8.2 Direct synthesis of O-glycosides.

because of the need for a more complicated process and because of cleavage of the inner glycosyl bonds during the long procedure. Direct conversion to glycosyl derivatives from free saccharides is a useful approach to solve these problems (Figure 8.1a). Many direct synthetic methods for glycosyl derivatives from free saccharides have been reported to date. The Fischer glycosylation reaction is the most classic method providing directly O-glycosides from free saccharides under acidic conditions and heat (Scheme 8.1) [1–3]. This method, however, can be applied only to monosaccharides due to cleavage of an inner glycosyl bond. Cation-exchange resin works as a catalyst to produce O-glycosides from free saccharides in alcohol solvent [4]. Bhattacharya et al. reported the activation of free saccharides to synthesize alkyl and aryl glycopyranosides using bismuth nitrate pentahydrate (Scheme 8.2a) [5]. The use of sulfamic acid increased the yields of the desired alkyl glycosides (Scheme 8.2b) [6]. Direct O-glycosylations under basic conditions have also been reported. Penadés et al. and Schmidt et al. reported direct anomeric O-arylation (Scheme 8.3a) [7, 8]. Mukherjee et al. reported the synthesis of alkyl glycosides using ammonium chloride as a catalyst with heating (Scheme 8.3b) [9].

The use of ionic liquids for direct O-glycosylation has been explored. 1-Ethyl-3-methylimidazolium benzoate was used in the presence of Amberlite IR-120 (H^+) resin or p-toluenesulfonic acid as promoter [10]. Augé and Sizun reported the use of 1-butyl-3-methylimidazolium trifluoromethanesulfonate ([BMIM][OTf]) for O-glycosylations catalyzed by Lewis acid salts such as indium(III) chloride, indium(III) trifluoromethanesulfonate, scandium(III) trifluoromethanesulfonate (Sc(OTf)$_3$), and ytterbium(III) trifluoromethanesulfonate (Scheme 8.4) [11]. A 74% yield ($\alpha/\beta = 75/25$) was obtained when D-glucose (Glc) reacted with n-octanol in [BMIM][OTf] in the presence of a catalytic amount of Sc(OTf)$_3$ at 80°C. Straightforward glycosylations of various alcohols with free monosaccharides were successfully performed by using scandium catalyst in ionic liquid solvent.

Pfaffe and Mahrwald reported ligand exchange acetalization of acetals in the presence of a catalytic amount of mandelic acid and titanium $tert$-butoxide at room temperature

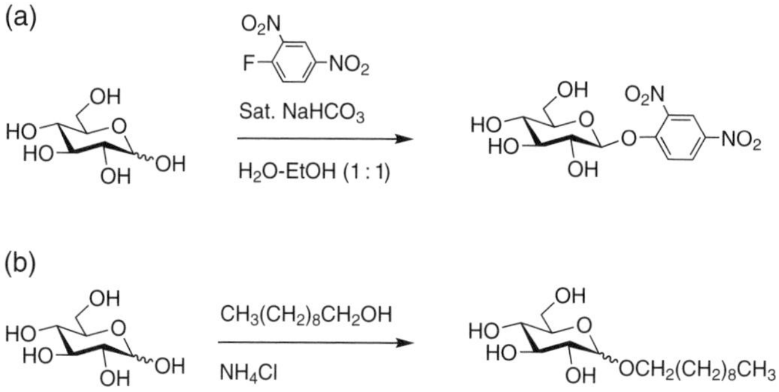

(a)

Scheme 8.3 Direct synthesis of O-glycosides under basic conditions.

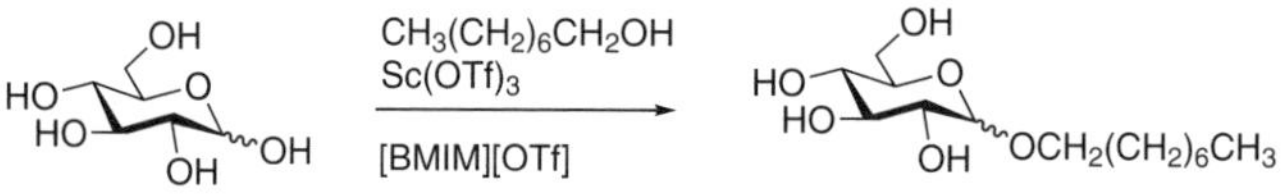

Scheme 8.4 Direct synthesis of O-glycosides in ionic liquids.

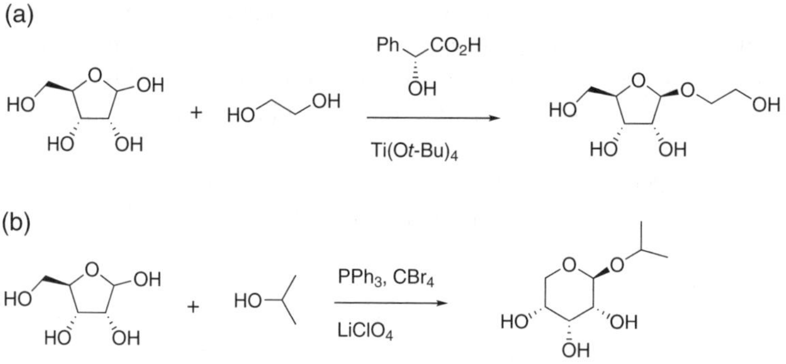

(a)

(b)

Scheme 8.5 Direct synthesis of O-glycosides from free furanose.

(Scheme 8.5a) [12]. Addition of lithium bromide enhanced the yield in acetonitrile media. The same group reported organocatalyzed β-selective direct glycosylation of free D-ribose using a catalytic amount of triphenylphosphine (PPh$_3$) and tetrabromomethane as the catalyst, with lithium perchlorate as an essential additive (Scheme 8.5b) [13].

Mitsunobu reaction conditions have also been employed for the conversion of glycosyl derivatives [14–16]. Aime et al. reported synthesizing 1-O-(but-2-ynoyl)-glycoside by using diisopropyl azodicarboxylate (DIAD) and PPh$_3$ in N,N-dimethylformamide (DMF) (Scheme 8.6a) [17]. Shoda et al. reported synthesizing p-nitrophenyl glycosides under similar Mitsunobu conditions using PPh$_3$ in N,N'-dimethylethylene urea (DMEU) (Scheme 8.6b) [18]. These products, however, are all mixtures of α- and β-anomeric isomers. Kawabata et al. demonstrated highly β-stereoselective glycosylation with gallic acid derivatives under Mitsunobu conditions (Scheme 8.6c) [19].

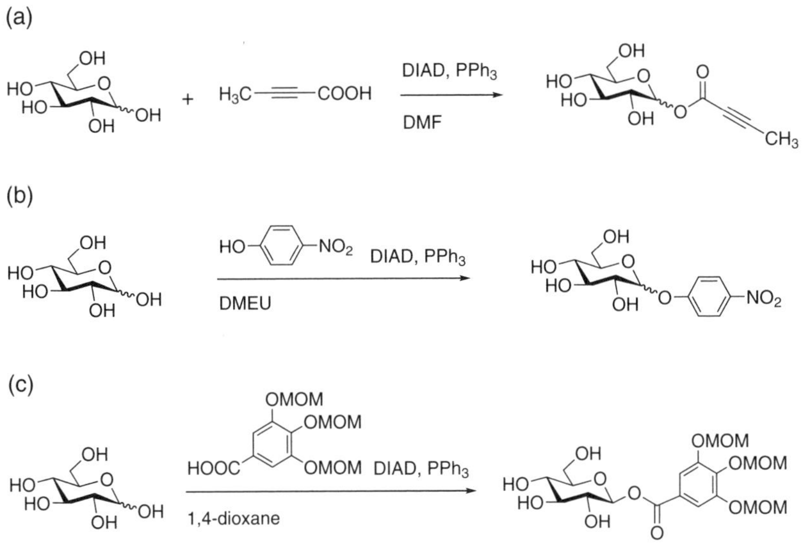

Scheme 8.6 Direct synthesis of O-glycosides under the Mitsunobu reaction conditions.

Nitz et al. reported the direct and β-selective synthesis of O-glycosides, glycosyl azides, and oxazoline derivatives via N'-glycopyranosyl sulfonohydrazides as glycosyl donors from free saccharides in polar organic solvent (Scheme 8.7a) [20, 21]. His group also developed a protecting-group-free two-step synthesis of glycosyl 1-phosphates by means of acid-catalyzed condensation of p-toluenesulfonyl hydrazide with various free mono- and disaccharides (Scheme 8.7b) [22].

Furthermore, N,O-dialkyloxyamine-N-glycoside, which can be directly synthesized from free saccharides, was shown to give primarily the equatorial glycosides in high yields under mildly acidic aqueous conditions [23]. Nitz's group reported that N,O-dimethylhydroxylamine-N-glycosides are stable to a variety of protecting-group manipulations, including acylation, alkylation, silylation, and acetal formation, and to subsequent glycosylation reactions (Scheme 8.8a) [24]. In addition, the alkoxyamine-N-glycosides can be cleaved selectively with N-chlorosuccinimide (NCS) to give the desired hemiacetals in excellent yields (Scheme 8.8b).

Recently, many dehydrative condensation reagents have become commercially available. Shoda's group, with which I was previously affiliated, reported many one-step conversions of free saccharides to various glycosyl derivatives using water-soluble condensing agents in water. His group was the first to report the direct formation of sugar oxazoline derivatives using 1-ethyl-3-(3-dimethylaminopropyl)carbodiimide hydrochloride (WSC-HCl) [25] from free N-acetyl-2-amino sugars in aqueous media (Scheme 8.9) [26]. Sugar oxazoline derivatives are very useful activated glycosyl donors for glycosidase-catalyzed glycosylations [27]. This method, however, requires a long reaction time (4 days) and provides a moderate yield of 28–40% due to the low reactivity of the carbodiimide reagent. The reactions proceed through a reactive intermediate formed by preferential attack of the anomeric hydroxy group toward the condensing agent, given that the pK_a value of the hemiacetal anomeric hydroxy group is much lower than that of the other hydroxy groups on saccharides (i.e. primary and secondary

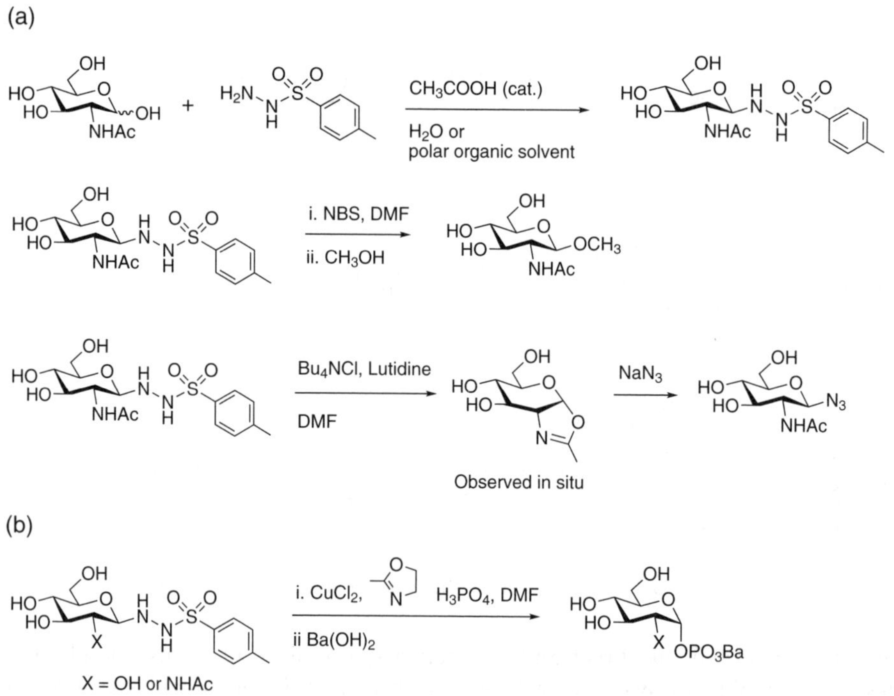

(a)

(b)

X = OH or NHAc

Scheme 8.7 Direct synthesis of glycosyl derivatives via *N'*-glycopyranosyl sulfonohydrazides and its utilization.

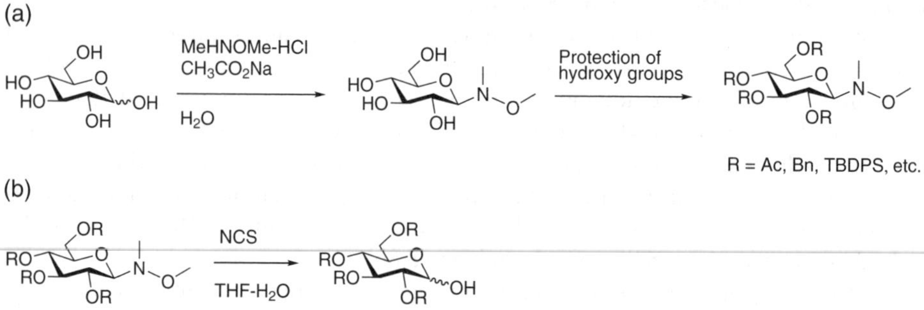

(a)

R = Ac, Bn, TBDPS, etc.

(b)

Scheme 8.8 Direct synthesis of *N,O*-dimethylhydroxylamine-*N*-glycoside and its utilization.

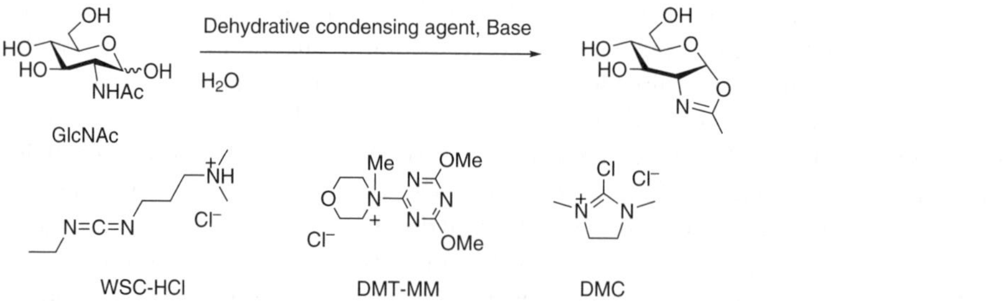

Scheme 8.9 Direct synthesis of sugar oxazoline derivative by using dehydrative condensing agents in water.

alcohols) and of water (Figure 8.2) [28]. Direct synthesis of sugar oxazoline derivatives using the triazine derivative 4-(4,6-dimethoxy-1,3,5-triazin-2-yl)-4-methylmorpholinium chloride (DMT-MM) [29] was also developed (Scheme 8.9) [30]. When the method was applied to N-acetyl-D-glucosamine (GlcNAc) with the intent of improving the yield, the required sugar oxazoline derivative was obtained in only 33% yield.

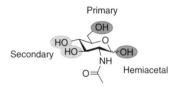

Primary

Secondary

Hemiacetal

pK_a : Hemiacetal < primary < secondary

Figure 8.2 Hydroxy groups on a free saccharide, GlcNAc.

A cation-type formamidinium salt, 2-chloro-1,3-dimethylimidazolinium chloride (DMC), was tested and identified as a more reactive reagent. DMC was found to convert quantitatively to the oxazoline derivative from free GlcNAc in water in the presence of base within 15 min [30]. This method is called "Shoda activation" [31] and is applicable to other monosaccharides, such as N-acetyl-D-galactosamine and N-acetyl-D-mannosamine, and longer chitooligosaccharides, from disaccharide to heptasaccharide. Furthermore, monosaccharides possessing a sulfuric acid moiety or phosphoric acid moiety could be converted to oxazoline derivatives without affecting the structures of these functional groups. It is noteworthy that a high mannose-type heptasaccharide and disialo-complex-type decasaccharide (N-glycan) having two N-acetylneuraminic acid moieties at the nonreducing ends could be converted to the corresponding oxazoline derivatives in high yields. These higher molecular weight biological oligosaccharides are difficult to derivatize by conventional methods that employ protecting groups. The transglycosylation reaction was achieved using the oxazoline derivative of N-glycan catalyzed by a mutant endo-β-N-acetylglucosaminidase from *Mucor hiemalis* (Endo-M) to obtain homogeneous glycopeptides and glycoproteins in high yields (Scheme 8.10) [32, 33].

DMC is a useful reagent for synthesizing other saccharide derivatives from free saccharides in aqueous solution. When DMC was reacted with free gluco-type saccharides, which have a hydroxy group at the 2-position in the presence of triethylamine (TEA) in water, the corresponding 1,6-anhydro sugars were directly obtained in excellent yields (Scheme 8.11a) [34]. This reaction is also applicable to oligosaccharides. A series of 1,6-anhydro formations using various starting free saccharides provided β-glycoside and 1,2-epoxide derivatives as intermediates, given that no product was produced when 2-deoxy-D-glucose and 2-fluoro-2-deoxy-D-glucose were used as starting materials (Figure 8.3). The presence of nucleophiles in the DMC reaction system resulted in the direct and stereoselective production of various saccharide derivatives. β-Glycosyl azides were directly produced in water in good yields by the reaction of free saccharides and sodium azide under basic conditions (Scheme 8.11b) [35]. The azide group has recently become a key functional group for click chemistry [36]. The direct azidation of free saccharides is applicable to not only monosaccharides such as Glc and GlcNAc but also to longer and higher molecular weight oligosaccharides such as malto-, cello-, and chitooligosaccharides, oligoxyloglucans, and N-glycans. An azide ion attaches to the anomeric carbon on the 1,2-epoxide and α-glycosyl intermediates to produce stereoselectively the corresponding glycosyl azide (Figure 8.3).

Thioglycosides are directly generated by the reaction of free saccharides with aryl thiol derivatives in acetonitrile aqueous solution using DMC (Scheme 8.11c) [37, 38]. According to this method, not only free monosaccharides but also free

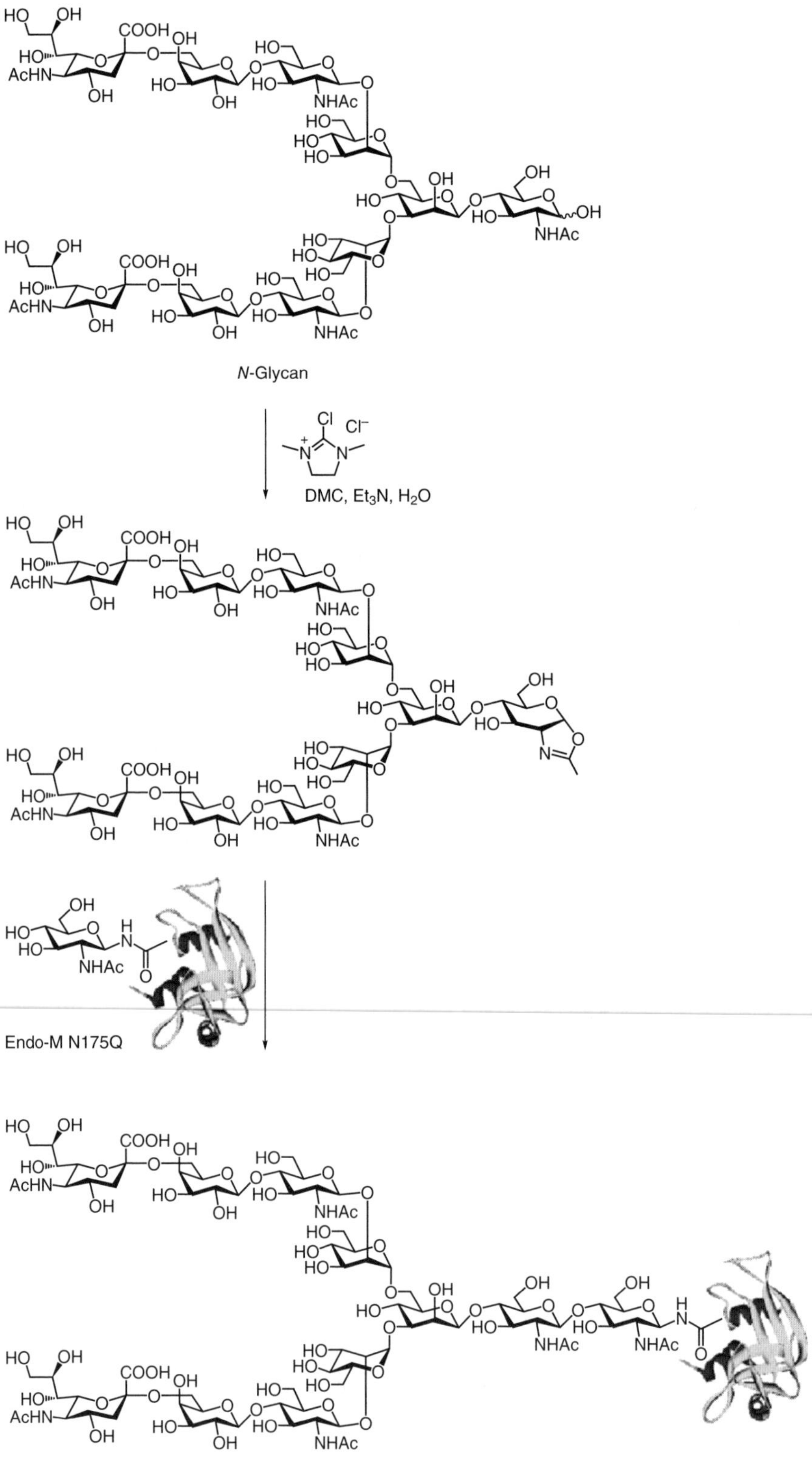

Scheme 8.10 Chemoenzymatic synthesis of glycoprotein via sugar oxazoline derivative.

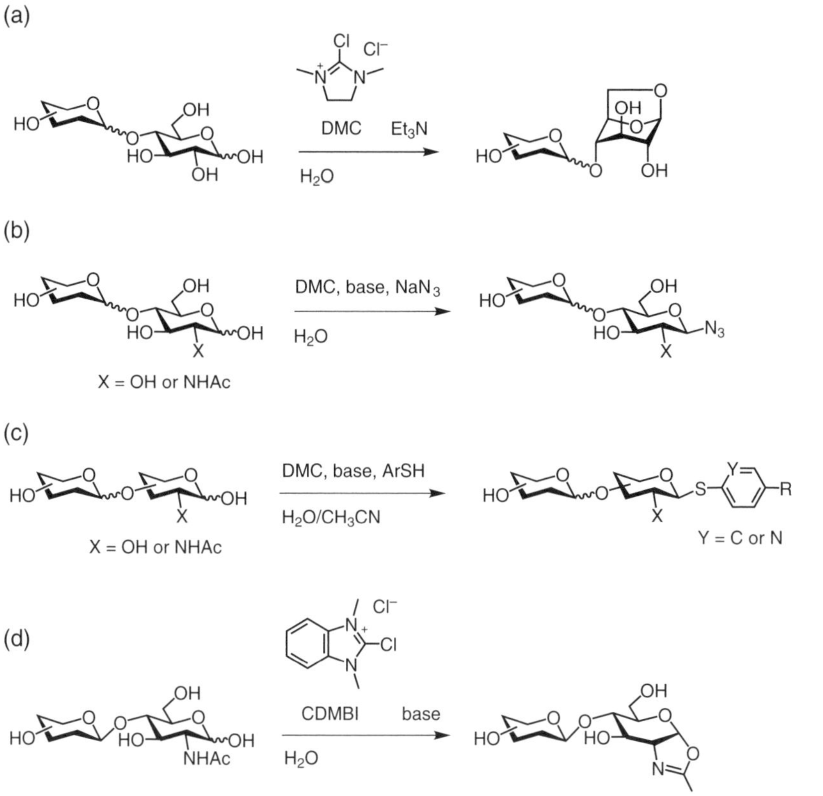

(a)

(b)

X = OH or NHAc

(c)

X = OH or NHAc

Y = C or N

(d)

Scheme 8.11 Direct synthesis of (a) 1,6-anhydro sugars, (b) glycosyl azides, and (c) thioglycosides by using DMC. (d) Direct synthesis of sugar oxazoline derivative by using CDMBI.

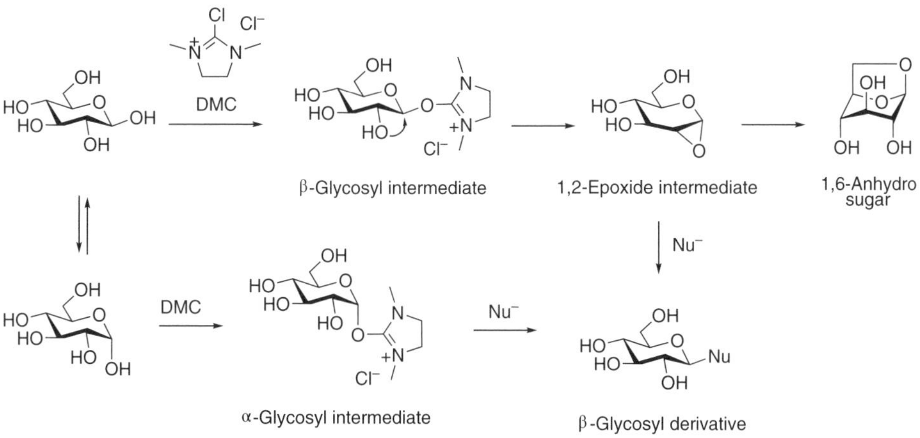

β-Glycosyl intermediate   1,2-Epoxide intermediate   1,6-Anhydro sugar

α-Glycosyl intermediate   β-Glycosyl derivative

Figure 8.3 Plausible reaction mechanism of direct synthesis of glycosyl derivatives by using DMC. Nu⁻: nucleophile (N_3, RS, etc.).

oligosaccharides, such as cello-, chito-, and maltooligosaccharides and GlcNAc oligomers, could be directly converted to the corresponding thioglycosides with β-configuration selectively. Glycosyl dithiocarbamates were also directly produced from free saccharides by using DMC in aqueous media [39]. The reaction was applied to direct labeling with a detachable fluorescent tag and the 4-methyl-7-thioumbelliferyl (MUS) group at the anomeric position of free oligosaccharides [40]. The resulting MUS-labeled saccharides showed high sensitivity for fluorescence detection and could be used for the quantification of oligosaccharide mixtures. The free saccharides could be successfully recovered by selective oxidative cleavage of the thioglycosidic bond under mild reaction conditions.

Shoda et al. also developed a practical one-step synthesis of sugar oxazoline derivatives from free *N*-acetyl-2-amino sugars using a similar form of amidinium salt-type agent, 2-chloro-1,3-dimethyl-1*H*-benzimidazol-3-ium chloride (CDMBI) (Scheme 8.11d) [41]. After the reaction, the hydrolysate of CDMBI, 3-dihydro-1,3-dimethyl-2*H*-benzimidazol-2-one (DMBI) precipitates from the aqueous mixture because of its higher hydrophobicity. The resulting DMBI can be removed simply by filtration, and the filtrate can be utilized for the subsequent enzymatic transglycosylation step without purification of the synthesized sugar oxazoline derivative.

Hindsgaul et al. used another formamidinium salt-type agent, 2-imidazolyl-1,3-dimethylimidazolinium chloride (ImIm), which was prepared using DMC and imidazole (Scheme 8.12a), to synthesize uridine diphosphate glucose (UDP-Glc) from nucleoside monophosphates in aqueous solution (Scheme 8.12b) [42]. ImIm reagent formed *in situ* can activate nucleoside 5'-monophosphate to UDP-sugars by reaction with D-glucose 1-phosphate (G-1-P).

Another use of the previous dehydrative condensing agent, DMT-MM, is the direct preparation of novel activated glycosyl donors for glycosidase-catalyzed transglycosylation. In general, glycosyl donors such as *p*-nitrophenyl glycosides are synthesized through a multistep process, which includes the protection–deprotection of hydroxy groups on saccharide moieties. When free saccharides were reacted with DMT-MM in

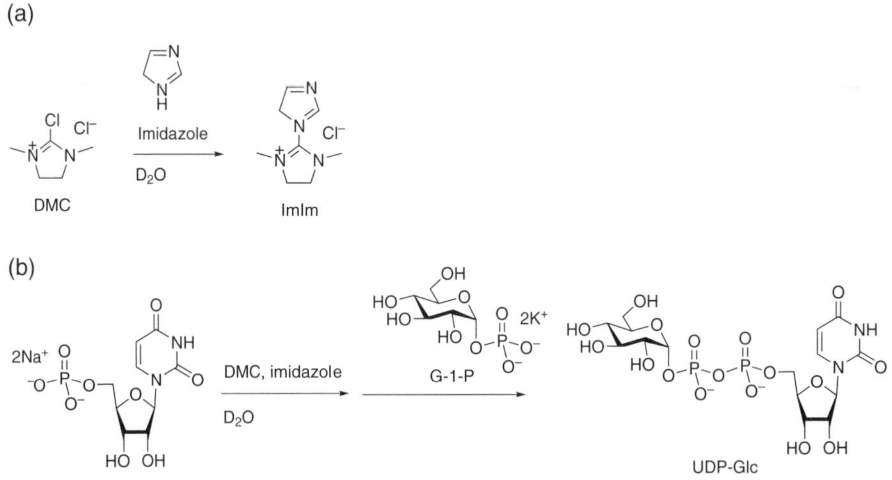

(a)

(b)

Scheme 8.12 Synthesis of UDP-Glc by using DMC.

the presence of a base reagent in water, the novel glycosyl compounds 4,6-dimethoxy-1,3,5-triazin-2-yl glycosides (DMT-glycosides) were directly produced without protection of the hydroxy groups on the saccharide moiety. The resulting DMT-glycosides were found to be efficient activated glycosyl donors for enzymatic glycosylations catalyzed by glycosidases. In the case of lactose (Lac), 4,6-dimethoxy-1,3,5-triazin-2-yl β-lactoside (DMT-β-Lac) was obtained from free Lac in good yield (Scheme 8.13a) [43–46]. In this reaction, a disubstituted α-lactoside derivative (DMT$_2$-α-Lac) resulting from derivatization at the 1- and 2-positions of the Lac moiety was formed as the by-product and could be completely removed by recrystallization. The formation of monosubstituted DMT-α-lactoside (DMT-α-Lac) was observed early in the reaction and was converted into DMT$_2$-α-Lac. The nucleophilicity of the 2-position hydroxy group of DMT-α-Lac may be intramolecularly enhanced by the lone pair on the N atom of the triazine ring, resulting in the formation of DMT$_2$-α-Lac. Efficient enzymatic transglycosylation catalyzed by *endo*-β-1,4-glucanase III from *Trichoderma reesei* (EGIII) was achieved using DMT-β-Lac as a glycosyl donor. On the other hand, when *N*-acetyl-2-amino sugars such as GlcNAc and its oligosaccharides were reacted with DMT-MM, α-configuration DMT-glycosides were produced as the main products. Enzymatic α-*N*-acetylglucosaminidation, resulting in the formation of α-1,4-linked *N*-acetylglucosaminyl galactose, was successfully performed using DMT-α-GlcNAc and α-*N*-acetylglucosaminidase (α-GlcNAcase) as a glycosyl donor and catalyst, respectively (Scheme 8.13b) [47]. DMT-glycosides of D-glucosamine (GlcN) and *N*-acetyl-D-galactosamine (GalNAc), DMT-β-GlcN and DMT-α-GalNAc, were also obtained from the corresponding free saccharides reacted with DMT-MM and were used as activated glycosyl donors for enzymatic glycosylations (Scheme 8.13c) [48, 49]. Furthermore, cellotetrose-backboned free heptasaccharide (XXXG) and nonasaccharide (XLLG) were directly converted to the corresponding DMT-β-glycosides, DMT-β-XXXG and DMT-β-XLLG, by the action of DMT-MM. The resulting DMT-β-glycosides were recognized as activated substrates and polymerized by EGIII, affording nonnatural polysaccharides having XXXG or XLLG repeating units in the main chain (Scheme 8.14) [50]. It is extremely difficult to construct such precise repeating structures via conventional synthetic routes that include protection and deprotection steps. The chemoenzymatic process, which consists of the one-step preparation of the glycosyl donor and subsequent enzymatic transglycosylation catalyzed by a glycosidase, will be a practical method in synthetic chemistry for various glycosides.

Triazinyl glycosides such as DMT-glycosides can be used for protecting-group-free chemical glycosylations. DMT-glycosides as glycosyl donors provided 1,2-*cis*-alkylglycosides by using a catalytic amount of metal compound, such as tetrakis(acetonitrile) copper(I), or silver(I) or palladium(II) catalysts. The use of tetrakis(acetonitrile) copper(I) as a catalyst provided a β-enriched mannoside from DMT-α-D-mannopyranoside (DMT-α-Man) as a glycosyl donor in alcohol solvent (Scheme 8.15a) [51]. The use of other metal catalysts, such as silver(I) or palladium(II), also provided the β-enriched mannoside products. When DMT-β-glycosides of Glc, maltose, melibiose, and maltopentaose, which have an equatorial hydroxy group at the 2-position, were reacted in the presence of the metal catalyst in alcohol solvent, α-enriched glycosides were produced. The preferential formation of 1,2-*cis*-glycosides from 1,2-*trans* DMT-glycosides suggests that the reaction proceeds through a stereospecific S$_N$2-type mechanism. It is assumed that the nitrogen atom in the triazine ring or the glycosidic oxygen

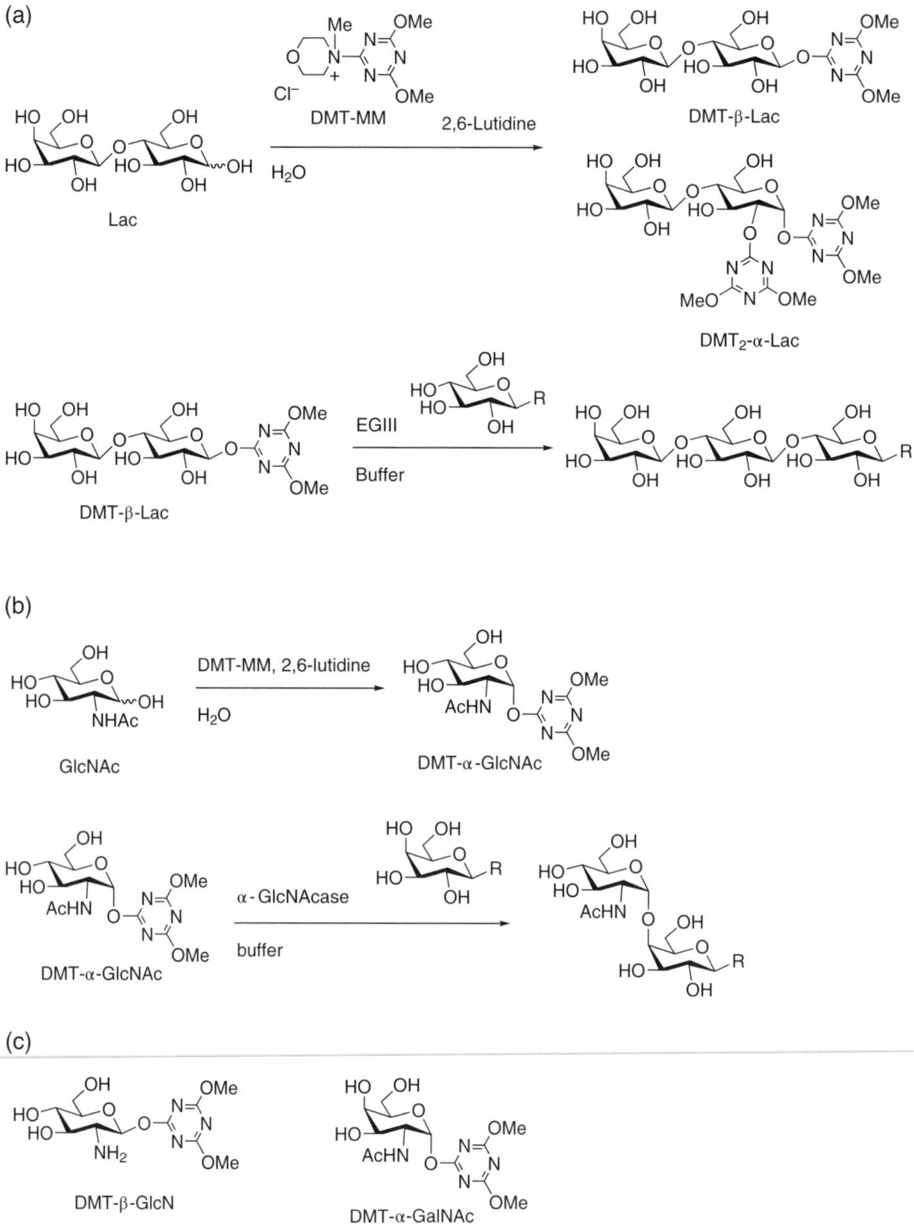

Scheme 8.13 Synthesis of DMT-glycosides and enzymatic transglycosylation by using DMT-glycosides as glycosyl donors.

is protonated by the metal and the anomeric center is activated. 4,6-Dibenzyloxy-1,3,5-triazin-2-yl glycosides (DBT-glycosides), which were prepared using 4,6-dibenzyloxy-2-chloro-1,3,5-triazine and free saccharides in the presence of *N*-methylmorpholine and ammonia in aqueous media, are also useful glycosyl donors (Scheme 8.15b) [52]. Hydrogenolytic conditions in the presence of palladium/carbon gave the corresponding

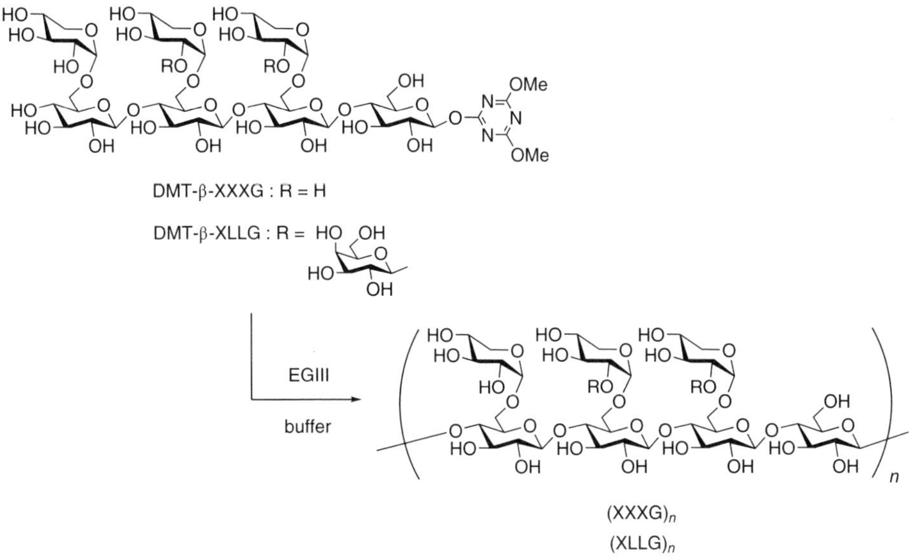

Scheme 8.14 Synthesis of nonnatural oligoxyloglucans by enzymatic polymerization using DMT-glycosides.

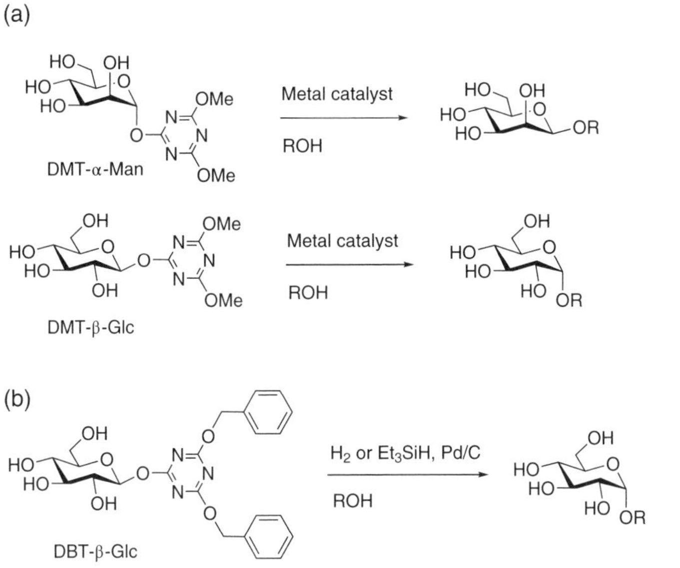

Scheme 8.15 Direct synthesis of O-glycosides using DMT- and DBT-glycosides.

glycosides in good yields without the addition of acid promoter. This glycosylation strategy can be successfully applied to not only monosaccharides but also to an acid-labile disaccharide, melibiose, and a longer-chain oligosaccharide, maltopentasaccharide. When primary alcohols were used as nucleophiles, the corresponding α-glycosides were formed preferentially, suggesting that the reactions proceeded through an

S_N2-type mechanism. On the other hand, the reaction in 1-octanol proceeded through an S_N1-type mechanism due to the poorer solubility of DBT-β-Glc in 1-octanol. The use of secondary alcohols also resulted in S_N1-type reactions, affording anomeric mixtures. When DBT-β-Glc was treated with alcohol in the presence of palladium/carbon without hydrogen gas, the corresponding glucoside was not obtained, clearly indicating that reductive cleavage of the benzyl ether is essential for glycosylation to occur. It is suggested that the resulting phenolic hydroxy group on the oxygen atom of the solvent alcohol becomes protonated. The nitrogen atom on the DBT derivative is effectively activated by the phenolic hydroxy group of the hydrogen-bonded alcohol molecule.

Although many protecting-group-free reactions at the anomeric position of free saccharides in polar organic solvents and ionic liquids have been reported, most of these methods are limited to low molecular weight saccharides because of the low solubility of high molecular weight oligosaccharides. On the other hand, reactions using water-soluble condensing agents such as DMC, DMT-MM, and their derivatives in aqueous solution have significant advantages. In particular, high molecular weight oligosaccharides can be reacted in aqueous solution, giving rise to various useful glycosyl derivatives without the use of protecting groups.

8.3 Protecting-Group-Free Synthesis of Glycopolymers

Saccharide–protein interactions are generally weak, but these interactions are amplified by multivalent forms of saccharides, giving rise to the "glycocluster effect," which underlies various biological processes [53, 54]. Glycopolymers, which comprise a synthetic polymer backbone with pendant saccharides, similarly give rise to artificial glycoclusters and amplify the saccharide signals [55–60]. Glycodendrimers [61–63], glyconanoparticles [64–67], and glycoconjugates [68–70] also act in a manner similar to glycopolymers. Although many synthetic methods and types of glycopolymers have been reported, most require the use of protecting groups for hydroxy and other functional groups on the saccharide moieties. A typical approach for the synthesis of glycopolymers involves the polymerization of glycomonomers. Another approach involves the post-attachment of glycosyl derivatives to a polymer backbone. However, the synthesis of glycomonomers and glycosyl derivatives is generally lengthy and requires multistep processes, including the protection and deprotection of hydroxy, carboxy, and amino groups on the saccharide moieties. Both approaches involve the laborious synthesis of glycosyl derivatives from free saccharides. Therefore, a simpler and more efficient synthetic method is desirable for the preparation of glycopolymers from free saccharides.

There have been few reports of the protecting-group-free synthesis of glycomonomers from free saccharides. The Fischer glycosylation method directly provides a glycomonomer from free monosaccharide by using an excess of hydroxyethyl acrylate in the presence of phosphomolybdic acid as a catalyst (Scheme 8.16a) [71]. However, this method cannot be applied to oligosaccharides or to control stereoselectivity at the anomeric position. Another approach, involving the synthesis of lactone derivatives, is well known (Scheme 8.16b) [72, 73]. Kobayashi et al. synthesized styrene derivatives carrying di- and trisaccharides by coupling vinyl-functionalized amines and lactone derivatives, which were prepared by the oxidation of free saccharides using iodine in methanol

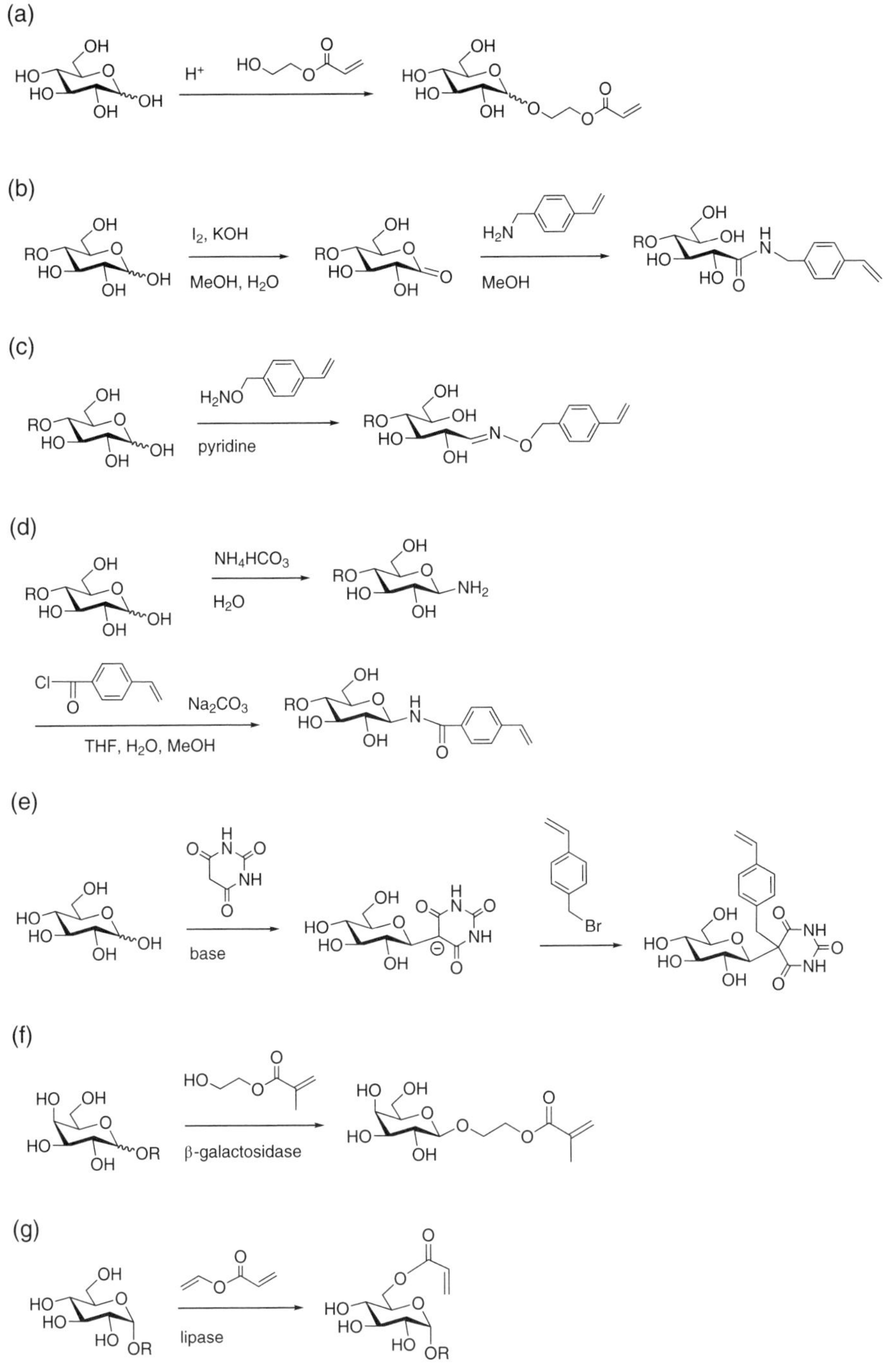

(a)

(b)

(c)

(d)

(e)

(f)

(g)

Scheme 8.16 Protecting-group-free synthesis of glycomonomers.

at elevated temperature. Kurth et al. directly synthesized styrene derivatives in organic solvent using cation-exchange resin and vinyl benzyl oxime while heating (Scheme 8.16c) [74, 75]. The reducing ends of the resulting glycomonomers formed ring-opened structures. Narumi et al. synthesized maltopentaosyl styrene monomers by using 4-aminoalkyl styrene in methanol [76]. The route to glycomonomers through glycosylamines, which were prepared using an excess amount of ammonium hydrogen carbonate in water, is also efficient [77–80]. Glycosylamines, however, are unstable and must be handled carefully when reacted with a carboxylic acid chloride such as acryloyl chloride or methacryloyl chloride, or with isocyanates or epoxides, to obtain the corresponding N-linked glycomonomers. Kobayashi et al. synthesized styrene derivatives carrying sialyllactoses by coupling glycosylamine and p-vinyl benzoyl chloride (Scheme 8.16d) [81]. Wulff et al. synthesized a C—C bond-linked glycomonomer using 1,3-dimethylbarbituric acid and 4-vinylbenzyl bromide with a saccharide derivative containing a barbiturate ring at the anomeric position (Scheme 8.16e) [82]. Enzymatic syntheses of glycomonomers using glycosidase or lipase have also been reported (Scheme 8.16 f and g) [83, 84]. All the previous methods can only be applied to linear and relatively low molecular weight saccharides such as mono-, di-, and trisaccharides because higher molecular weight oligosaccharides are poorly soluble in organic solvents and are susceptible to cleavage of the glycosyl linkage, resulting in low yields. Therefore, a simpler and more efficient synthetic method that is applicable to various free oligosaccharides is required for the preparation of glycomonomers and glycopolymers.

We recently developed protecting-group-free synthetic routes toward glycomonomers from free saccharides through glycosyl azides or thioglycosides, which were directly prepared using DMC as described previously. Triazole-linked glycomonomers were synthesized from free saccharides without the use of protecting groups in two steps through direct synthesis of glycosyl azide and subsequent copper-catalyzed azide–alkyne cycloaddition (CuAAC) with N-propargylacrylamide (Scheme 8.17) [85]. Click chemistry, such as CuAAC, is a powerful tool for synthesizing glycomonomers, glycopolymers, and other glycoclusters, as described in several reports on the application of click chemistry to glycotechnology [86–88]. Our protecting-group-free method is applicable to not only disaccharides such as Lac but also to higher molecular weight oligosaccharides bearing N-acetylneuraminic acid, including α-2,6-sialyllactose (6'SALac) and N-glycan. The resulting glycomonomers were subjected to copolymerization with acrylamide (AAm) using a reversible addition–fragmentation chain transfer (RAFT) technique [89, 90] in dimethyl sulfoxide (DMSO) at 35°C. 2,2-Azobis(4-methoxy-2,4-dimethylvaleronitrile) (V-70) and 2-(benzylsulfanylthiocarbonylsulfanyl) ethanol (BTSE) were used as an initiator and chain transfer agent, respectively, giving rise to glycopolymers bearing triazole-linked oligosaccharides. The glycopolymers were obtained from free saccharides without the use of protecting groups. The saccharide unit ratios in the product polymers were found to be slightly lower than the glycomonomer ratios in the feed because of the lower reactivity of the glycomonomer compared with that of AAm. Di- and trisaccharide-containing monomers provided glycopolymers with low dispersity, whereas N-glycan-containing monomer provided glycopolymer with wide dispersity due to steric hindrance of the larger oligosaccharide. Although there have been many reports of the synthesis of glycopolymers, the present synthetic method, utilizing direct anomeric activation of free saccharide by DMC, is the first to incorporate a large biologically relevant oligosaccharide such as N-glycan without the use of protecting groups.

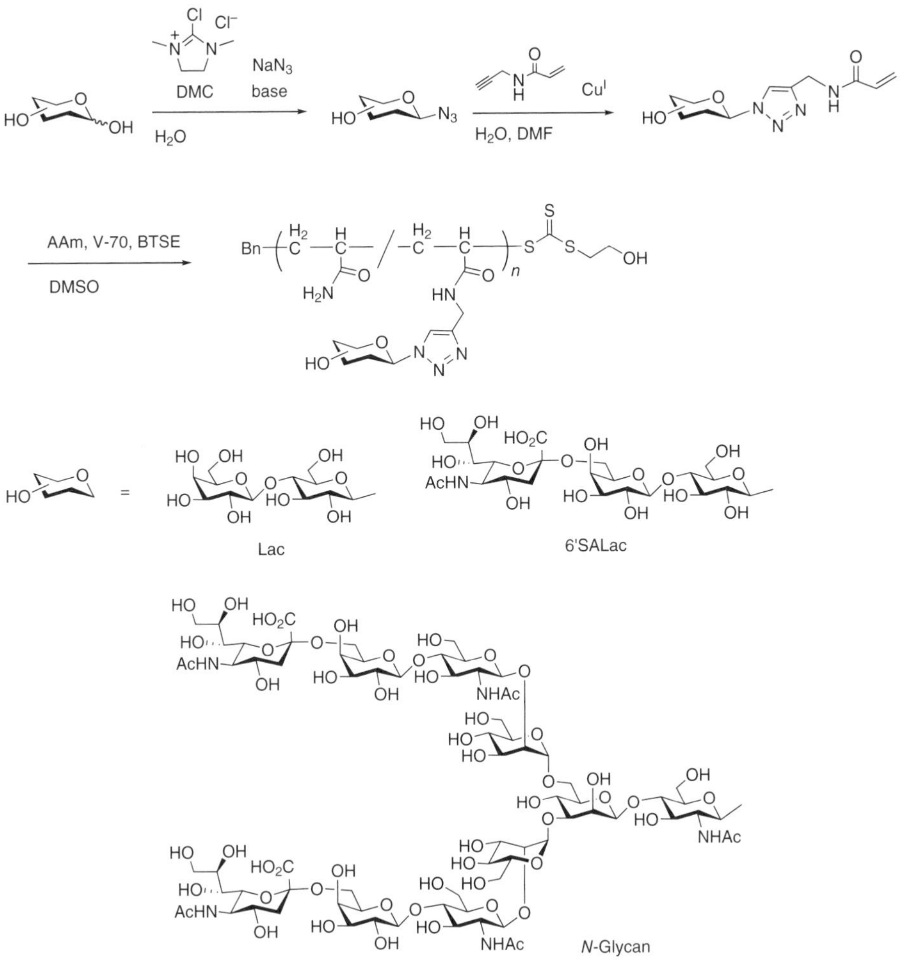

Scheme 8.17 Protecting-group-free synthesis of glycomonomers and glycopolymers via glycosyl azides.

An enormous number of post-attachment methods by CuAAC for the preparation of glycopolymers have been reported. CuAAC is one of the most useful methods for attaching saccharide moieties efficiently to side chains on the polymer backbone (Scheme 8.18) [91–94]. The direct azidation method using DMC, starting from free saccharides, is widely used to synthesize glycopolymers. However, protection of the alkyne group of alkyne-containing monomer compounds with the trimethylsilyl (TMS) group, followed by deprotection, is necessary before and after the polymerization reaction, respectively.

Our protecting-group-free method toward glycomonomers is realized using thioglycosides. Thioglycoside monomers were synthesized from free saccharides by a one-pot method through 4-aminophenyl-1-thioglycosides (Scheme 8.19) [95]. The products of the first reaction step, 4-aminophenyl-1-thioglycosides, were coupled with acryloyl chloride to obtain the glycomonomers. The reaction mixture was washed with

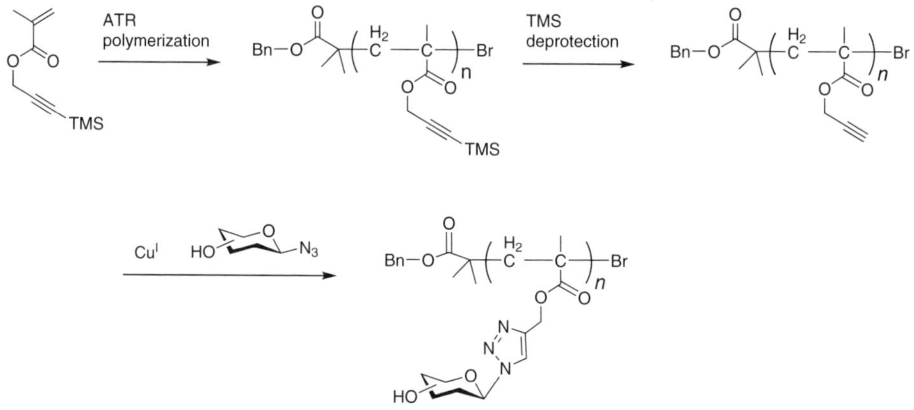

Scheme 8.18 Synthesis of glycopolymers by post-attachment CuAAC using glycosyl azides.

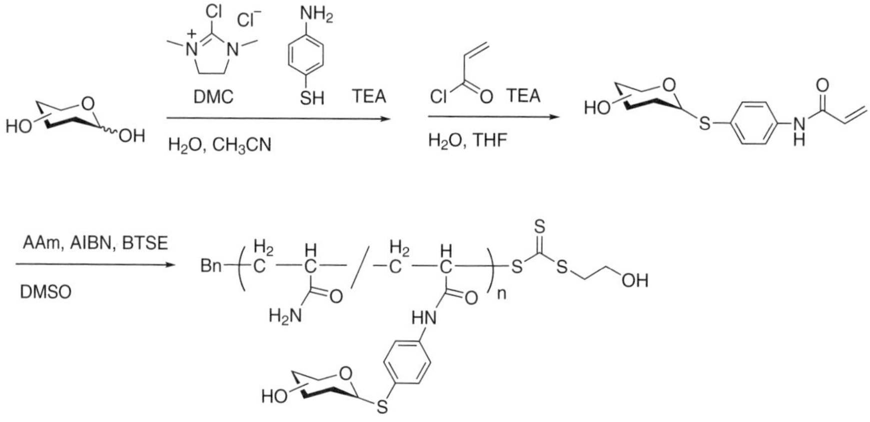

Scheme 8.19 Protecting-group-free synthesis of glycopolymers by the one-pot glycomonomer synthesis.

chloroform to thoroughly extract excess 4-aminobenzenthiol, and then the mixture was directly reacted with acryloyl chloride in the presence of TEA in aqueous THF solution to yield thioglycoside monomers without isolation of the 4-aminophenyl-1-thioglyco-sides. Removal by preparative high-performance liquid chromatography of the small amount of stereoisomeric product provided the glycomonomers in good yield. The thioglycosidic monomers were subjected to RAFT copolymerization with AAm to obtain glycopolymers bearing thioglycosides. The polymerization reaction was performed in DMSO at 70°C using α,α-azobisisobutyronitrile (AIBN) and BTSE. The glycopolymers were obtained from free saccharides using a one-pot monomer synthesis method without the use of protecting groups. The saccharide-containing monomers provided glycopolymers with low dispersity, and the saccharide unit ratios in the product were consistent with the monomer ratios in the feed. This one-pot approach from free saccharides is facile and effective for the synthesis of thioglycosidic monomers and polymers.

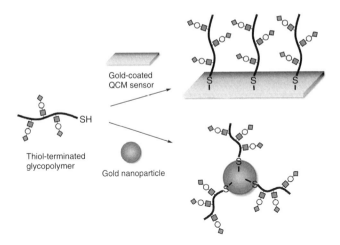

Figure 8.4 Immobilization of glycopolymers on gold surfaces.

Glycopolymers synthesized by the RAFT polymerization method possess a trithio-carbonate group, resulting from the BTSE, at the terminus of the polymer backbone. The aforementioned triazole-linked and thioglycosidic glycopolymers were easily converted to thiol-terminated polymers in the presence of sodium borohydride and were then immobilized on a gold-coated quartz crystal microbalance (QCM) sensor via Au—S bond formation (Figure 8.4). When subjected to lectin-binding tests in phosphate buffered saline (PBS), both triazole-linked and thioglycosidic glycopolymers strongly bound with the corresponding lectins: Man with concanavalin A (ConA) from *Canavalia ensiformis*, Lac with peanut lectin (PNA) from *Arachis hypogaea*, 6'SALac with *Sambucus sieboldiana* agglutinin (SSA), and *N*-glycan with SSA. Bovine serum albumin (BSA) did not interact with the glycopolymers. The association constants (K_a) for the lectin–saccharide interactions were estimated to be in the order of $10^7\,M^{-1}$. The K_a value for the binding of lectin and free saccharide is reported to be in the order of $10^3\,M^{-1}$ [96]. Furthermore, when lectin-binding tests were conducted using glycopolymer-immobilized gold nanoparticles (GNPs) in a UV/Vis spectrometer, the interaction between the saccharide and the corresponding lectin was clearly detected. The absorbance of GNP gradually decreased as the amount of lectin added increased, suggesting that the lectin-induced aggregation of the glycopolymer-modified GNP due to lectin–saccharide interactions. These results from lectin-binding tests indicate that the glycopolymers strongly interact with the corresponding lectin, suggesting that the glycopolymers exhibit the glycocluster effect.

8.4 Protecting-Group-Free Synthesis of Glycoconjugates

The combination of the direct synthesis of glycosyl azides and subsequent click chemistry is efficient for preparing glycoconjugates such as saccharide-displayed particles, glycosylated peptides and polyamino acids, glycodendrimers, and glycoproteins, as well as glycopolymers. We recently reported the facile synthesis of saccharide-terminated poly-L-lactides (Sac-PLLAs) (Scheme 8.20a) [97]. Various Sac-PLLAs were synthesized readily through direct azidation using DMC, followed by CuAAC with

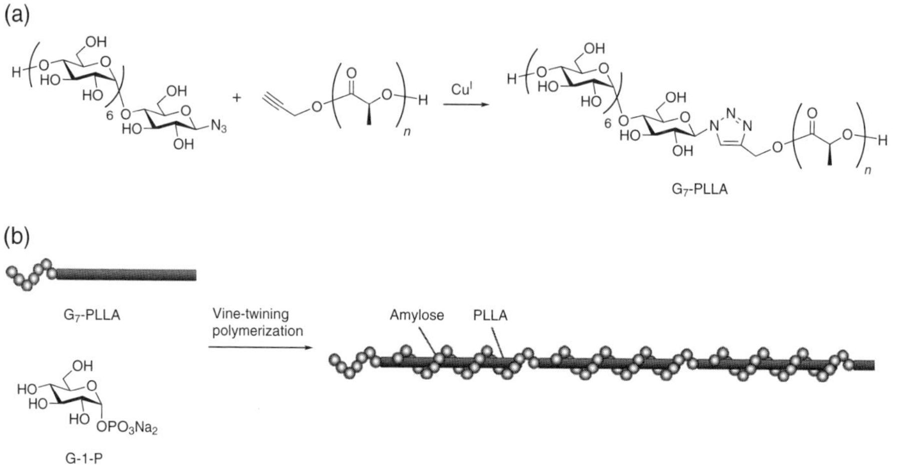

(a)

(b)

G$_7$-PLLA   Vine-twining polymerization   Amylose   PLLA

G-1-P

Scheme 8.20 Synthesis of (a) saccharide-terminated PLLA by CuAAC and (b) amylose–PLLA inclusion supramolecular polymer by vine-twining polymerization.

alkyne-terminated polylactides. The Sac-PLLAs were found to form aggregates with the saccharide moieties, with the saccharides forming a core and the PLLA segment forming a shell around the aggregates in nonpolar solvents. In contrast, in polar solvents, the aggregates comprised a PLLA core and a saccharide shell. Furthermore, enzymatic polymerization catalyzed by a phosphorylase, using a maltoheptaose-terminated poly-L-lactide (G$_7$-PLLA) and G-1-P as a primer–guest polymer conjugate and a monomer, respectively, was conducted via vine-twining polymerization [98–100] to provide novel inclusion supramolecular polymers consisting of amylose and PLLA (Scheme 8.20b) [101–103]. Vine-twining polymerization using a maltoheptaose-terminated poly(tetrahydrofuran) and G-1-P provided a novel inclusion supramolecular polymer consisting of amylose and poly(tetrahydrofuran) [104].

Fairbanks et al. developed a protecting-group-free and one-pot synthesis of triazole-linked glycopeptides by the combined use of 2-azido-1,3-dimethylimidazolinium hexafluorophosphate (ADMP), a derivative of DMC, and CuAAC with alkyne-bearing peptides in aqueous solution (Scheme 8.21a) [105]. This method may be applied to the synthesis of glycopeptides bearing not only monosaccharides but also larger oligosaccharides such as N-glycan and the core tetrasaccharide of N-glycan. The same group reported application of this DMC-mediated reaction to the direct synthesis of alkyl thioglycosides. The addition of mercaptoacetonitrile to the DMC-mediated reaction mixture produced the corresponding cyanomethyl thioglycosides, which were converted into 2-imino-2-methoxyethyl thioglycosides for protein glycosylation (Scheme 8.21b) [106]. The resulting thioglycosides were directly attached to the surface protein lysine residues. Winssinger et al. applied this DMC-mediated reaction to the solid-phase synthesis of glycopeptides (Scheme 8.22a) [31]. The thiol group of the cysteine residue on the peptide was linked to the anomeric carbon of free saccharides, producing S-linked glycopeptides. Rademann et al. reported the one-pot conversion of free GlcNAc and its derivatives to glycosyl thiol in water (Scheme 8.22b) [107]. In this reaction, sugar oxazoline derivatives were first produced using DMC under basic conditions.

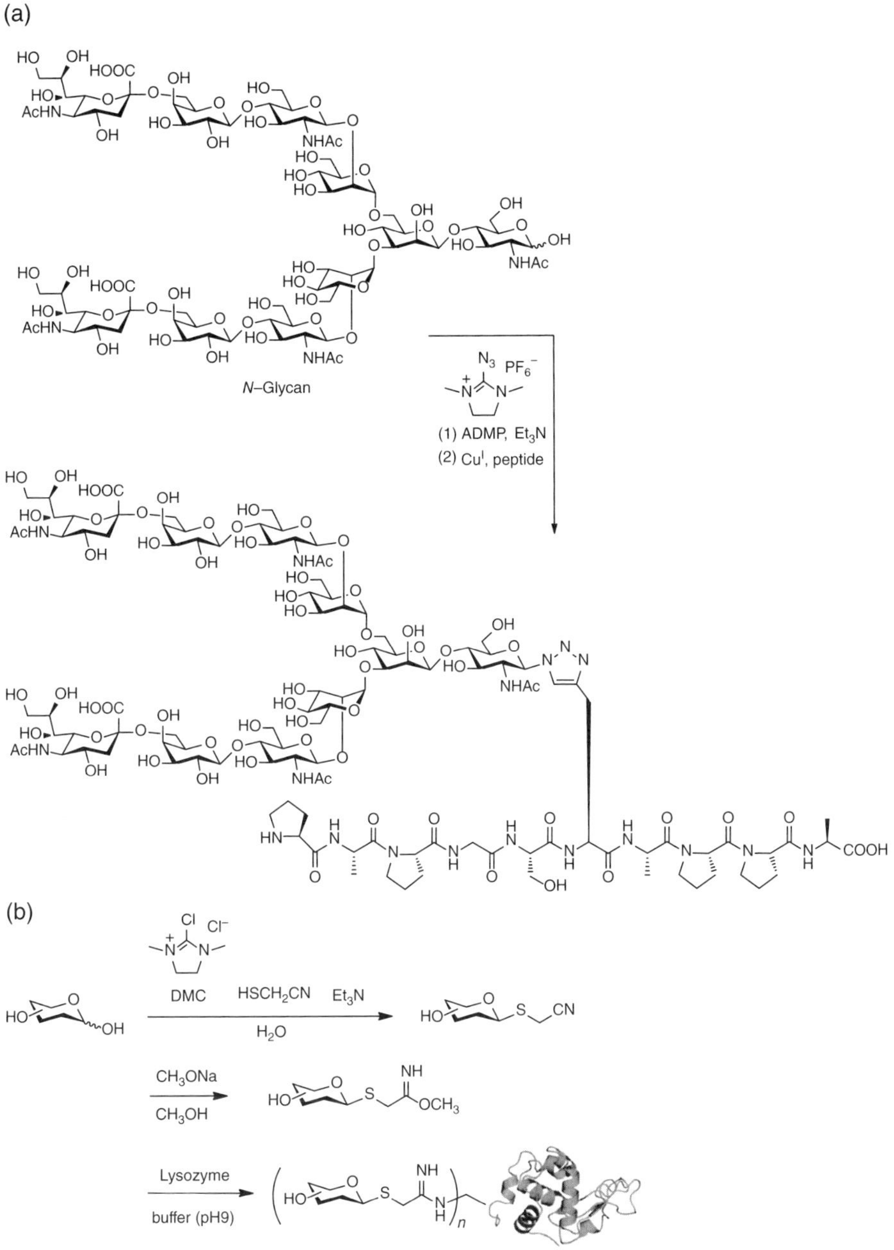

(a)

(b)

Scheme 8.21 Protecting-group-free synthesis of glycopeptides and glycoproteins using glycosyl azides.

The oxazoline derivatives were converted into 1-glycosyl thioesters through treatment with 1-thioacids. The conjugation of amino acids and proteins (*S*-glycosidic amino acids and *S*-glycoproteins) was accomplished using the thiol–ene reaction with terminal olefins.

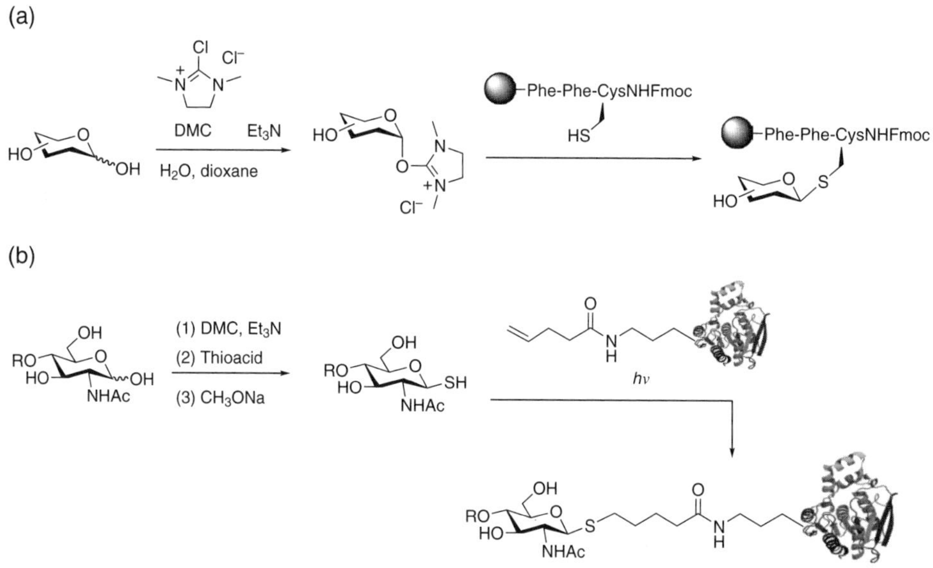

(a)

(b)

Scheme 8.22 Protecting-group-free synthesis of thio-linked glycopeptides and glycoproteins.

Tanaka et al. reported using CuAAC to synthesize glycodendrimers and glycoclusters on a protein carrying a number of N-glycans. An alkyne-terminated polylysine-based dendrimer was modified by azide-containing N-glycans [63]. The resulting glycodendrimers were useful glycocluster probes *in vivo* due to the multivalent cluster of oligosaccharides. His group also reported synthesizing homo- and heterogeneous N-glycan clusters on albumin using a combination of CuAAC and the 6π-azaelectrocyclization protocol [108].

Chemoenzymatic routes are also simple methods toward glycoconjugates from free saccharides. Usui et al. reported the chemoenzymatic synthesis of artificial glycopeptides containing multivalent sialyloligosaccharides with a polyglutamic acid (PGA) backbone (Scheme 8.23) [109–112]. p-Aminophenyl-β-D-N-acetyllactosaminide (LacNAc-β-pAP), which was enzymatically prepared from p-nitrophenyl β-D-N-acetyl-glucopyranoside (GlcNAc-β-pNP) and Lac using β-galactosidase and subsequent reduction of a nitro group, was attached to the carboxy groups on PGA using the dehydrative condensation agents 1H-benzotriazol-1-yloxytris(dimethylamino)phosphonium hexafluorophosphate (BOP) and 1-hydroxybenzotriazole (HOBt), giving rise to LacNAc-carrying PGA (LacNAc-PGA). The attached saccharide moieties on PGA were modified by sialic acids via catalysis by a sialyltransferase to produce sialyl LacNAc-carrying PGA (Neu5AcLacNAc-PGA). An addition reaction of an amino group to an aldehyde group is also a simple approach to attach saccharide moieties onto prepared glycoconjugates. Umemura et al. chemoenzymatically synthesized chitosan derivatives bearing N-glycans from chitosan and hen egg yolk sialylglycopeptide (SGP) (Scheme 8.24) [113]. The sialyloligosaccharide transglycosylation product (SGTG) with an aldehyde group, which was prepared through a transglycosylation reaction catalyzed by Endo-M using SGP, was reacted with amino groups on chitosan to provide complex disialooligosaccharide-chitosan (CDO-chitosan). Both

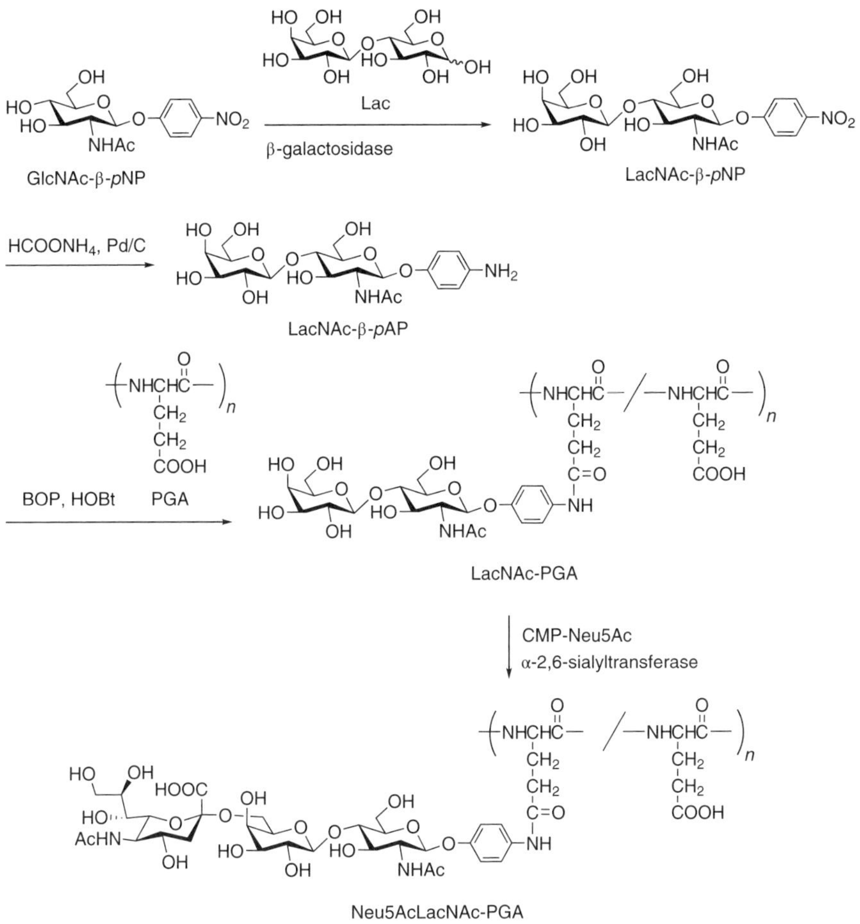

Scheme 8.23 Chemoenzymatic synthesis of artificial glycopeptides.

sialylglycoconjugates, Neu5AcLacNAc-PGA and CDO-chitosan, inhibited infection by influenza viruses.

8.5 Conclusions

This chapter provided an overview of the protecting-group-free synthesis of various glycosyl derivatives from free saccharides. Several direct methods for the conversion of free saccharides at the anomeric position, including classical methods and recent strategies arising from our work and that of others, were described. Most of the reactions were performed using simple and low molecular weight saccharides in polar organic solvents because of the low solubility of free saccharides in low-polarity organic solvents. However, little has been reported regarding the direct synthesis of glycosyl derivatives in water. Anomeric activation methods using dehydrative condensing agents in water are simple and efficient approaches for preparing various glycosyl derivatives

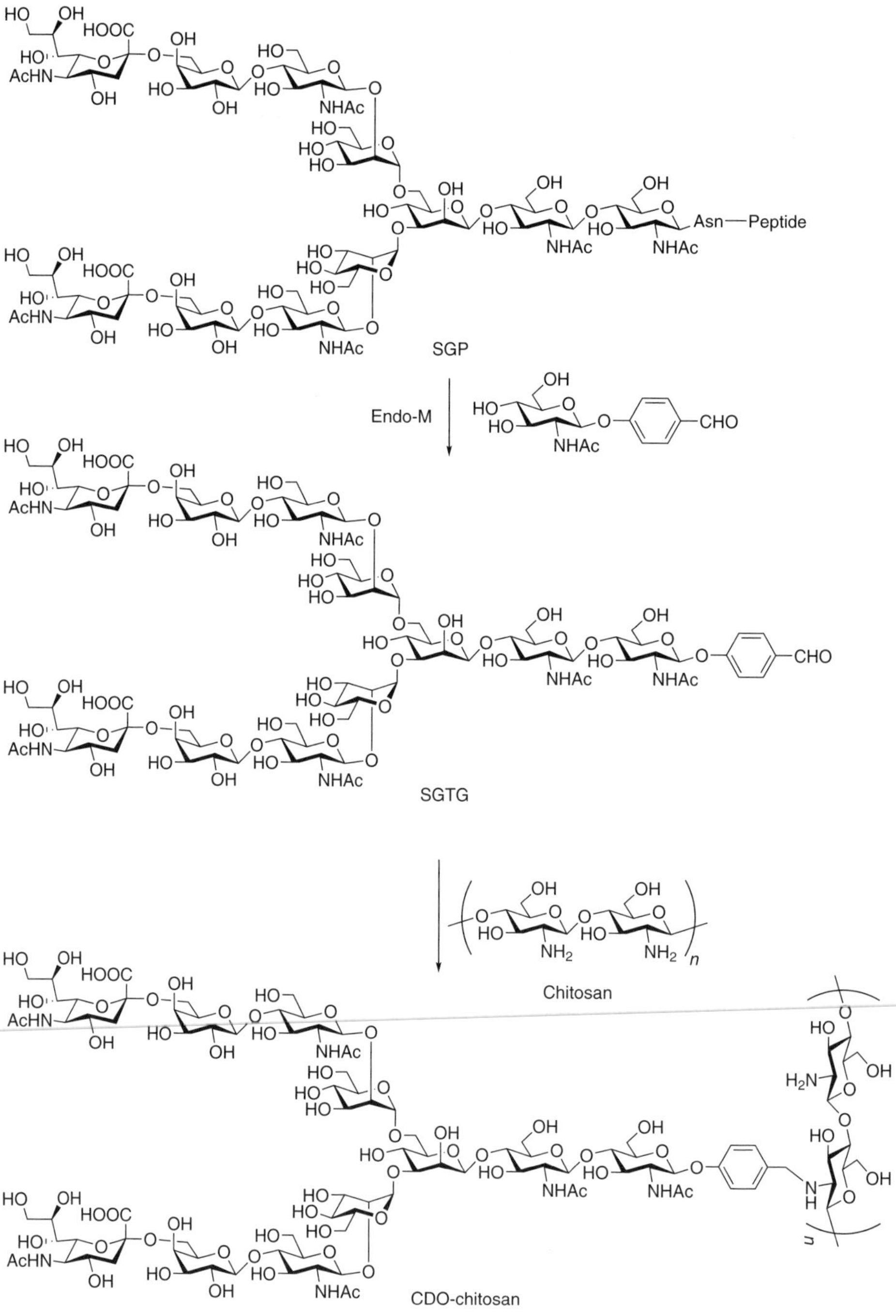

Scheme 8.24 Chemoenzymatic synthesis of *N*-glycan-bearing chitosan.

from free saccharides. The use of formamidinium salt-type reagents such as DMC directly and stereoselectively provided various glycosyl derivatives such as sugar oxazoline derivatives, glycosyl azides, and thioglycosides in aqueous solution. The strategy of anomeric activation in water is based on the anomeric hemiacetal being more acidic than other hydroxy groups on saccharides and on water. This method is applicable to not only mono- and disaccharides but also to various higher molecular weight oligosaccharides such as *N*-linked oligosaccharides typically found on cell surfaces, since the hemiacetal moiety is only found at the anomeric position of free saccharides. Furthermore, the reaction of free saccharide with triazine-type reagents such as DMT-MM produced novel activated glycosyl donors for glycosidase-catalyzed glycosylation. In addition, subsequent reaction with the synthesized glycosyl derivatives, such as by click chemistry, can relatively easily produce glycomonomers, glycopolymers, and glycoconjugates without the use of protecting groups. Protecting-group-free reactions in water have a low environmental impact because they can reduce the number of synthetic steps and the use of organic solvents, since no protecting groups are required and the reactions can be conducted in water. In the near future, protecting-group-free synthesis using free saccharides will be the preferred method for obtaining various glycosyl derivatives and glycomaterials. In addition, glycomaterials prepared by a protecting-group-free method will likely play significant roles in providing efficient biomaterials such as pathogen inhibitors, cell cultivation agents for tissue engineering, and biosensors for the detection of viruses and toxins.

References

1 Fischer, E. (1983). *Ber. Dtsch. Chem. Ges.* 26: 2400–2412.
2 Fischer, E. and Beensch, L. (1894). *Ber. Dtsch. Chem. Ges.* 27: 2478–2486.
3 Fischer, E. (1895). *Ber. Dtsch. Chem. Ges.* 28: 1145–1167.
4 Cadotte, J.E., Smith, F., and Spriestersbach, D. (1952). *J. Am. Chem. Soc.* 74: 1501–1504.
5 Polanki, I.K., Kurma, S.H., and Bhattacharya, A.K. (2015). *J. Carbohydr. Chem.* 34: 196–205.
6 Guchhait, G. and Misra, A.K. (2011). *Catal. Commun.* 14: 52–57.
7 Sharma, S.K., Corrales, G., and Penadés, S. (1995). *Tetrahedron Lett.* 36: 5627–5630.
8 Huchel, U., Schmidt, C., and Schmidt, R.R. (1995). *Tetrahedron Lett.* 36: 9457–9460.
9 Sharma, D.K., Lambu, M.R., Sidiq, T. et al. (2013). *RSC Adv.* 3: 11450–11455.
10 Park, T.-J., Weïwer, M., Yuan, X. et al. (2007). *Carbohydr. Res.* 342: 614–620.
11 Augé, J. and Sizun, G. (2009). *Green Chem.* 11: 1179–1183.
12 Pfaffe, M. and Mahrwald, R. (2012). *Org. Lett.* 14: 792–795.
13 Schmalisch, S. and Mahrwald, R. (2013). *Org. Lett.* 15: 5854–5857.
14 Mitsunobu, O. (1981). *Synthesis* 1981: 1–28.
15 Mitsunobu, O., Yamada, M., and Mukaiyama, T. (1967). *Bull. Chem. Soc. Jpn.* 40: 935–939.
16 Mitsunobu, O. and Yamada, M. (1967). *Bull. Chem. Soc. Jpn.* 40: 2380–2382.
17 Reineri, F., Santelia, D., Viale, A. et al. (2010). *J. Am. Chem. Soc.* 132: 7186–7193.
18 Shoda, S., Kobayashi, A., and Takahashi, S. (2006). Process for producing glycoside derivative. PCT Int. Appl., WO 2006038440.
19 Takeuchi, H., Mishiro, K., Ueda, Y. et al. (2015). *Angew. Chem. Int. Ed.* 54: 6177–6180.

20 Gudmundsdottir, A.V. and Nitz, M. (2008). *Org. Lett.* 10: 3461–3463.

21 Williams, R.J., Paul, C.E., and Nitz, M. (2014). *Carbohydr. Res.* 386: 73–77.

22 Edgar, L.J.G., Dasgupta, S., and Nitz, M. (2012). *Org. Lett.* 14: 4226–4229.

23 Peri, F., Dumy, P., and Mutter, M. (1998). *Tetrahedron* 54: 12269–12278.

24 Dasgupta, S. and Nitz, M. (2011). *J. Org. Chem.* 76: 1918–1921.

25 Newman, M. and Boden, H. (1961). *J. Org. Chem.* 26: 2525.

26 Kadokawa, J., Mito, M., Takahashi, S. et al. (2004). *Heterocycles* 63: 1531–1535.

27 Kobayashi, S., Kiyosada, T., and Shoda, S. (1996). *J. Am. Chem. Soc.* 118: 13113–13114.

28 Thamsen, J. (1952). *Acta Chem. Scand.* 6: 270–284.

29 Kunishima, K., Kawachi, C., Monta, J. et al. (1999). *Tetrahedron* 55: 13159–13170.

30 Noguchi, M., Tanaka, T., Gyakushi, H. et al. (2009). *J. Org. Chem.* 74: 2210–2212.

31 Novoa, A., Barluenga, S., Serba, C., and Winssinger, N. (2013). *Chem. Commun.* 49: 7608–7610.

32 Yamamoto, K. (2013). *Biotechnol. Lett.* 35: 1733–1743.

33 Umekawa, M., Higashiyama, T., Koga, Y. et al. (2010). *Biochim. Biophys. Acta* 1800: 1203–1209.

34 Tanaka, T., Huang, W.C., Noguchi, M. et al. (2009). *Tetrahedron Lett.* 50: 2154–2157.

35 Tanaka, T., Nagai, H., Noguchi, M. et al. (2009). *Chem. Commun.* 3378–3379.

36 Kolb, H.C., Finn, M.G., and Sharpless, K.B. (2001). *Angew. Chem. Int. Ed.* 40: 2004–2021.

37 Tanaka, T., Matsumoto, T., Noguchi, M. et al. (2009). *Chem. Lett.* 38: 458–459.

38 Yoshida, N., Noguchi, M., Tanaka, T. et al. (2011). *Chem. Asian. J.* 6: 1876–1885.

39 Li, G., Noguchi, M., Kashiwagura, H. et al. (2016). *Tetrahedron Lett.* 57: 3529–3531.

40 Yoshida, N., Fujieda, T., Kobayashi, A. et al. (2013). *Chem. Lett.* 42: 1038–1039.

41 Noguchi, M., Fujieda, T., Huang, W.C. et al. (2012). *Helv. Chim. Acta* 95: 1928–1936.

42 Tanaka, H., Yoshimura, Y., Jørgensen, M.R. et al. (2012). *Angew. Chem. Int. Ed.* 51: 11531–11534.

43 Tanaka, T., Noguchi, M., Kobayashi, A., and Shoda, S. (2008). *Chem. Commun.* 2016–2018.

44 Tanaka, T., Kobayashi, A., Noguchi, M. et al. (2009). *J. Appl. Glycosci.* 56: 83–88.

45 Tanaka, T., Noguchi, M., Watanabe, K. et al. (2010). *Org. Biomol. Chem.* 8: 5126–5132.

46 Kobayashi, A., Tanaka, T., Watanabe, K. et al. (2010). *Bioorg. Med. Chem. Lett.* 20: 3588–3591.

47 Noguchi, M., Nakamura, M., Ohno, A. et al. (2012). *Chem. Commun.* 48: 5560–5562.

48 Tanaka, T., Wada, T., Noguchi, M. et al. (2012). *J. Carbohydr. Chem.* 31: 634–646.

49 Kiyohara, M., Nakatomi, T., Kurihara, S. et al. (2012). *J. Biol. Chem.* 287: 693–700.

50 Tanaka, T., Noguchi, M., Ishihara, M. et al. (2010). *Macromol. Symp.* 297: 200–209.

51 Tanaka, T., Kikuta, N., Kimura, Y., and Shoda, S. (2015). *Chem. Lett.* 44: 846–848.

52 Ishihara, M., Takagi, Y., Li, G. et al. (2013). *Chem. Lett.* 42: 1235–1237.

53 Lee, Y.C. and Lee, R.T. (1995). *Acc. Chem. Res.* 28: 321–327.

54 Mammen, M., Choi, S., and Whitesides, G.M. (1998). *Angew. Chem. Int. Ed.* 37: 2754–2794.

55 Narain, R. (2011). *Engineered Carbohydrate-Based Materials for Biomedical Applications: Polymers, Surfaces, Dendrimers, Nanoparticles, and Hydrogels.* Wiley.

56 Le Droumaguet, B. and Nicolas, J. (2010). *Polym. Chem.* 1: 563–598.

57 Slavin, S., Burns, J., Haddleton, D.M., and Becer, C.R. (2011). *Eur. Polym. J.* 47: 435–446.

58 Miura, Y. (2012). *Polym. J.* 44: 679–689.

59 Sunasee, R. and Narain, R. (2013). *Macromol. Biosci.* 13: 9–27.

60 Ahmed, M., Wattanaarsakit, P., and Narain, R. (2013). *Eur. Polym. J.* 49: 3010–3033.

61 Woller, E.K. and Cloninger, M.J. (2002). *Org. Lett.* 4: 7–10.

62 Chabre, Y.M. and Roy, R. (2008). *Curr. Top. Med. Chem.* 8: 1237–1285.

63 Tanaka, K., Siwu, E.R.O., Minami, K. et al. (2010). *Angew. Chem. Int. Ed.* 49: 8195–8200.

64 de la Fuente, J.M., Barrientos, A.G., Rojas, T.C. et al. (2001). *Angew. Chem. Int. Ed.* 40: 2257–2261.

65 de la Fuente, J.M. and Penadés, S. (2005). *Tetrahedron: Asymmetry* 16: 387–391.

66 Spain, S.G., Albertin, L., and Cameron, N.R. (2006). *Chem. Commun.* 4198–4200.

67 Housni, A., Cai, H., Liu, S. et al. (2007). *Langmuir* 23: 5056–5061.

68 Ejaz, M., Ohno, K., Tsujii, Y., and Fukuda, T. (2000). *Macromolecules* 33: 2870–2874.

69 Yang, Q., Xu, Z., Hu, M. et al. (2005). *Langmuir* 21: 10717–10723.

70 Guo, T., Liu, P., Zhu, J. et al. (2006). *Biomacromolecules* 7: 1196–1202.

71 Kitazawa, S., Okumura, M., Kinomura, K., and Sakakibara, T. (1990). *Chem. Lett.* 1733–1736.

72 Kobayashi, K., Sumitomo, H., and Ina, Y. (1985). *Polym. J.* 17: 567–575.

73 Narain, R. and Armes, S.P. (2003). *Biomacromolecules* 4: 1746–1758.

74 Zhou, W., Wilson, M.E., Kurth, M.J. et al. (1997). *Macromolecules* 30: 7063–7068.

75 Zhou, W., Naik, S.S., Kurth, M.J. et al. (1998). *J. Polym. Sci. A Polym. Chem.* 36: 2971–2978.

76 Togashi, D., Otsuka, I., Borsali, R. et al. (2015). *J. Polym. Sci. A Polym. Chem.* 53: 1671–1679.

77 Marcotte, E.M., Monzingo, A.F., Ernst, S.R. et al. (1996). *Nat. Struct. Biol.* 3: 155–162.

78 Klein, J. and Herzog, D. (1987). *Macromol. Chem. Phys.* 188: 1217–1232.

79 Klein, J. and Begli, A.H. (1989). *Macromol. Chem. Phys.* 190: 2527–2534.

80 Klein, J., Begli, A.H., and Engelke, S. (1989). *Macromol. Rapid Commun.* 10: 629–636.

81 Tsuchida, A., Kobayashi, K., Matsubara, N. et al. (1998). *Glycoconj. J.* 15: 1047–1054.

82 Wulff, G. and Clarkson, G. (1994). *Macromol. Chem. Phys.* 195: 2603–2610.

83 Gill, I. and Valivety, R. (2000). *Angew. Chem. Int. Ed.* 39: 3804–3808.

84 Degoede, A., Vanoosterom, M., Vandeurzen, M.P.J. et al. (1993). *Heterogeneous Catalysis and Fine Chemicals III*, vol. 78 (ed. M. Guisnet, J. Barbier, J. Barranult, C. Bouchoule, D. Duprez, G. Perot and C. Montassier), 513–520. Elsevier Science.

85 Tanaka, T., Ishitani, H., Miura, Y. et al. (2014). *ACS Macro Lett.* 3: 1074–1078.

86 Witczak, Z.J. and Bielski, R. (2013). *Click Chemistry in Glycoscience: New Developments and Strategies*. Wiley.

87 Dondoni, A. (2007). *Chem. Asian. J.* 2: 700–708.

88 Binder, W.H. and Sachsenhofer, R. (2007). *Macromol. Rapid Commun.* 28: 15–54.

89 Moad, G., Rizzardo, E., and Thang, S.H. (2005). *Aust. J. Chem.* 58: 379–410.

90 Lowe, A.B., Sumerlin, B.S., and McCormick, C.L. (2003). *Polymer (Guildf)* 44: 6761–6765.

91 Vinson, N., Gou, Y.Z., Becer, C.R. et al. (2011). *Polym. Chem.* 2: 107–113.

92 Gou, Y.Z., Richards, S.J., Haddleton, D.M., and Gibson, M.I. (2012). *Polym. Chem.* 3: 1634–1640.

93 Nagao, M., Kurebayashi, Y., Seto, H. et al. (2016). *Polym. J.* 48: 745–749.

94 Nagao, M., Kurebayashi, Y., Seto, H. et al. (2016). *Polym. Chem.* 7: 5920–5924.

95 Tanaka, T., Inoue, G., Shoda, S., and Kimura, Y. (2014). *J Polym. Sci. Part A Polym. Chem.* 52: 3513–3520.

96 Miura, Y., Ikeda, T., and Kobayashi, K. (2003). *Biomacromolecules* 4: 410–415.

97 Tanaka, T., Fukuhara, H., Shoda, S., and Kimura, Y. (2013). *Chem. Lett.* 42: 197–199.

98 Kadokawa, J.-I. (2012). *Polymers (Basel)* 4: 116–133.

99 Kadokawa, J.-I. (2013). *Biomolecules* 3: 369–385.

100 Kadokawa, J.-I. (2013). *Trends Glycosci. Glycotechnol.* 25: 57–69.

101 Tanaka, T., Sasayama, S., Nomura, S. et al. (2013). *Macromol. Chem. Phys.* 214: 2829–2834.

102 Tanaka, T., Sasayama, S., Yamamoto, K. et al. (2015). *Macromol. Chem. Phys.* 216: 794–800.

103 Tanaka, T., Gotanda, R., Tsutsui, A. et al. (2015). *Polymer. (United Kingdom)* 73: 9–16.

104 Tanaka, T., Tsutsui, A., Gotanda, R. et al. (2015). *J. Appl. Glycosci.* 62: 135–141.

105 Lim, D., Brimble, M.A., Kowalczyk, R. et al. (2014). *Angew. Chem.* 126: 12101–12105.

106 Alexander, S.R. and Fairbanks, A.J. (2016). *Org. Biomol. Chem.* 14: 6679–6682.

107 Köhling, S., Exner, M.P., Nojoumi, S. et al. (2016). *Angew. Chem. Int. Ed.* 55: 15510–15514.

108 Ogura, A., Tahara, T., Nozaki, S. et al. (2016). *Sci. Rep.* 6: 21797–21813.

109 Totani, K., Kubota, T., Kuroda, T. et al. (2003). *Glycobiology* 13: 315–326.

110 Ogata, M., Murata, T., Murakami, K. et al. (2007). *Bioorg. Med. Chem.* 15: 1383–1393.

111 Ogata, M., Hidari, K.I.P.J., Murata, T. et al. (2009). *Bioconjug. Chem.* 20: 538–549.

112 Ogata, M., Uzawa, H., Hidari, K.I.P.J. et al. (2014). *J. Appl. Glycosci.* 61: 1–7.

113 Umemura, M., Itoh, M., Makimura, Y. et al. (2008). *J. Med. Chem.* 51: 4496–4503.

9

Latent Functionality: A Tactic Toward Formal Protecting-Group-Free Synthesis

Rodney A. Fernandes

Department of Chemistry, Indian Institute of Technology Bombay, Mumbai, India

9.1 Introduction

The synthesis of a natural product or any complex organic molecule needs (i) identification of a molecule to be synthesized, (ii) a synthetic strategy or general plan, (iii) selection of specific reaction conditions for each step, and (iv) the bench execution of the synthetic plan. These steps are interlinked making the task of synthesizing a molecule a formidable challenge at each stage. This is mainly because of the enormous diversity and a gallery of reactions available for synthesis and the unpredictable nature and severe limits on the applicability of any given reaction. This makes the designed synthetic route hypothetical and of a general nature, which can be neither binding nor decisive. Many times, the synthetic route may fail in the first step or even the last step, jeopardizing the strategy especially in the latter case. Hence total synthesis is an unpredictable task, and many synthetic efforts have gone down the drain, nevertheless with long lasting impressions and the realization that "nature" is still an unbeatable force when it comes to achieving total synthesis of complex molecules. However, the rational design for synthesis of a target molecule, as formulated by Corey through the concept of "retrosynthesis," is still desirable and is followed explicitly today in any venture of a total synthesis.

For a given molecule, many synthetic routes can be proposed with the involvement of intermediates, which can be directly synthesized, rather than the target molecule. Thus the starting point for any route is a readily available material, most often a commercially available and relatively cheap material. The synthetic routes involve distinct structural moieties of the molecule that can be derived through possible synthetic operations. These units are called synthons. Since it is possible to devise many routes, there could be equal number or more synthons. The choice needs to be judiciously made and none could be in practice, superior to other. However, with general knowledge and practice of total synthesis, an expert would use his/her genius in selecting a particular synthon

Protecting-Group-Free Organic Synthesis: Improving Economy and Efficiency, First Edition.
Edited by Rodney A. Fernandes.
© 2018 John Wiley & Sons Ltd. Published 2018 by John Wiley & Sons Ltd.

for a particular task. This then relates to availability of that synthon or its ready preparation using easily attainable reaction conditions. The concept of *"synthons"* or *"functional group equivalents"* has spurred research in organic synthesis. This then led to new ideas and concepts like *"umpolung"* chemistry and *"latent functionality"* While umpolung relates to reversal of reactivity, the latter concept of latent functionality relates to masked groups or synthetic equivalents not appearing directly related to a functional group but when unmasked result in the desired functional group or structural unit whose introduction would be difficult otherwise following conventional reactivity or use of direct functional group manipulations. The two concepts of synthetic equivalent and latent functionality appear close knit. In this chapter the use of latent functionality will be dealt in relation to avoiding protecting groups. However, in total synthesis involving many steps, some protecting groups may still be involved; in that case the focus will be on other parts of the synthesis where a latent functionality is used, avoiding any further requirement of protecting-group manipulations. It is rather a visionary choice how a latent functionality is identified at first instance in a given complex molecule and used or carried forward over one or several steps and then unmasked to get the desired structural unit or functional group. Thus latent functionality in total synthesis has many times saved the use of masking group as this functionality would be in general stable to a variety of reactions and can be distinctly unmasked when required. The chapter consists of four sections covering the use of latent functionality for direct conversions, which may relate to one to two steps, silicon-centered latent functionality, latent functionality in total synthesis that is carried through three or more steps, and symmetry-based latent functionality considerations. The full chapter is not explicit literature on the titled topic, but most common latent functionalities are covered with examples.

9.2 Latent Functionality for Direct Conversions Using Short-Term Latent Groups

Many syntheses have been possible due to availability of synthons with different reactivity and reaction conditions that can alter their reactivity in a predictable manner. Functional groups that cannot be directly derived can be obtained indirectly through latent groups. In this section direct use of different synthetic equivalents for latent functional groups is discussed through representative examples that are documented in the literature. The latent groups are called short-term latent groups as they are carried over fewer reactions or steps before unmasking. These groups can also be visualized for use over multistep synthesis and bear potential to be exploited in future.

Xu and coworkers in their synthesis of S-(+)-XJP (**1a**) and R-(–)-XJP (**1b**) employed indirect acyloin equivalent to prevent possible isomerization at α-alkoxy position (Scheme 9.1) [1]. The earlier synthesis reported on these molecules observed substantial racemization [2]. (±)-XJP, an isochromanone, was isolated from the peel of *Musa sapientum* L. [3]. It exhibits antioxidative, anti-inflammatory, antihypertensive, and cardioprotective effects [4]. In their synthesis [1], compound **4** was prepared from commercially available **2** in five steps using **3a** or **3b**. An intramolecular Heck reaction under optimized conditions provided the desired methyleneisochroman **5** in 78–80% yields. The presence of methylene group prevented any isomerization at the alkoxy

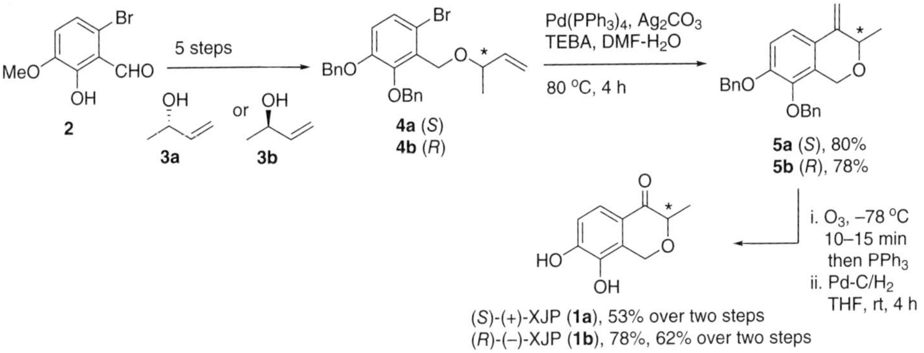

Scheme 9.1 Synthesis of *S*-(+)-XJP (**1a**) and *R*-(−)-XJP (**1b**) with allyl alcohol as acyloin equivalent.

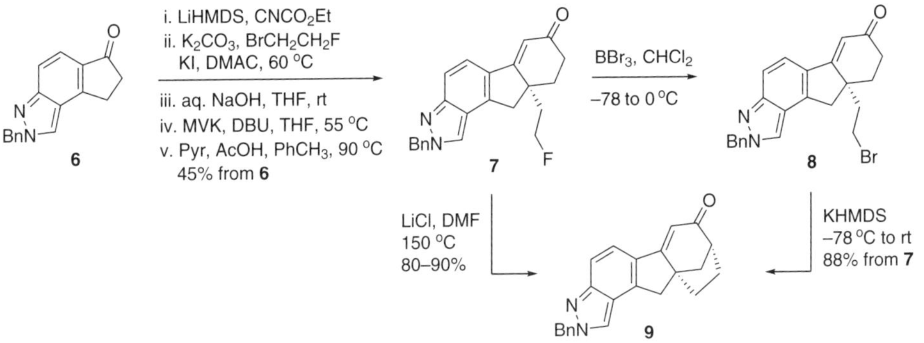

Scheme 9.2 Cyclization of fluoroalkyl pyrazolotetrahydrofluorenone **7**.

carbon and is the latent functionality to the desired acyloin moiety in the target molecule. The acyloin group was put in place through ozonolytic cleavage of the exomethylene group at −78 °C, and further debenzylation provided **1a** or **1b**. The enantiopurity of the final products was same as that of the chiral pool material **3a** or **3b**, indicating no racemization during introduction of the acyloin moiety.

Use of alkyl halides for alkylation normally involves I, Br, or Cl in decreasing order of reactivity. It is quite uncommon to use alkyl fluorides. These are less reactive and quite stable to a variety of conditions. They are, however, good alternatives to protected alkoxides, which would require deprotection and subsequent conversion to a leaving group for alkylation. The alkyl fluoride can be activated using boron trihalides when required [5]. Parker et al. [6] employed an alkyl fluoride unit for an intramolecular cyclization toward the synthesis of bridged fluorenones (Scheme 9.2). Pyrazoloindanone **6** upon alkylation with fluoroethyl bromide and then Robinson annulation provided tetrahydrofluorenone **7** in 45% overall yield. Halide exchange with BBr_3 gave **8**, and intramolecular cyclization furnished the bridged fluorenone **9**. Alternatively, treatment of **7** with LiCl/DMF at 150 °C directly provided **9** presumably through alkyl chloride as intermediate. Thus the fluoroalkyl group served as a latent bridging ring. It was less reactive during introduction as fluoroethyl bromide but was later activated at the appropriate time for intramolecular alkylation.

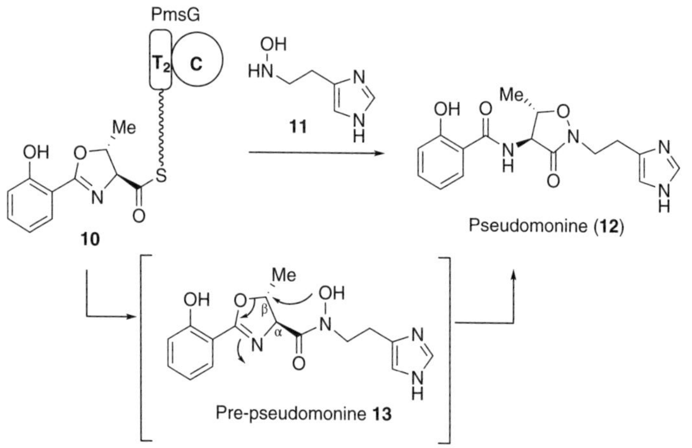

Scheme 9.3 Latent oxazoline electrophile in pseudomonine biosynthesis.

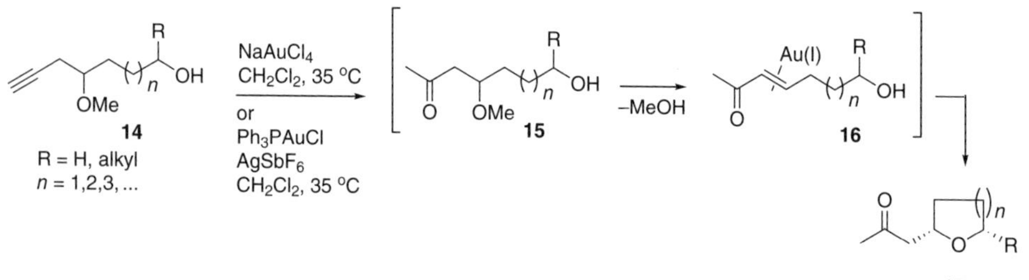

Scheme 9.4 Homopropargylic ethers as latent Michael acceptors.

In the *Pseudomonas* metabolite pseudomonine (**12**), the hydroxamate moiety is found uncommonly as an isoxazolidinone ring (Scheme 9.3). A latent oxazoline electrophile **13** is implicated for N—O—C bond formation wherein the oxygen of a hydroxamate intermediate serves as nucleophile [7]. Thus the isoxazolidinone of pseudomonine (**12**) is accessed through the rearrangement of intermediate named pre-pseudomonine **13** (an initial oxazoline hydroxamate condensation product). S$_N$2 attack on the β-carbon of Thr-derived oxazoline results in N—O—C bond formation, leading to oxazolidinone unit of pseudomonine (**12**). This mode of reactivity is supported by the relative stereochemistry inversion at the β-carbon of L-Thr-derived oxazoline; the pseudomonine isoxazolidinone bears a modified L-*allo*-Thr residue with substituents placed with *trans*-relative stereochemistry about the heterocycle.

Homopropargylic ethers have been efficiently utilized by Jung and Floreancig [8] as latent electrophiles (Michael acceptors). A pendant hydroxyl group is used for intramolecular Michael addition to the *in situ* generated α,β-unsaturated methyl ketones providing the oxygen heterocycles (Scheme 9.4). Thus tetrahydrofurans, tetrahydropyrans, and higher homologs **17** were synthesized by gold-mediated hydration of alkynes **14** to methyl ketones **15**, followed by MeOH elimination generating the α,β-unsaturated methyl ketones **16** for subsequent intramolecular Michael addition with the pendant hydroxyl group. The cyclic ethers produced are important moieties present in many natural products [9].

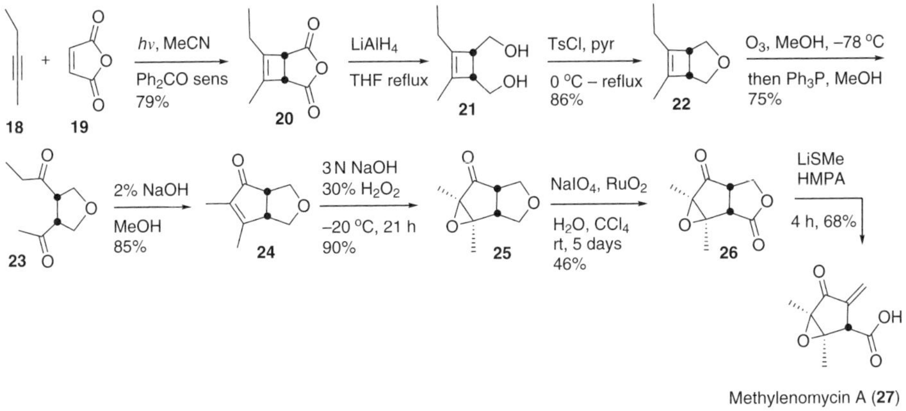

Scheme 9.5 Cyclobutene as latent functionality for cyclopentenone synthesis.

Smith et al. [10] employed cyclobutene obtained by [2+2] cycloaddition as a latent functionality to unmask it as highly substituted cyclopentenone in the total synthesis of methylenomycin A (27;Scheme 9.5) without involvement of any protecting groups. Methylenomycin A is isolated from the culture broth of *Streptomyces violaceoruber* [11] and is active against Gram-positive and Gram-negative bacteria *in vitro*. A [2+2] cycloaddition of butyne 18 with maleic anhydride 19 gave the adduct 20. Further the LiAlH₄ reduction provided the cyclobutene diol 21. Tetrahydrofuran formation (22) and cleavage of olefin gave the diketone 23. The latter on intramolecular aldol condensation furnished the desired cyclopentenone 24. The reaction was regioselective in giving the tetrasubstituted olefin rather than the other possible trisubstituted isomer. Stereoselective epoxidation of the less hindered convex face led to epoxide 25, and subsequent introduction of carbonyl group regioselectively using RuO₂/NaIO₄ afforded a single lactone 26. Opening of the lactone with lithium thiomethoxide in HMPA directly gave methylenomycin A (27) in 68% yield. Thus the desired latent cyclopentenone was easily derived through the use of cyclobutene intermediate.

A suitably placed cyclobutene ring served as a latent furan moiety in the Koft and Smith synthesis of hibiscone C (32, gmelofuran; Scheme 9.6) [12]. This is opposite to

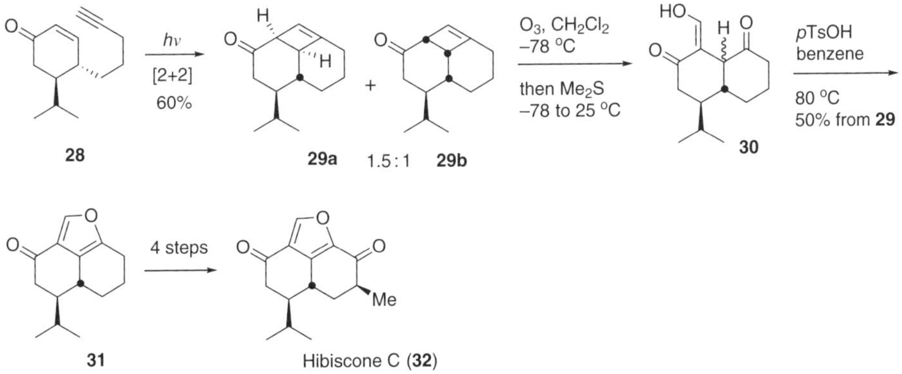

Scheme 9.6 Cyclobutene as latent furan moiety in the synthesis of hibiscone C (32).

the use of furan ring as 1,4-dicarbonyl equivalent in many synthesis (discussed in a later part of this chapter, Section 9.4). An intramolecular [2+2] cycloaddition of enone moiety with alkyne in **28** led to a mixture of cyclobutenes **29a** and **29b** (1.5 : 1). Ozonolytic cleavage and reduction led to a mixture of epimers **30**. Cyclization–dehydration provided the furan compound **31**. This was converted in few steps to hibiscone C. Thus the cyclobutene obtained by [2+2] cycloaddition served as latent furan moiety in hibiscone C (**32**).

The well-known Ramberg–Bäcklund rearrangement [13] of dialkylsulfones can be viewed as latent functionality for olefinic bonds. The method is powerful to generate sterically hindered, tetrasubstituted, and strained double bonds in cyclic systems. An excellent example of this chemistry is the synthesis of β-carotene (**35**; Scheme 9.7) [14a]. Retinol (**33**) in two steps was converted into sulfone **34** as a mixture of olefin isomers. This on LDA treatment gave the dianion, which on reaction with iodine gave a mixture of stereoisomeric β-carotenes. Thermal- or iodine-catalyzed isomerization led to all *trans*-β-carotene (**35**).

Paquette and coworkers [14b] had employed the previous rearrangement in the synthesis of strained propellane-type compounds. For example, the sulfone **36** or **38** on Ramberg–Bäcklund rearrangement gave the [4.4.2]propella-3,11-diene **37** in 56 and 47% yields, respectively (Scheme 9.8). The cyclobutene ring formation was independent of the configuration of **36** or **38**.

There are many such examples of synthesis of strained systems reported in the literature [13e, f]. The use of Ramberg–Bäcklund rearrangement in the end steps of the total synthesis of (+)-eremantholide A (**45**) is noteworthy (Scheme 9.9) [15]. The synthesis started from lactolide **39**, which was converted into lactone **40** in eight steps. Further seven steps were required to install the third ring, leading to **41**. The free hydroxyl was converted into mesylate and displaced with iodide. Further allylic bromination provided bromoiodide **42**. This was cyclized using TMS₂S to give 10-membered cyclic sulfide that was further oxidized to the sulfone **43**. The Ramberg–Bäcklund rearrangement

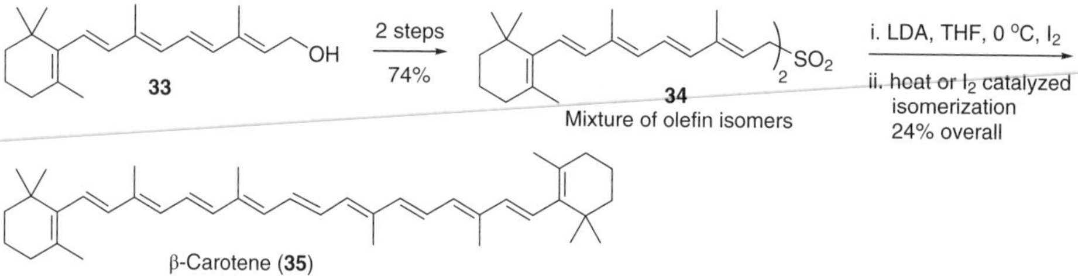

Scheme 9.7 Dialkylsulfones as latent olefin functionality in β-carotene (**35**) synthesis.

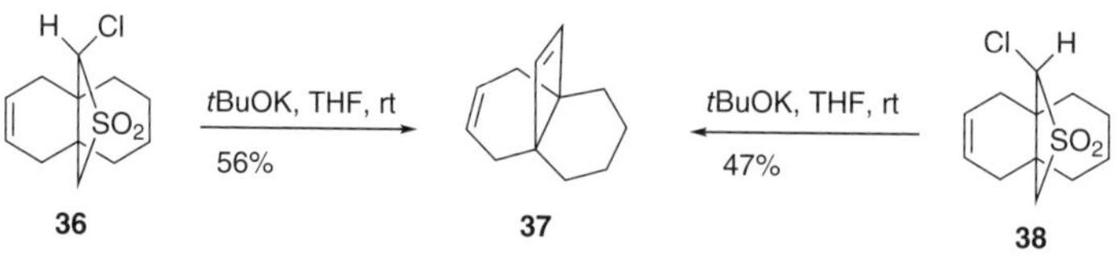

Scheme 9.8 Cyclic sulfones to generate strained cyclobutene.

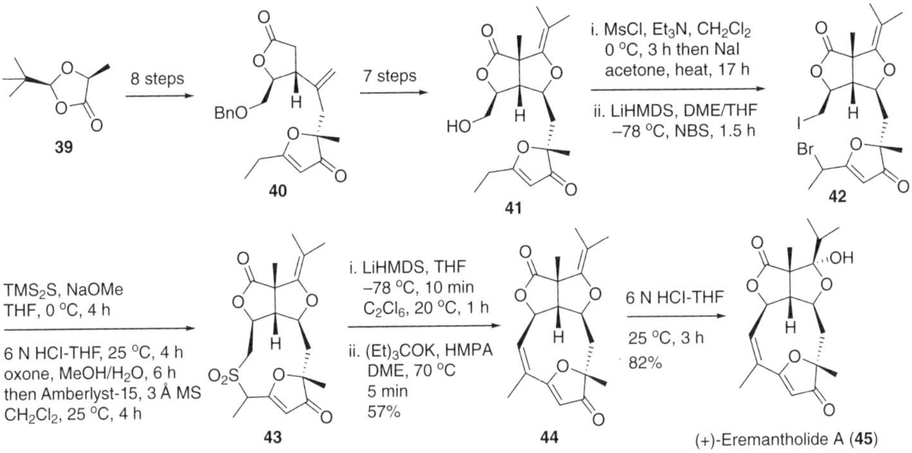

Scheme 9.9 Cyclic sulfones in ring-contracting olefination for eremantholide (**45**) synthesis.

was initiated on **43** by first kinetic α-chlorination at the more substituted side of the sulfone followed by extrusion of sulfur, giving the ring-contracted olefin **44** in 57% yield. The latter on treatment with 6 N HCl delivered eremantholide A (**45**) in 82% yield.

9.3 Silicon-Centered Latent Functionalities

Silicon-tethering and silicon-based latent functionalities are widely used [16]. These could be short term or long term and carried through several steps due to the stability of silicon center to many reactions. The tethering, a well-established strategy, offers intramolecular reactions to be carried out for various synthetic transformations. This has advantageous implications on both reactivity and selectivity in organic synthesis. The tether in many instances is stable, can be carried forward through a wide range of reactions, and is also removed or converted into other functional groups under specific conditions. This constitutes the concept of latent functionality involving unmasking of silicon tether at appropriate stage, giving diverse products. Hence we considered a separate section for their applications.

A γ-silane on allylic halides serves as latent proton equivalent enabling stereospecific alkylation, which otherwise is prone to competing S_N2' reactions especially with strong nucleophiles [17]. If the silyl group could be removed or replaced by other groups, a range of other functionalities can be achieved [18]. Kang and coworkers [17] developed this chemistry as shown in Scheme 9.10. Reaction of vinylsilane **46** with dibutylcuprate gave **47** as the only product. Similarly the reaction with diethylmalonate led to **48**. However the reaction with butylcopper reagent gave exclusively S_N2' product **49**. Thus there was a sharp distinction in reaction path with dialkylcuprates and alkylcopper reagents. The silyl center in **47**, **48**, or **49** can be converted to various other functionalities or replaced with hydrogen [18].

Ehlinger and Magnus [19] employed the (trimethylsilyl)allyl anion **51** as latent β-acyl anion equivalent having γ-mode of ambient reactivity (Scheme 9.11). This was efficiently added to aldehydes or ketones to give δ-hydroxy vinylsilane adducts. The latter can be converted into lactones through epoxidation, $BF_3 \cdot OEt_2$-mediated lactol

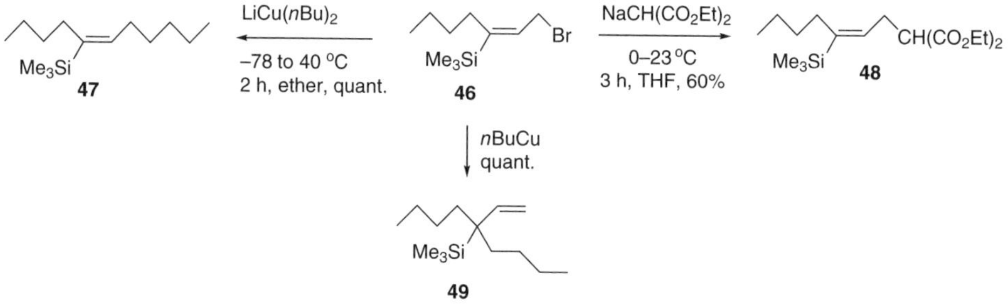

Scheme 9.10 γ-Silane on allylic halides as latent proton equivalent in stereospecific alkylation.

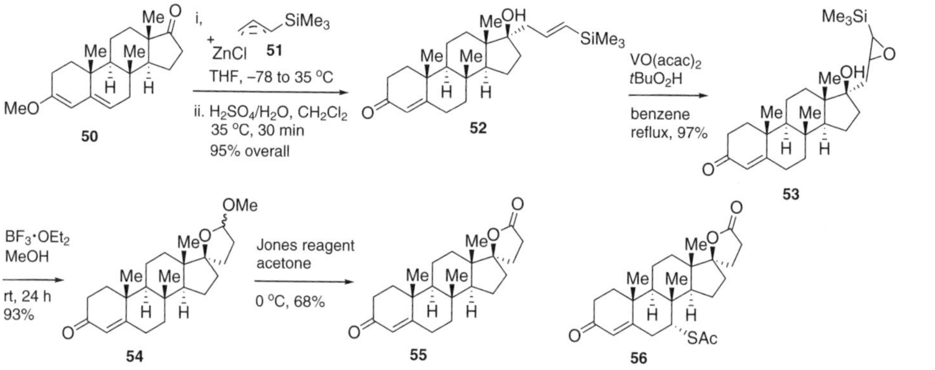

Scheme 9.11 (Trimethylsilyl)allyl anion **51** as latent β-acyl anion equivalent.

formation, and oxidation. This chemistry is exemplified by the addition of (trimethylsilyl)allyl anion **51** to 3-methoxyandrosta-3,5-dien-17-one **50** to provide the adduct **52** (Scheme 9.11). The usual mCPBA reaction for epoxidation failed in this case due to competing Baeyer–Villiger oxidation of ring A enone. The Sharpless procedure using VO(acac)$_2$/tBuO$_2$H in benzene gave the epoxide **53**. This upon reaction with BF$_3$·OEt$_2$ in MeOH gave the lactol O-methyl ether **54**, and subsequent Jones oxidation provided the lactone **55**. The latter is an important precursor to spironolactone **56**, an aldosterone blocker [20]. Thus the (trimethylsilyl)allyl anion worked efficiently as latent β-acyl anion equivalent, and the spirolactone could be easily assembled through easy manipulation from the starting ketone.

Paquette and Han [21] utilized similar chemistry as before using vinylsilane as latent aldehyde functionality in the efficient synthesis of (±)-gymnomitrol (**62**; Scheme 9.12). This avoided the use of protected hydroxyl compound as aldehyde precursor. Addition of Grignard reagent form (E)-2-bromovinyltrimethylsilane under Cu catalysis to compound **57** followed by α-methylation provided compound **58**. MCPBA treatment gave the epoxide **59**, which on acid-catalyzed hydrolysis provided the aldehyde **60** along with small amount of **61**. Intramolecular aldol reaction of **60** with 2% KOH in MeOH gave **61**. Further synthetic manipulation of **61** over four steps provided gymnomitrol (**62**). Thus vinylsilane served as latent aldehyde functionality, which was unmasked in the presence of keto carbonyl without the need of any blocking procedures.

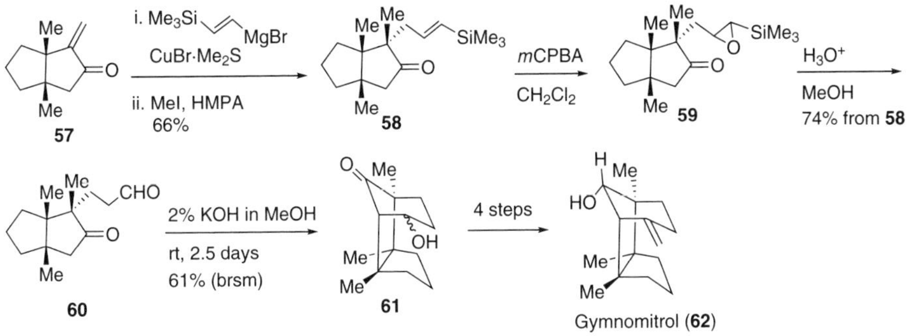

Scheme 9.12 Vinylsilane as latent aldehyde group in the synthesis of gymnomitrol (**62**).

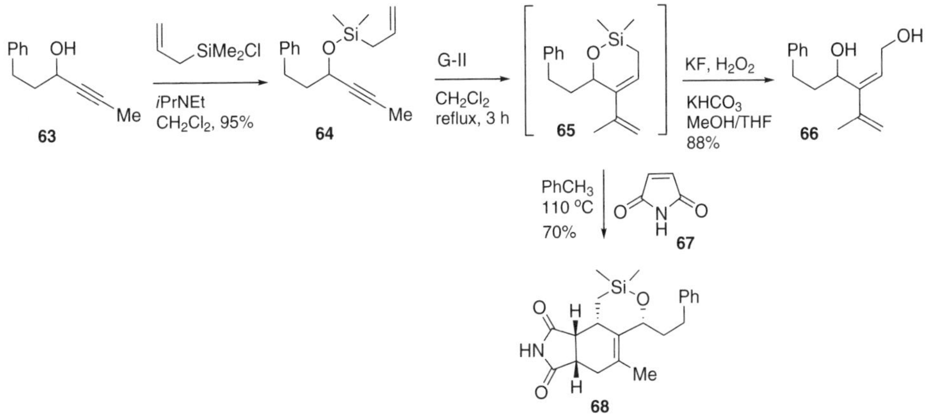

Scheme 9.13 Silicon-tethered enyne metathesis for latent diene synthesis.

Yao explored the ring-closing metathesis of silicon-tethered enynes to give intermediate cyclic siloxanes, which upon Tamao oxidation afforded conjugated dienes with 1,4-dihydroxyl groups [22]. The tethered silicon served as latent hydroxyl group and the enyne as conjugated diene (Scheme 9.13). Thus the alkynol **63** on silicon tethering provided the enyne **64**. The latter on ring-closing metathesis provided the siloxane **65**, which on Tamao oxidation with KF and H_2O_2 gave the diene **66**, with 1,4-dihydoxyl groups in 88% yield. It was also possible to employ the intermediate diene–siloxane **65** for Diels–Alder reaction to furnish the adduct **68** in 70% yield. The siloxane **68** could be further functionalized similar to **66**.

Kim and Lee [23] employed a tandem ring-closing metathesis of silaketal-tethered dienyne **75** to construct the C1–C21 linear carbon skeleton **77** of tartrolon B based on the intermediate bicyclic siloxane **76** as latent functionality to unmask the stereodefined E,Z-1,3-diene unit in **77** (Scheme 9.14). The fragment **70** was prepared from aldehyde **69** in nine steps. Further an optimized aldol reaction using (+)-DIPCl with 4-pentenal gave **71** as 6 : 1 epimer mixture at C-11. The silaketal formation with compound **71** led to β-hydroxy elimination, giving α,β-unsaturated ketone. Hence the β-hydroxy ketone **71** was converted into β-acetoxy alcohol **72** and then into C-11 hydroxy compound **73**. Silaketal formation with compound **74** gave silicon-tethered compound

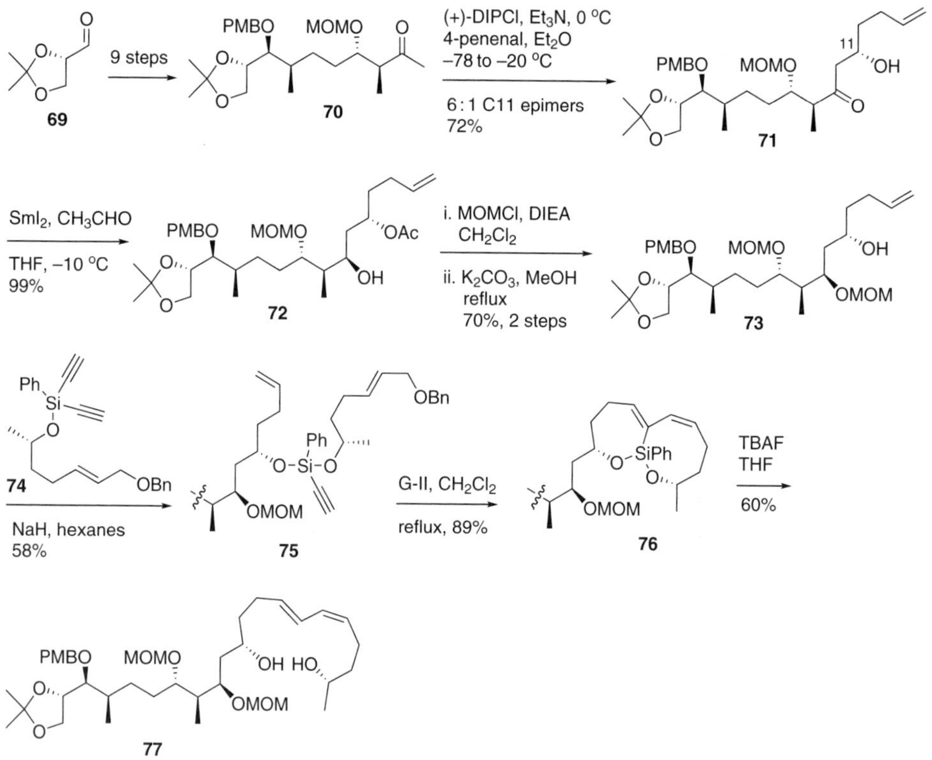

Scheme 9.14 Silicon-tethered dienynes for latent diene synthesis.

75. The tethered olefin fragments were chosen that the most accessible terminal alkene would react for enyne ring closure, giving a seven-membered ring followed by the eight-membered ring closure to provide bicycle **76**. Removal of the silicon tether unmasked the stereodefined *E,Z*-1,3-diene unit with 1,10 placement of desired hydroxyl groups and completing the synthesis of C1–C21 skeleton **77** of dimer molecule tartrolon B.

Intramolecular hydrosilylation of alkenes is a technique to introduce hydroxyl functionality. The suitably placed olefin serves as latent hydroxyl functionality. Tamao et al. [24a] introduced hydromethylsilyl group into homoallyl alcohol **78** using 1,3-dihydro-tetramethyldisilazane to form silyl ether **79** (Scheme 9.15). The intramolecular hydrosilylation of olefin **79** in the presence of $H_2PtCl_6 \cdot 6H_2O$ gave the five-membered siloxane, which was then oxidized with H_2O_2 to form 1,3-diol **81** in high yields. The same was extended to allyl alcohol **82**, which gave *endo*-ring closure (to **83**) unlike *exo* for homoallyl alcohol, though the final product is 1,3-diol in both cases [24b]. A repeat reaction on diol **84** provided the triol **85** with good *erythro* selectivity. In the case of allyl amine **86**, a similar reaction gave 1,2-amino alcohol **89** via the intermediate four-membered 1-aza-2-silacyclobutane compound **88** [24c].

The silicon tethering has been explored by Leighton's group for an efficient generation of 1,5-dihydroxyl moiety with intramolecular transfer of allyl group to the intermediate aldehyde obtained by tandem alkyne silylformylation (Scheme 9.16) [25a]. Thus, homopropargyl alcohols could serve as latent 1,5-diols to be obtained with good

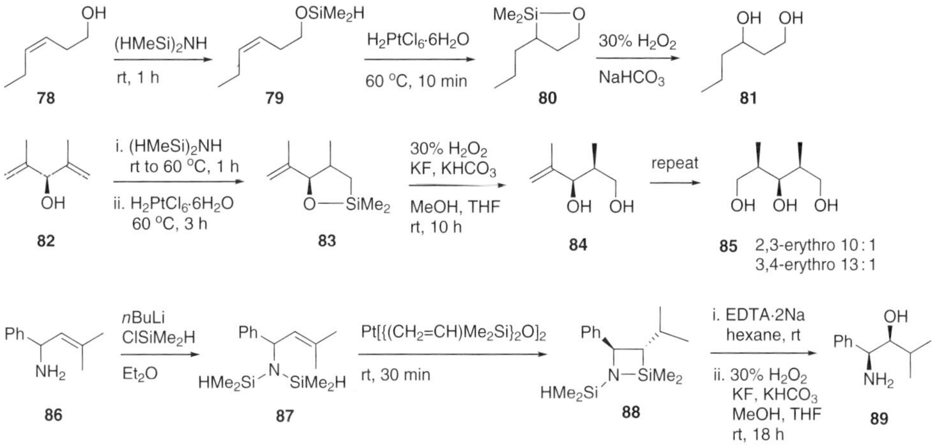

Scheme 9.15 Hydrosilylation to introduce latent hydroxyl group.

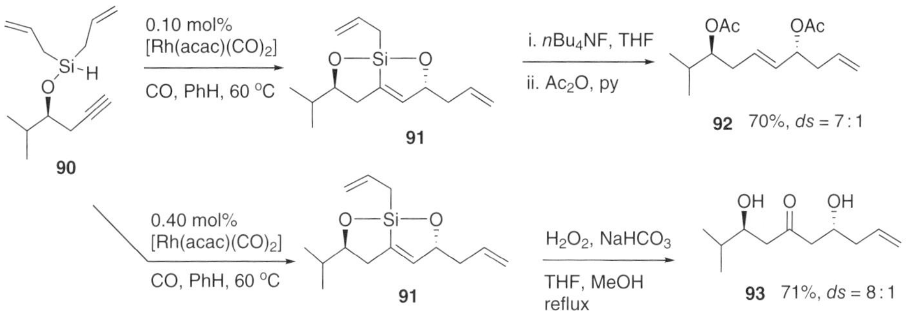

Scheme 9.16 Tandem alkyne silylformylation/allylation.

anti-1,5-diastereoselectivity. For example, the tandem alkyne silylformylation and allylation of **90** provided **91**. Protodesilylation and acetylation gave the 1,5-diacetate **92** in 70% overall yield in good 7 : 1 *anti*-diastereoselectivity. Tuning of protodesilylation of **91** led to 3-keto-1,5-diol **93** in 8 : 1 diastereoselectivity [25b].

An alkyl/aryl silyl carbon-centered group can be progressed through a series of reactions due to its stability under various conditions and can be unveiled as latent hydroxyl group [26a]. In the total synthesis of decahydroquinoline alkaloids, Polniaszek and Dillard [26b] employed dimethylphenylsilyl group for latent hydroxyl functionality (Scheme 9.17). The 4-methoxypyridine **94** was progressed to **95** in five steps. Hydrolysis of ketal and silylvinyl lithium addition stereoselectively gave alcohol **96**. The anionic oxy-Cope rearrangement of **96** provided **97**, which upon hydrogenation gave **98**. During the hydrogenation the C2 center isomerized to the more stable compound **98**. The fluoroboric acid-promoted desilylation gave the labile fluorosilane, which underwent rapid hydrolysis giving silanol. This on H_2O_2 treatment gave the hydroxyl compound **99**. Thus, the stable silyl group was unmasked as hydroxyl functionality in **99**, which also bears a keto group, whose introduction would otherwise require protecting-group manipulations.

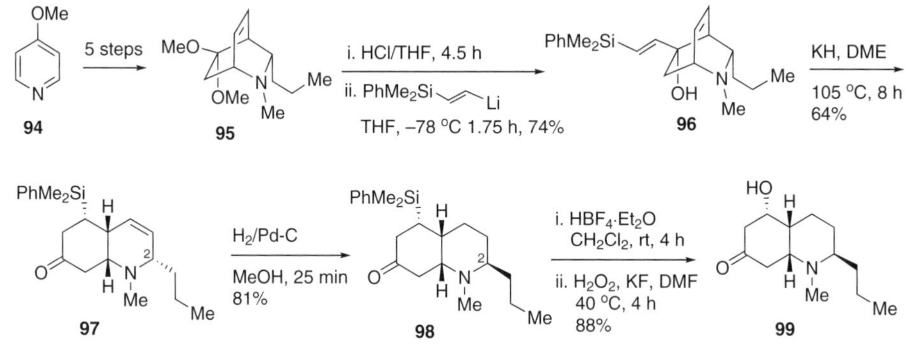

Scheme 9.17 Alkyl/aryl silyl as latent hydroxyl group.

9.4 Latent Functionality in Total Synthesis (Long-Term Latent Groups)

The use of latent functionality carried through multistep sequence comes through a strategic design in total synthesis. The latent group masked in some form should be stable to multiple reactions and easily unmaskable to the desired function at the appropriate stage in the synthesis. This section unfolds some of the latent groups carried through longer sequences and normally uncovered at end stages of the total synthesis.

In the synthesis of (±)-coronafacic acid (**109**), the direct Diels–Alder approach for the bicyclic structure has been difficult due to the need to construct the labile diene and dienophile moieties (Scheme 9.18). Ichihara and coworkers [27] developed a unique strategy based on thermal reaction of latent diene–dienophile moieties, which were

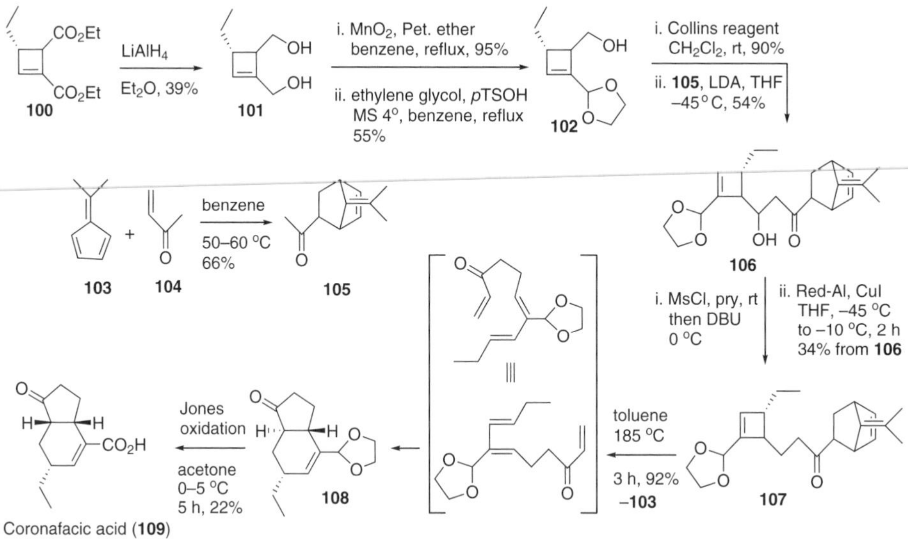

Scheme 9.18 Synthesis of coronafacic acid (**109**) using latent diene–dienophile for Diels–Alder reaction.

masked as thermally labile cyclobutene and methyl ketone Diels–Alder adduct (**107**; Scheme 9.18). The LiAlH$_4$ reduction of diester **100** provided diol **101**, which on selective oxidation gave monoaldehyde. This was converted into acetal **102**. Methyl ketone **105** was prepared by Diels–Alder reaction of dimethyl fulvene **103** and methyl vinyl ketone **104**. The aldol condensation of **105** with the aldehyde from **102** delivered **106**. The elimination of hydroxyl and selective reduction of double bond gave the latent diene–dienophile compound **107**. Heating this in toluene beautifully unmasked the diene by conrotatory opening of the cyclobutene ring and the dienophile by retro-Diels–Alder opening with extrusion of fulvene **103**. The immediate intramolecular Diels–Alder (IMDA) reaction furnished single product **108** in 92% yield. Further Jones oxidation resulted in deacetalization, isomerization, and oxidation to give coronafacic acid (**109**).

An early example of using furan ring as latent 1,4-diketone was elegantly demonstrated by Büchi and Wüest [28] in the synthesis of *cis*-jasmone (**114**, Scheme 9.19). Addition of 2-lithio-5-methyl furan **110** to (*Z*)-olefin bromide **111** gave **112**. Further, furan ring hydrolysis led to the desired 1,4-diketone **113**. This on intramolecular aldol reaction delivered *cis*-jasmone (**114**) in overall 40–45% yield.

Kishi et al. used a furan moiety that served as latent ester group that was unveiled in the end steps of the synthesis [29a]. In their work on monensin [29b, c], a fragment **121** was synthesized as shown in Scheme 9.20. 2-(2-Furyl)-propionaldehyde **115** on Wittig

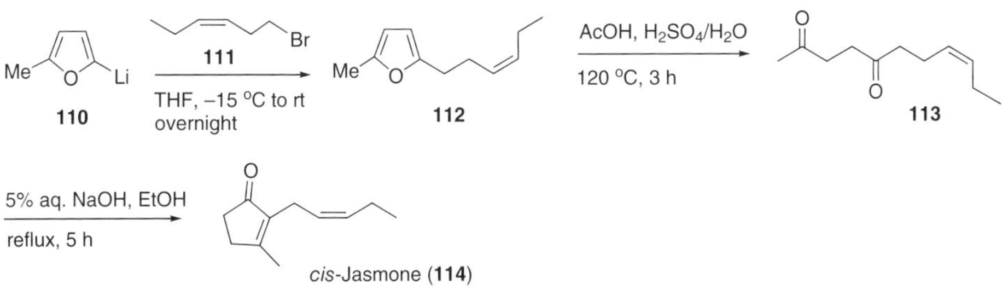

Scheme 9.19 Furan ring as latent 1,4-diketone **113**.

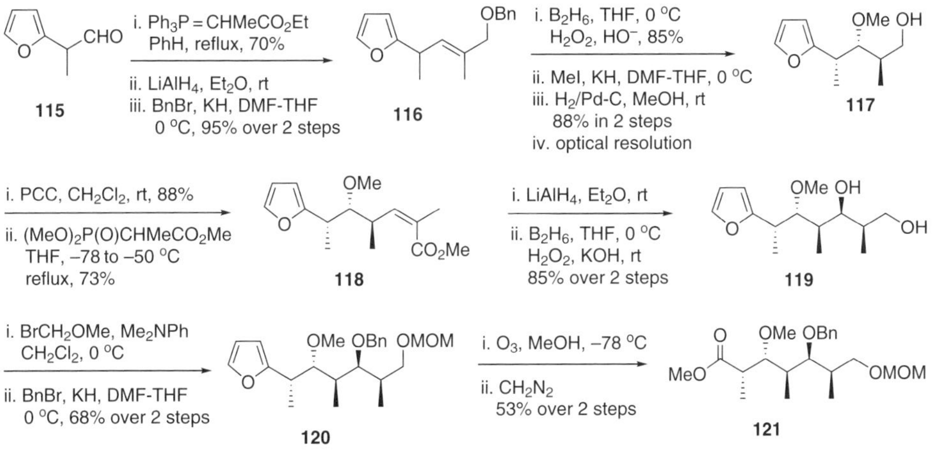

Scheme 9.20 Furan as latent ester functionality.

olefination, LiAlH$_4$ reduction, and benzyl protection gave **116**. Further the sequence of hydroboration–oxidation, methylation, debenzylation, and optical resolution of diastereomers provided **117**. Oxidation and *cis*-Wittig olefination gave **118**. Further ester reduction and hydroboration–oxidation stereoselectively led to diol **119**. Orthogonal protection of both hydroxyls gave **120**. The ester group was now unveiled by ozonolytic cleavage of furan moiety and esterification, giving the fragment **121** for monensin synthesis [29a]. Thus the ester moiety was carried forward in masked furan form without the necessity of using protecting groups to ester precursors like alcohol or acid.

The furan ring is stable under various reaction conditions and can be oxidized at the right stage to unmask as 1,4-dicarbonyl equivalent, as discussed earlier. In the total synthesis of macrosphelides A and B, Kobayashi et al. employed 2-substituted furan as latent 4-oxo-2-alkenoic acid [30]. Esterification of **122** with alcohol **123** gave **124** (Scheme 9.21). This was used again to esterify acid **125** (also prepared by furan oxidation from **123**) to produce bis-ester **126**. This was converted in three steps to **127** and then subjected to furan oxidation to lead to 4-oxo-2-alkenoic acid **128**. The latter on macrocyclization and MOM deprotection gave macrosphelide B (**129**). Similarly, **128** was converted into macrosphelide A (**130**) in five steps. Thus, the required 4-oxo-2-alkenoic acid moiety in **128** could be derived through furan oxidation in the presence of other hydroxyl groups.

Similarly, the oxazole moiety is carried through several steps before unmasking as latent carboxyl function toward the synthesis of oasomycin A by Evans and coworkers [31]. The aldehyde **131** obtained from 4,5-diphenyloxazole was carried through 13 steps to obtain **132** (Scheme 9.22). This sulfone on Kocienski–Julia olefination reaction

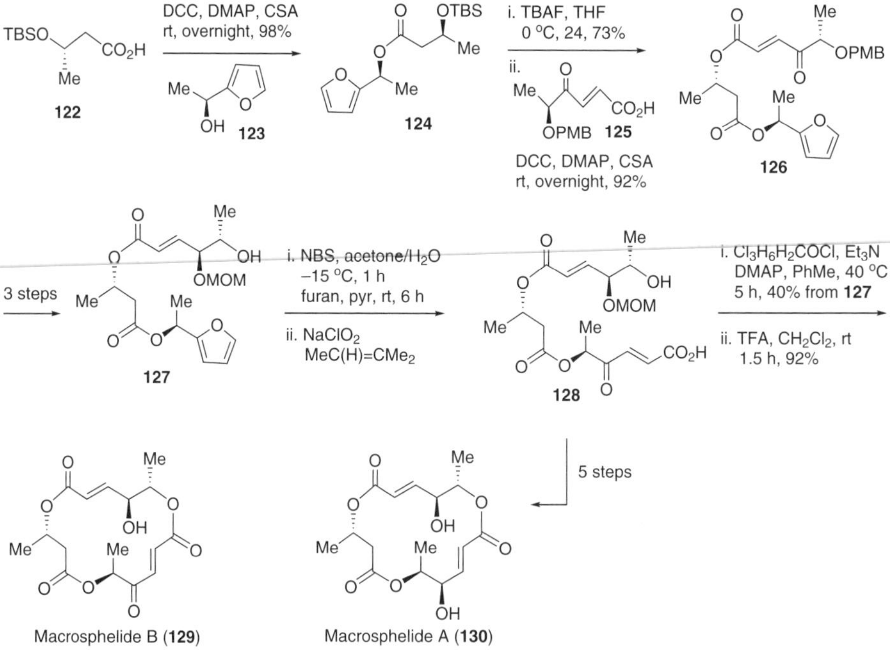

Scheme 9.21 Furan as latent 4-oxo-2-alkenoic acid in total synthesis of macrosphelides A and B.

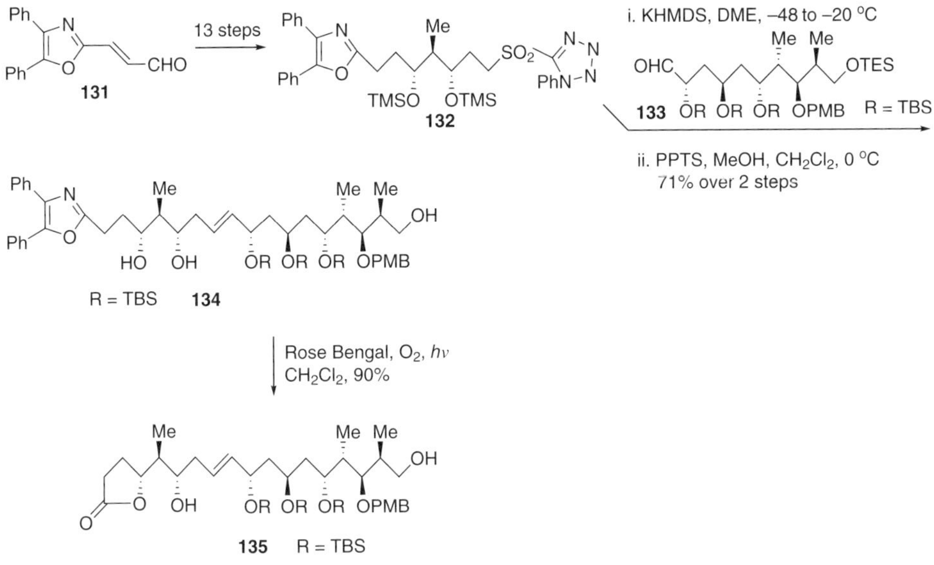

Scheme 9.22 Oxazole as latent carboxyl function.

with aldehyde fragment **133** under Barbier conditions followed by TMS and TES deprotection gave olefin compound **134**. At this stage the latent carboxyl function was unmasked by singlet-oxygen oxidation of the 4,5-diphenyloxazole moiety to give the lactone **135** through simultaneous lactonization. It is remarkable that the diphenyloxazole was stable under several reaction conditions (over 15 steps) and carried the masked carboxyl function. The compound **135** is an advanced intermediate in the synthesis of oasomycin A.

A suitably substituted methoxy phenyl group has been used to generate latent *p*-benzoquinone dienophile for IMDA reaction in the elegant total synthesis of elisapterosin B by Rawal and coworkers [32]. The coupling of acyl chloride **136** with the Grignard reagent followed by ketalization led to compound **137** (Scheme 9.23). This was converted over seven steps to the vinyl bromide **138**. Cross-coupling with (*E*)-2-bromopropene generated the desired diene **139**. Further DIBAL-H reduction and Wittig olefination gave triene **140**. Demethylation and salcomine-mediated oxidation provided the latent benzoquinone **141** for IMDA, generating the angular tricyclic compound **142**. This was converted in two steps to elisapterosin B (**143**). Thus the reactive benzoquinone dienophile was carried masked as methoxy phenyl group in the synthesis.

In many steroid-type compounds the enone moiety in A ring is carried in masked form as methoxy phenyl group. This is unmasked at the end to generate the latent cyclohexenone unit. The methoxy phenyl group is robust enough to be carried untouched through many reaction steps. This is exemplified by the synthetic work on alnusenone by Ireland and coworkers (Scheme 9.24) [33]. The β-tetralone compound **144** on Robinson's annulation with vinyl ketone **145** gave **146** in 72% yield. The latter was progressed to pentacyclic diether **147** through four steps. Selective demethylation followed by Birch reduction and *in situ* remethylation gave cyclohexenone **148**. This was converted over seven steps to the tetramethylphosphorodiamide derivative **149**.

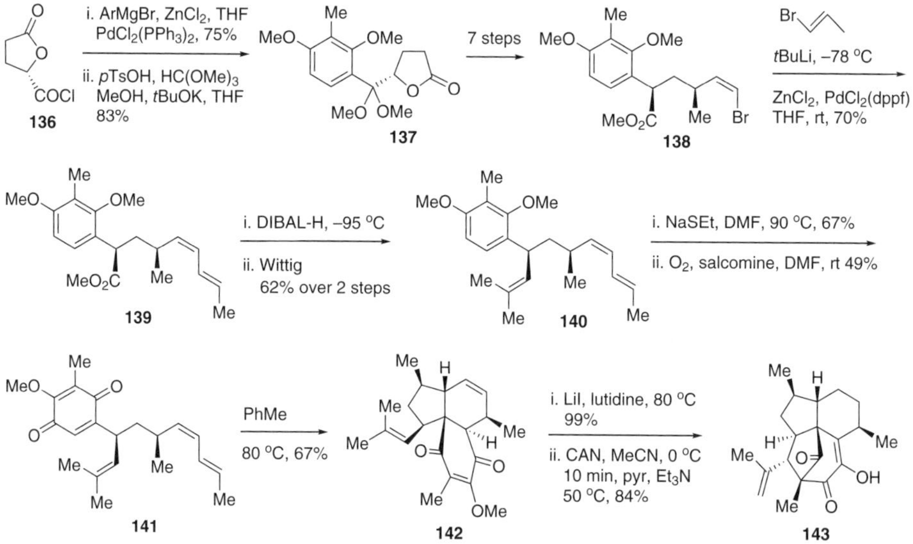

Scheme 9.23 Methoxy phenyl group as latent *p*-benzoquinone.

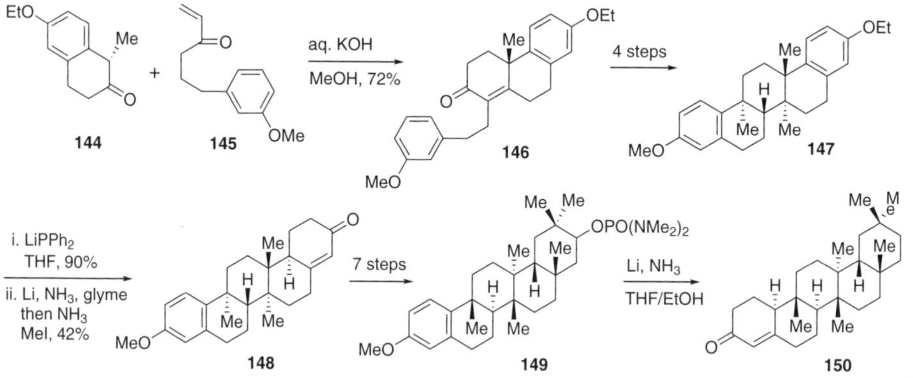

Scheme 9.24 Methoxy phenyl as latent A ring cyclohexanone in steroids.

The complete reduction by removing the phosphorodiamide and unmasking of cyclohexenone A ring was achieved under Birch reduction conditions, giving compound **150**. This was then converted in one step to alnusenone.

Similarly, the α-picolyl group serves as a latent cyclohexenone unit (Scheme 9.25). In the total synthesis of (±)-D-homosterone, Danishefsky et al. [34] carried out Michael addition of enone **152** to methyl-2-vinylpyridine **151** to give **153** after acidic work-up (Scheme 9.25). Further ketone and olefin reduction and ketalization of remaining keto group led to **154**. This on Birch-type reduction and treatment with aq. NaOH gave the cyclohexenone **156** after acidic work-up for deketalization. The reaction proceeds through the two-enamine intermediate **155**, which upon hydrolysis gave the 1,5-diketone and after intramolecular aldol reaction provided the cyclohexenone **156**. This was then oxidized to triketone and subjected to annulation reaction giving dienone **157**,

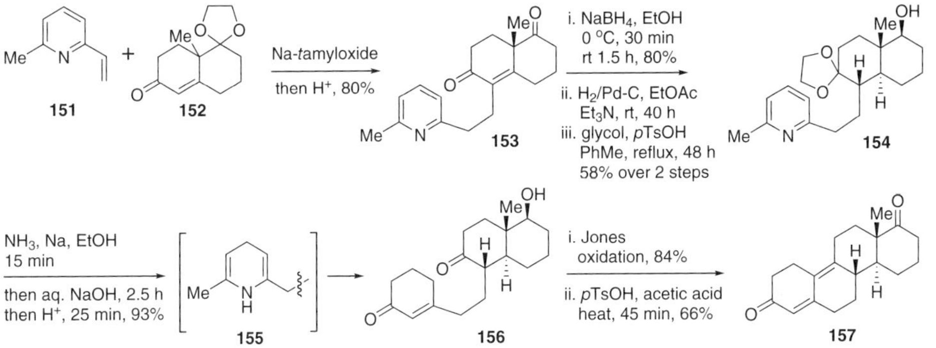

Scheme 9.25 α-Picolyl group as latent A ring cyclohexenone.

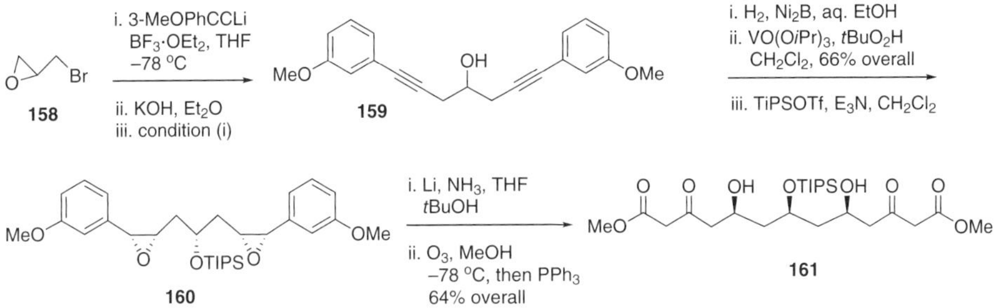

Scheme 9.26 Methoxy phenyl as latent β-ketoester functionality.

which was then used to prepare (±)-D-homosterone. It is notable that the methyl pyridine unit gives the 1,5-diketone and this undergoes regioselective aldol condensation to give cyclohexenone.

An excellent use of *o*-methoxy phenyl group to unravel the latent β-ketoester functionality was demonstrated by Wang and Deschênes [35]. Sequential homologation of epibromohydrin **158** with Li-3-MeO-phenylacetylide provided the carbinol **159** (Scheme 9.26). Hydrogenation followed by stereoselective epoxidation gave bisepoxide, which upon TIPS protection led to **160**. The latter on dissolving metal–NH$_3$ reduction and ozonolysis beautifully unmasked the β-ketoester functionality, giving **161**. This can be further elaborated to afford the all *syn*-pentanol compound by chelation controlled keto reduction.

The indoline ring is present in many *Strychnos* alkaloids (A and B rings) and can be arrived at by reductive cyclization of an aminophenyl compound with a suitably placed keto group. Carrying a free amine over several steps involving construction of other rings is a daunting task and therefore not explored. However use of the nitrophenyl group is gaining importance as a latent aminophenyl moiety as the indoline ring can be unmasked through catalytic hydrogenation of the nitro group and subsequent reductive cyclization with appropriate keto group. Bonjoch and coworkers employed this strategy in the total synthesis of many *Strychnos* alkaloids [36] (Scheme 9.27). The prochiral diketone **162** on ozonolysis and reductive amination gave a mixture of *cis*- and *trans*-**163**. The latter was converted into *cis*-**163** by a three-step sequence. This was

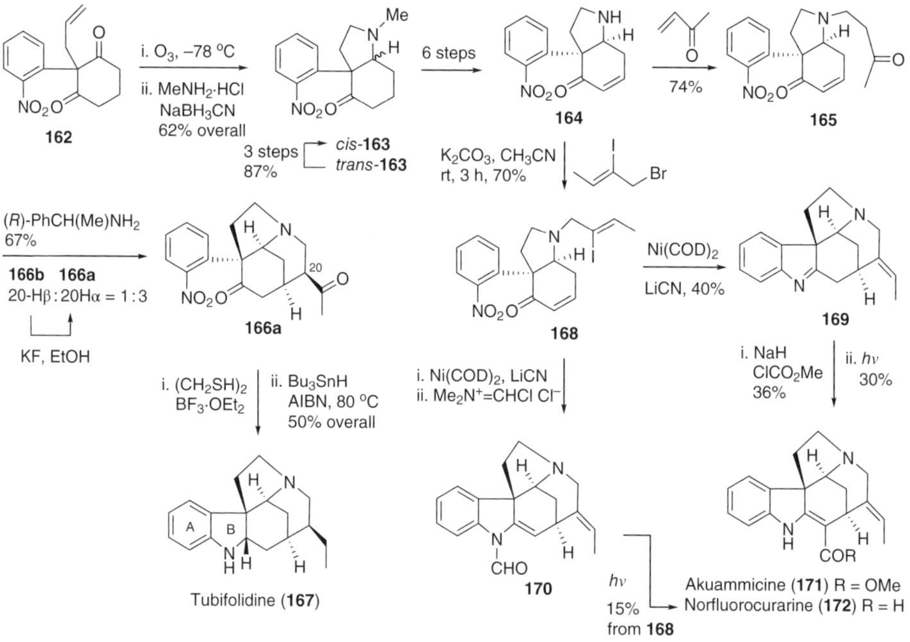

Scheme 9.27 Nitrophenyl ketone as latent indoline or AB ring of *Strychnos* alkaloids.

then carried forward over six steps, leading to enone **164**. Michael addition of **164** to methyl vinyl ketone gave **165**. This on second conjugate addition using α-methylbenzylamine provided a mixture of **166**, 20-Hβ : 20Hα = 1 : 3. The minor epimer **166b** was converted into **166a** using KF-EtOH. Further conversion of **166a** to dithioacetal and Bu₃SnH-mediated reduction led to tubifolidine (**167**). This conversion involved reductive removal of dithiane, nitro to amino group, and intramolecular reductive amination to form the B ring. Similarly, the alkylation of **164** with (Z)-1-bromo-2-iodo-2-butene gave **168**. A tandem Ni(0)-promoted cyclization of the vinyl iodide on the C=C and the reductive cyclization led to the imine compound **169**. Treatment of this with methyl chloroformate gave the *N*-methoxycarbonyl enamine, which on photoisomerization gave akuammicine (**171**). A direct reaction of **168** with Ni(0)/LiCN and trapping of the intermediate with (chloromethylene)dimethyliminium chloride provided the *N*-formyl enamine **170**. This on photoisomerization gave norfluorocurarine (**172**).

A concise synthesis of cephalosporolides E and F was reported by us [37a] using olefin unit as latent keto function. A similar strategy at the same time was explored by Du et al. [37b] toward the synthesis of cephalosporolides H and I. The keto function was unveiled through regioselective Wacker-type oxidation of olefin bond. In the synthesis of cephalosporolides E and F, the lactone **174** was prepared in a one-pot procedure from L-mannonic-γ-lactone (**173**; Scheme 9.28). The olefin alcohol **175** was prepared in one step from commercially available (R)-propylene oxide and allylmagnesium chloride. The cross-metathesis of **174** with **175** gave **176** in 78% yield. The keto group was unveiled through regioselective Wacker-type oxidation of **176**, wherein simultaneous ketal formation by free hydroxyl groups led efficiently to separable diastereomers of cephalosporolides E (**177**) and F (**178**). Thus the keto group could be carried in masked

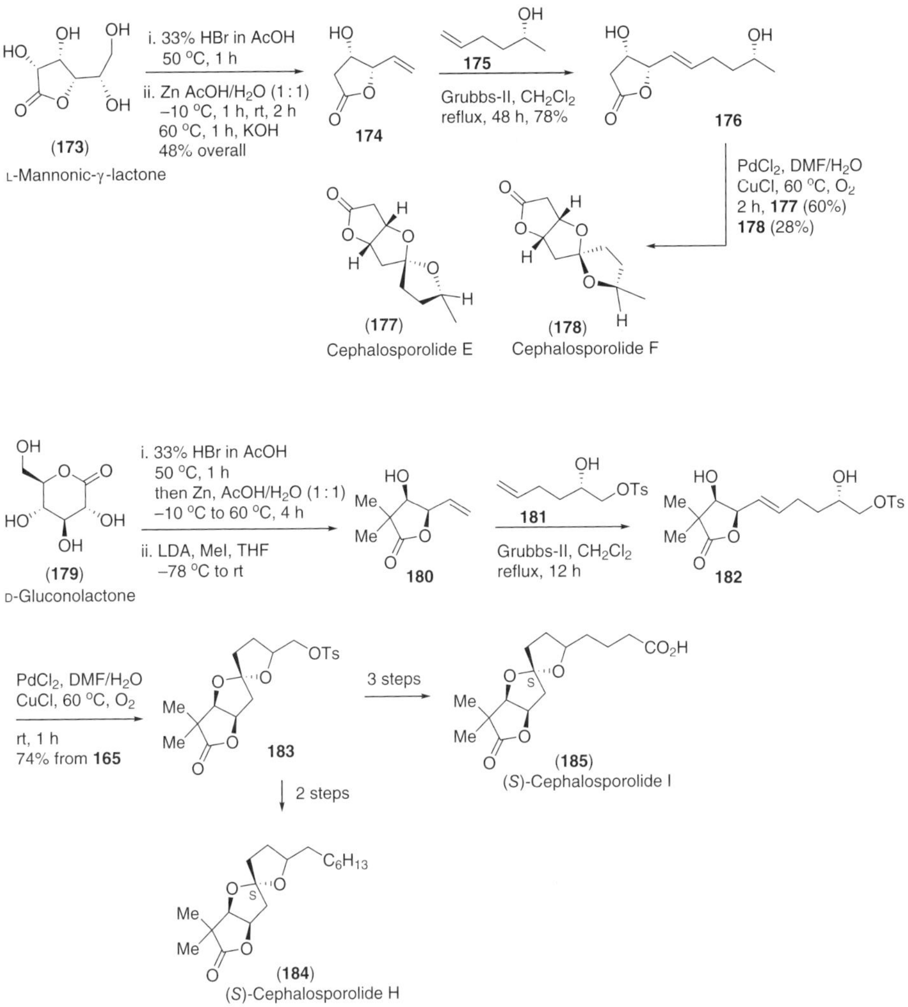

Scheme 9.28 Olefin as latent keto function with no protecting groups involved.

olefin form and required no protection. Had it been carried as OH group, selective oxidation would have been difficult in the presence of other hydroxyl groups.

In the synthesis of cephalosporolides H and I, Du and coworkers [37b] employed D-gluconolactone (**179**) to make *ent*-**174** [37c], which on α-dimethylation gave **180** (Scheme 9.28). Further the cross-metathesis of the latter with **181** provided olefin **182**. This was subjected to Wacker-type oxidation and *in situ* ketalization to furnish **183** as a single spiro-diastereomer. They observed that if the OTs group in **182** was mesyl or chloride, the spiro-diastereomer mixture was obtained. Finally the compound **183** was converted by attaching the required side chains in two steps to (*S*)-cephalosporolide H (**184**) and in three steps to (*S*)-cephalosporolide I (**185**).

A unique case of hydroxyl protecting group in the form of *ortho*-bromobenzyl serves as blocking group and, during removal, affects the hydroxyl oxidation as well, leading

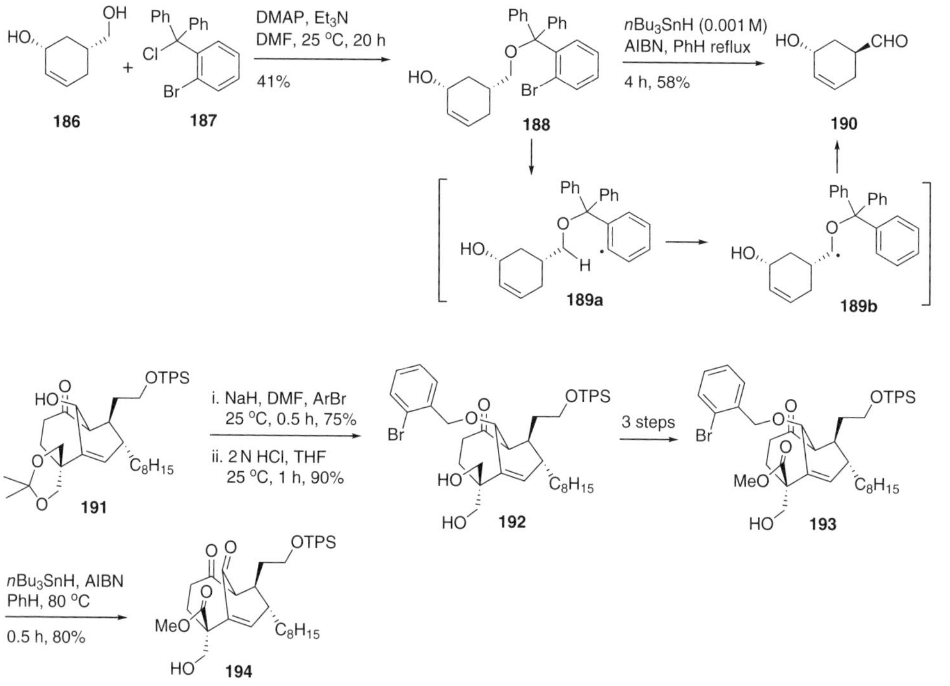

Scheme 9.29 Self-oxidizing *o*-bromobenzyl protecting group for latent carbonyl functionality.

directly to the keto functionality. This is advantageous when the molecule contains other hydroxyl groups, where regioselective oxidation would be impossible. Thus the hydroxyl protecting group serves as a latent keto functionality upon unmasking and avoids carrying the keto function as protected alcohol, which would require deprotection and selective oxidation. This is exemplified by the work of Curran and Yu [38a] as shown in Scheme 9.29. The selective oxidation of 1°-hydroxy group in compound **186** to the aldehyde **190** was achieved via the benzyl-protected compound **188** (using **187**). The radical-mediated deprotection–oxidation involves first generation of aryl radical **189a**, followed by 1,5-hydrogen transfer, giving benzylic or alkoxy radical **189b**. Fragmentation of this produces the aldehyde **190**.

A similar approach was utilized by Nicolaou et al. in their seminal synthetic work on CP molecules [38b]. Thus compound **191** was protected as benzyl ether and acetonide deprotection gave **192** (Scheme 9.29). Convex alcohol oxidation and esterification gave **193**. The latter on benzyl ether deprotection by radical chemistry underwent self-oxidation, giving the ketone **194** in good yield.

9.5 Symmetry-Based Latent Functionality Considerations

Symmetrical objects indeed look beautiful. Alternatively, "beauty lies in the eye of the beholder." An asymmetric molecule as a whole may have hidden symmetrical units. It depends upon the eye of the beholder to visualize the hidden symmetry. Many syntheses of complex molecules have been simplified based on hidden symmetry

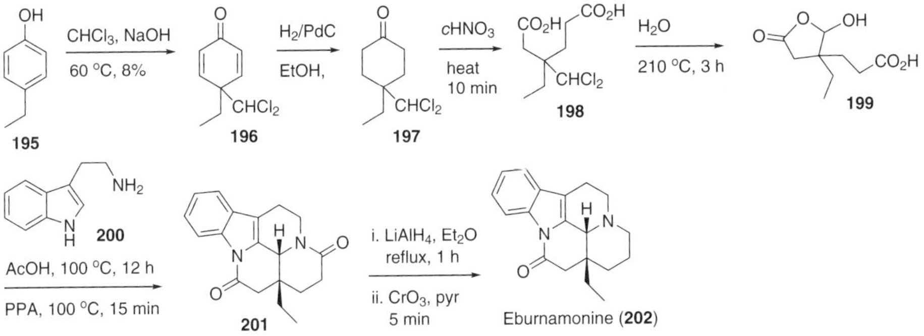

Scheme 9.30 Latent symmetry considerations in eburnamonine (**202**) synthesis.

considerations and retrosynthetically have led to simple molecules. The starting molecules can then be subjected to select reactions for rapid generation of molecular complexity. Thus hidden or otherwise symmetric considerations can be visualized as latent functionalities, which help avoid lengthy sequences or even use of protecting groups in total synthesis. A symmetric starting material can be desymmetrized, leading to the desired complex asymmetric target molecule. This may require reversal of oxidation states on either side of the symmetric molecule or generation of unequal carbon chains whereby functional groups or structural carbons (or heteroatoms) of an otherwise symmetric dimeric molecule get equally divided, yet in a unsymmetrical way generating the molecular complexity which is, as a whole unsymmetric. This can be illustrated by an elegant synthesis of eburnamonine (**202**) in early 1960 by Bartlett and Taylor [39]. The molecule **202** does not appear symmetric but is traced back to simple symmetric dienone **196** obtained through unusual Reimer–Tiemann reaction on *p*-ethylphenol **195** (Scheme 9.30). Reduction of double bonds in **196** led to symmetric cyclohexanone **197**. HNO$_3$-mediated oxidation gave the diacid **198** with different chain lengths from the branching point. This gave the desymmetrized molecule, which on hydrolysis provided the lactol anhydride **199**. Condensation of this with tryptamine (**200**) efficiently delivered the pentacyclic compound **201** with distinct reactions of both nitrogens on the acids **199**. This is uniquely designed such that at the expense of one lactol ring, three rings of **201** have been synthesized. The final steps involved LiAlH$_4$-based reduction of **201** and CrO$_3$ oxidation, furnishing eburnamonine (**202**).

Another example was reported by Tambar et al. in their total synthesis of (±)-trigonoliimine C (**209**) [40] (Scheme 9.31). An unsymmetric indoxyl intermediate **208** was desirable for the success of the synthesis. The nearly symmetric bis-indole compound **206** was synthesized starting from 6-methoxytryptamine **203**. N-formylation, Boc-protection, and C2 stannylation gave compound **204**. This on Stille cross-coupling with bromoindole **205** provided the bis-indole **206** with concomitant Boc deprotection. After a series of optimization, a very selective oxidation of **206** using PhI(TFA)$_2$ resulted in hydroxyindolenine **207**. A selective Wagner–Meerwein [1,2]-shift using HCl in dimethylacetamide gave the desired indoxyl compound **208**. Further deprotection of the phthalimide group and Ti(O*i*Pr)$_4$-mediated imine formation furnished the strained 7-membered (±)-trigonoliimine C (**209**). This could be a general route to many bis-indole natural products with latent structural symmetry.

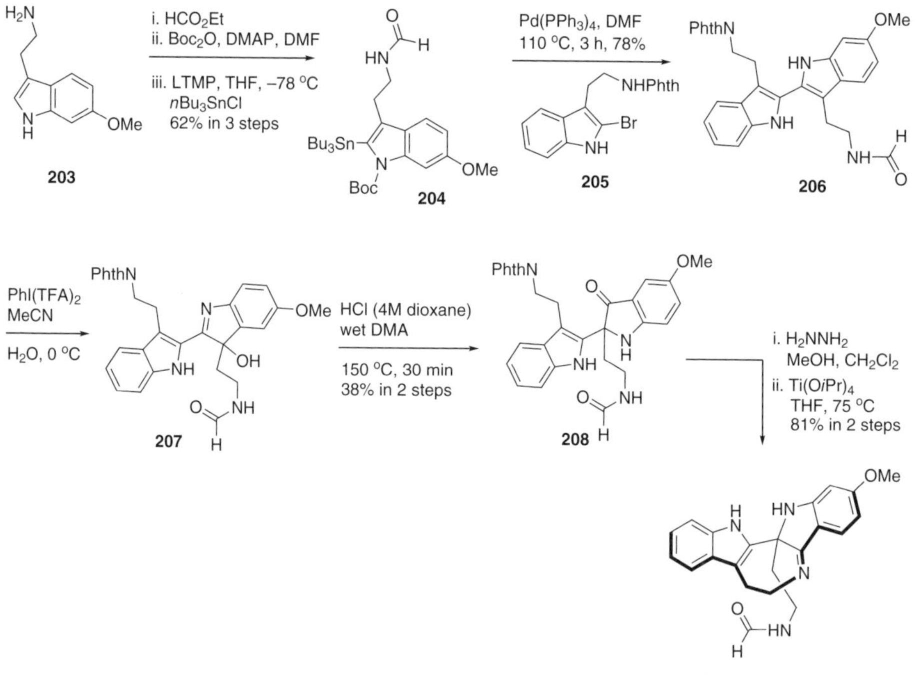

Scheme 9.31 Selective generation of indoxyl in a near symmetric bis-indole.

Somfai et al. explored a similar bis-indole chemistry in their work on (±)-dehaloperophoramidine (**217**) synthesis [41]. The disconnection of all C—N bonds revealed a latent symmetry in the molecule, and this was retrosynthetically traced back to bis(oxindole) compound **212**. The synthesis is shown in Scheme 9.32. The SmI₂-mediated Overman's reductive dialkylation on isoindigo **210** with 1,4-dichloro-2-butene followed by monoimidate formation with Meerwein's reagent and subsequent tosylation of the adjacent NH group gave **211**. The one-pot ozonolytic cleavage of the olefin, reduction to alcohols, and Mitsunobu reaction gave the bis-azide **212**. The differentiation of aminoethyl side chains was achieved by hydrogenation, which initiated a domino process, resulting in the hexacyclic ortho-amide **215**. The process involved reduction of diazide to diamine **213**, lactam formation and amidine ring closure by second aminoethyl group to **214**, and subsequent trapping by sulfonamide to form compound **215**. This reaction beautifully crafted the differentiation of two aminoethyl groups and formed the B and D rings. The final steps involved treatment of ortho-amide **215** to reductive amination under acidic conditions (paraformaldehyde, NaBH(OAc)₃, AcOH, CF₃CH₂OH), resulting in the formation of (±)-dehaloperophoramidine (**217**). This domino process involved conversion of *ortho*-amide **215** to the D/E amidine and 1,3-sulfur shift giving **216a**. This was followed by the A ring formation with the free amine and iminium ion formed with paraformaldehyde leading to **216b**. The subsequent reduction of iminium ion followed by the A/B amidine formation and hydrolysis of sulfonamide gave the target molecule **217**. The support for sulfonamide migration was obtained by isolation and X-ray characterization of the *N*-benzyl compound similar to

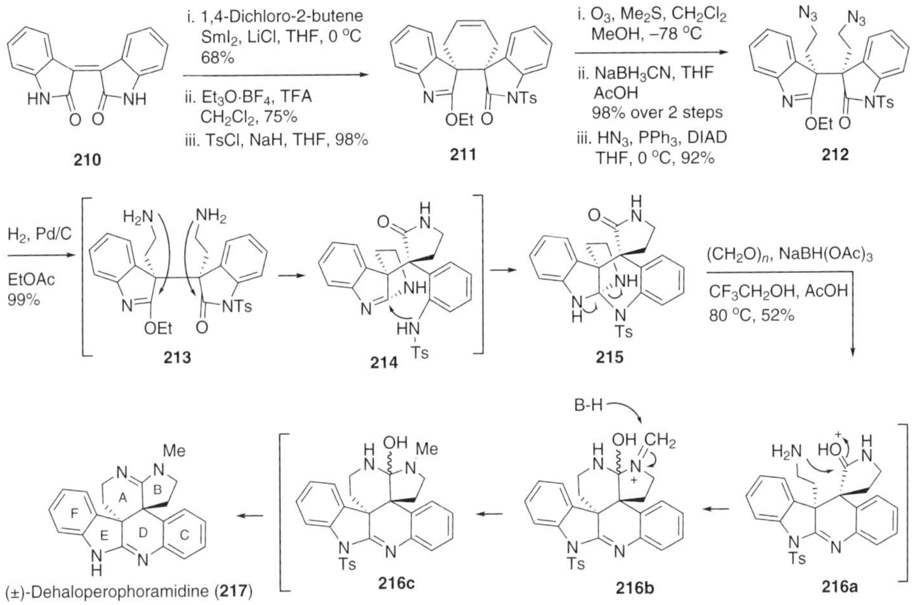

Scheme 9.32 Efficient domino processes in the synthesis of (±)-dehaloperophoramidine (**217**).

216c (instead of *N*-Me) when the reaction was run with benzaldehyde. The two domino processes (**212** to **215** and **215** to **217**) made the synthetic strategy highly step economic and efficient in generating the molecular complexity based on the hidden or latent symmetry in the molecule. The thermodynamic preferences for the hexacyclic ring system with distinct reactivity of the two aminoethyl groups and stepwise generation of the two amidine rings were crucial for success of this strategy.

Anderson and coworkers achieved the total synthesis of the core structure of popolohuanone E (**224**) [42]. From the substituents present on the aryl and quinone moiety, the latent symmetry in the molecule was quite obvious (Scheme 9.33). The biomimetic synthesis of the core structure **223** started from 1,2,4-trimethoxybenzene **218**, which, on lithiation, addition to pivalaldehyde and deoxygenation of benzylic hydroxyl gave **219**. Oxidative coupling of the latter provided the dimer **220**. The global demethylation and oxidation afforded biquinone **221**. This was subjected to a mild rearrangement under basic condition to afford the dibenzofuran-1,4-dione core **223** of popolohuanone E (**224**). The reaction proceeds via the intermediates **222a** and **222b**.

Snyder et al. demonstrated an efficient oxidative dearomatization/Diels–Alder cascade to rapidly access the bicyclo[2.2.2]octene systems and elaborate the intermediates toward the synthesis of rufescenolide and yunnaneic acids C and D [43]. The latent symmetry present in these molecules was tactfully unveiled, and two similar compounds with groups in different oxidation states were considered for distinct generation of the latent diene and its reaction with the dienophile provided by the other partner. As shown in Scheme 9.34, the phenol **226** on reaction with Pb(OAc)$_4$ gave the desired diene, which reacted with excess of acid **225** to give the bicyclo[2.2.2]octene adduct **228** in 69% yield. This upon NaBH(OAc)$_3$ reduction gave a diastereomeric mixture **229** (dr = 1.2 : 1). Reduction of the acetal using TMSOTf/Et$_3$SiH conditions delivered the

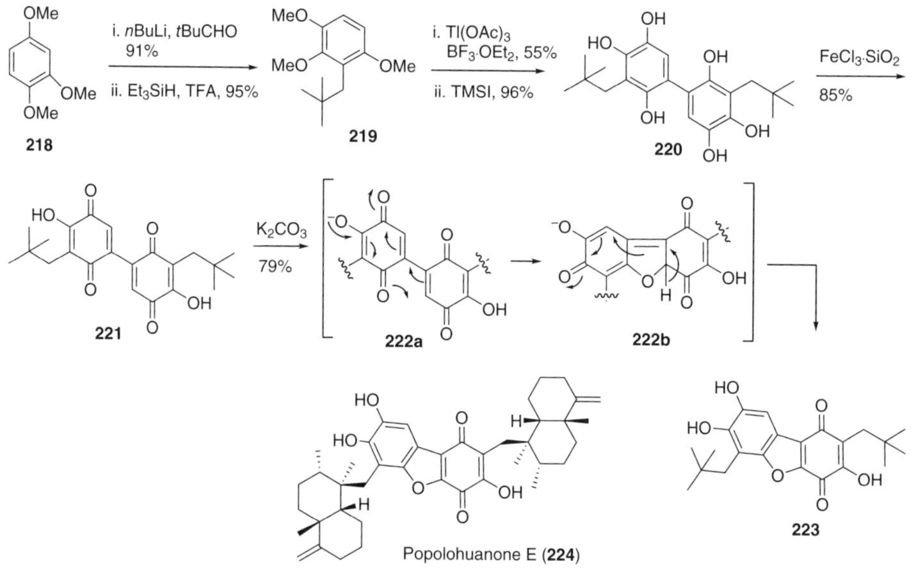

Scheme 9.33 Synthesis of popolohuanone E core structure **223** based on inherent symmetry.

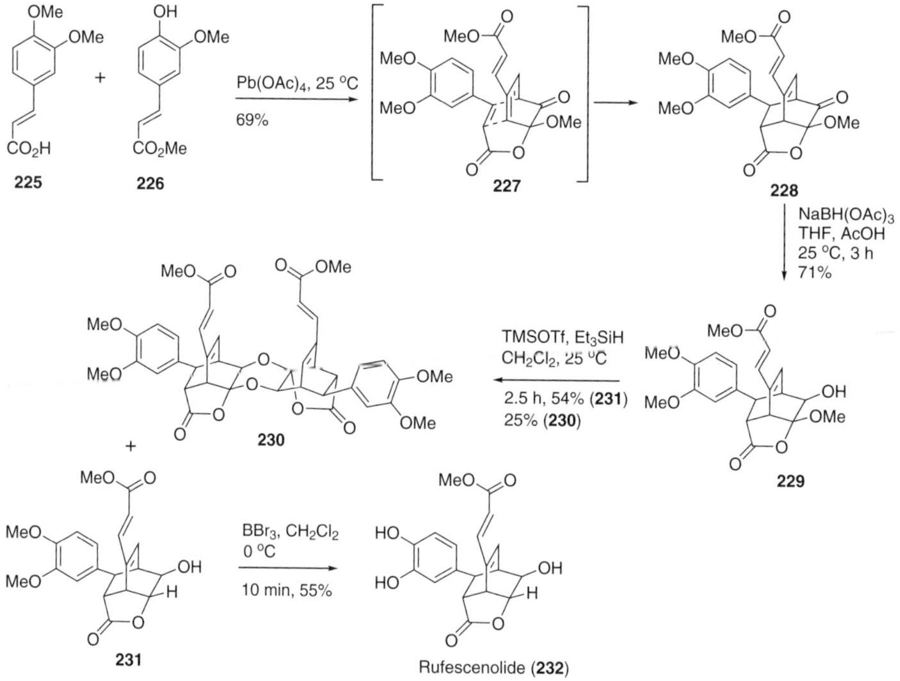

Scheme 9.34 Synthesis of rufescenolide (**232**) based on latent symmetry.

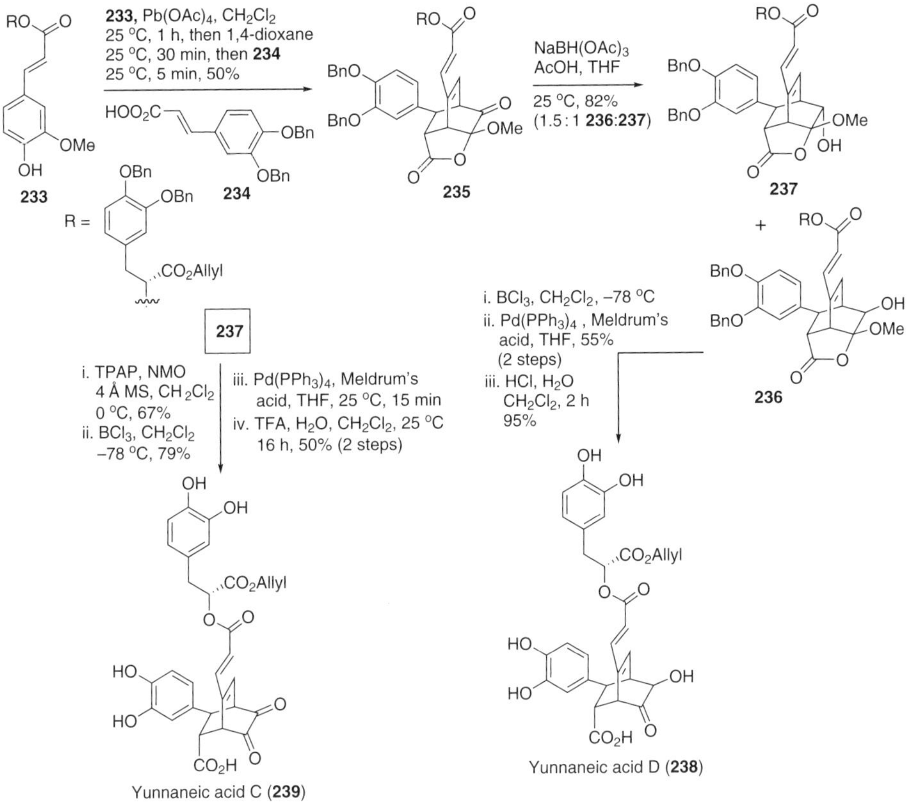

Scheme 9.35 Synthesis of yunnaneic acids C (**239**) and D (**238**) based on latent symmetry.

lactone **231** in 54% yield along with the dimer **230** (25%). The separated lactone **231** was demethylated to rufescenolide (**232**) in 55% yield.

A similar strategy was utilized for the synthesis of yunnaneic acids C and D (Scheme 9.35) [43]. The reaction of phenol **233** with Pb(OAc)$_4$ and then with excess of acid **234** gave the bicyclo[2.2.2]octene adduct **235** in 50% yield as inseparable 1 : 1 diastereomeric mixture. This upon NaBH(OAc)$_3$ reduction gave four diastereomers consisting of two *exo*-alcohols and two *endo*-alcohols that were separated by preparative TLC. The two *endo*-alcohols **237** were oxidized to the ketone, debenzylated with BCl$_3$, deallylated with Pd(PPh$_3$)$_4$, and hydrolyzed to the acid giving yunnaneic acid C (**239**) and the epimer of its enantiomer (not shown). Similarly the *exo*-alcohols **236** on debenzylation, deallylation, and hydrolysis provided yunnaneic acid D (**238**).

Zou and coworkers [44] in 2009 isolated (+)-angiopterlactone B from the rhizome of *Angiopteris caudatiformis*. A hidden symmetry could be visualized in angiopterlactone B that it could arise by intramolecular Michael addition from angiopterlactone A (**246**). In that case the former molecule would involve a δ- to γ-lactone isomerization of type **243** to give **244**, and these can get involved in double Michael addition cascade to give angiopterlactone B. Lawrence and coworkers designed this strategy as shown in Scheme 9.36 [45]. The synthesis started with enantioselective Noyori transfer hydrogenation of **240** to give **241** in 95% yield and 96% ee. The latter on Achmatowicz

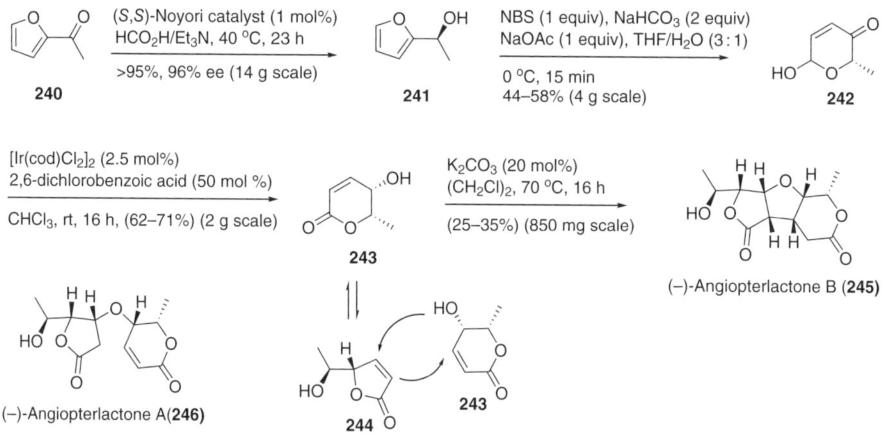

Scheme 9.36 Total synthesis of (−)-angiopterlactone B (**245**).

rearrangement [46] using *N*-bromosuccinimide (NBS) gave pyranone **242** as an incon-sequential mixture of diastereomers. Dynamic kinetic isomerization of pyranone **242** using tandem Brönsted acid and iridium catalysis gave δ-lactone **243** in 62–71% yield on a multigram scale. Further, **243** when treated with potassium carbonate in dichloro-ethane delivered the desired natural product **245** in 25% isolated yield. This reaction presumably involves the isomerization of **243** to **244** and the subsequent double Michael addition. The synthesized molecule **245** had opposite optical rotation to that of natural (+)-angiopterlactone B, indicating the need of revision of its absolute configuration.

A similar chemistry was explored at the same time by Bhattacharya et al. [47] for the synthesis of (−)-angiopterlactone B (**245**) and its antipode and diastereomers and other analogs (Scheme 9.37). Both lactones **248** and **243** were prepared from 3,4-di-*O*-acetyl-L-rhamnal **247**. These were subjected to intermolecular oxa-Michael addition using NaH as base, giving (−)-angiopterlactone B (**245**) in 32% yield. The lactone **248** was recovered unreacted. This indicated the rearrangement of **243** to **244** and subsequent reactions, giving **245**. Screening of various bases for this reaction led to TBAF as best option, wherein **243** gave **245** in 62% yield. The other enantiomer *ent*-**243** provided (+)-angiopterlactone B (*ent*-**245**). From the analytical data it was concluded that the natural product is levorotatory. The other diastereomeric lactones **249** and *ent*-**249** provided other diastereomers **250** and *ent*-**250**, respectively. Various other analogs (not shown here) were also synthesized using TBAF as base.

9.6 Conclusions

The use of latent functionality is beneficial in many syntheses where the actual functional groups are either labile to many reaction conditions or need protection–deprotection sequences. Many syntheses of natural products have been rendered highly efficient, and molecular complexity together with simplicity in execution has been pos-sible by considering the derivation of functional groups by carrying them in masked

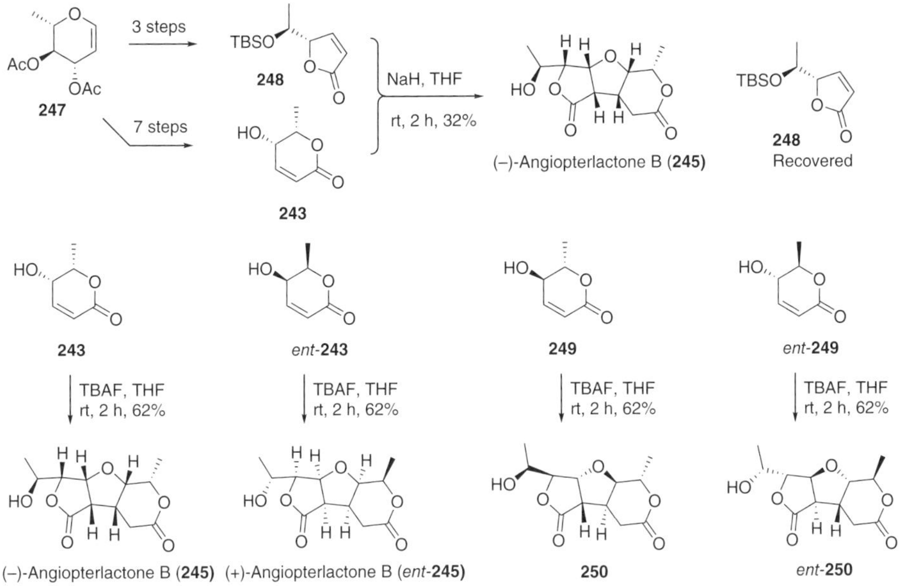

Scheme 9.37 Synthesis of (−)- and (+)-angiopterlactone B and diastereomers.

form and unveiling them into desired functional groups at appropriate stages. The use of furans and benzyl groups to unmask them as oxidized functional groups is noteworthy. The silyl tethering has allowed exploration of various functional groups by employing different reactions for their manipulations. Lastly, the use of latent symmetry consideration has been well executed to construct complex molecules by analyzing the hidden symmetry in otherwise nonsymmetric molecules. Understanding the biogenetic paths has enabled many practitioners to adopt replicable laboratory reactions in their synthesis.

References

1 Wang, W., Wei, G., Yang, X. et al. (2014). *Org. Biomol. Chem.* 12: 7338.
2 Bai, R., Liu, J., Zhu, Y. et al. (2012). *Bioorg. Med. Chem. Lett.* 22: 6490.
3 Qian, H., Huang, W.L., Wu, X.M. et al. (2007). *Chin. Chem. Lett.* 18: 1227.
4 (a) Fu, R., Chen, Z., Wang, Q. et al. (2011). *Atherosclerosis* 219: 40. (b) Fu, R., Yan, T., Wang, Q. et al. (2012). *Vasc. Pharmacol.* 57: 105. (c) Fu, R., Wang, Q., Guo, Q. et al. (2013). *Vasc. Pharmacol.* 58: 78.
5 (a) Olah, G.A., Narang, S.C., and Field, L.D. (1981). *J. Org. Chem.* 46: 3727. (b) Rozov, L.A., Lessor, R.A., Kudzma, L.V., and Ramig, K. (1998). *J. Fluor. Chem.* 88: 51.
6 Parker, D.L. Jr., Fried, A.K., Meng, D., and Greenlee, M.L. (2008). *Org. Lett.* 10: 2983.
7 Sattely, E.S. and Walsh, C.T. (2008). *J. Am. Chem. Soc.* 130: 12282.
8 Jung, H.H. and Floreancig, P.E. (2006). *Org. Lett.* 8: 1949.
9 (a) Faulkner, D.J. (1999). *Nat. Prod. Rep.* 16: 155. (b) Risi, R.M., Maza, A.M., and Burke, S.D. (2015). *J. Org. Chem.* 80: 204.

10 Scarborough, R.M. Jr., Toder, B.H., and Smith, A.B. III (1980). *J. Am. Chem. Soc.* 102: 3904.

11 (a) Haneishi, T., Kitahara, N., Takiguchi, Y. et al. (1974). *J. Antibiot.* 27: 386. (b) Haneishi, T., Terahara, A., Arai, M. et al. (1974). *J. Antibiot.* 27: 393.

12 Koft, E.R. and Smith, A.B. III (1982). *J. Am. Chem. Soc.* 104: 5568.

13 (a) Ramberg, L. and Bäcklund, B. (1940). *Ark. Kemi. Mineral. Geol.* 13A, 27: 1 Chem. Abstr. 1940, 34, 4725.(b) Bordwell, F.G. and Cooper, G.D. (1951). *J. Am. Chem. Soc.* 73: 5184. (c) Conant, J.B., Kirner, W.R., and Hussey, R.E. (1925). *J. Am. Chem. Soc.* 47: 488. (d) Meyers, C.Y., Malte, A.M., and Mathews, W.S. (1969). *J. Am. Chem. Soc.* 91: 7510. (e) Paquette, L.A. (1977). *Org. React.* 25: 1. (f) Taylor, R.J.K. and Casy, G. (2003). *Org. React.* 62: 357.

14 (a) Büchi, G. and Freidinger, R.M. (1974). *J. Am. Chem. Soc.* 96: 3332. (b) Paquette, L.A., Philips, J.C., and Wingard, R.E. Jr. (1971). *J. Am. Chem. Soc.* 93: 4516.

15 Boeckman, R.K. Jr., Yoon, S.K., and Heckendorn, D.K. (1991). *J. Am. Chem. Soc.* 113: 9682.

16 (a) Bols, M. and Skrydstrup, T. (1995). *Chem. Rev.* 95: 1253. (b) Bracegirdle, S. and Anderson, E.A. (2010). *Chem. Soc. Rev.* 39: 4114.

17 Kang, J., Cho, W., and Lee, W.K. (1984). *J. Org. Chem.* 49: 1838.

18 (a) Hudrlik, P.F., Peterson, D., and Rona, R.J. (1975). *J. Org. Chem.* 40: 2263. (b) Hudrlik, P.F., Hudrlik, A.M., Rona, R.J. et al. (1977). *J. Am. Chem. Soc.* 99: 1993. (c) Hirao, T., Yamada, N., Ohshiro, Y., and Agawa, T. (1982). *Chem. Lett.* 1997.

19 Ehlinger, E. and Magnus, P. (1980). *J. Am. Chem. Soc.* 102: 5004.

20 (a) Cella, J.A. and Kagawa, C.M. (1957). *J. Am. Chem. Soc.* 79: 4808. (b) Dodson, R.M. and Tweit, R.C. (1959). *J. Am. Chem. Soc.* 81: 1224. (c) Cella, J.A., Brown, E.A., and Burtner, R.R. (1959). *J. Org. Chem.* 24: 743. (d) Cella, J.A. and Tweit, R.C. (1959). *J. Org. Chem.* 24: 1109. (e) Brown, E.A., Muir, R.D., and Cella, J.H. (1960). *J. Org. Chem.* 25: 96. (f) Atwater, N.W., Bible, R.H., Brown, E.A. et al. (1961). *J. Org. Chem.* 26: 3077. (g) Lenz, G.R. and Schulz, J.A. (1978). *J. Org. Chem.* 43: 2334. (h) Heusler, K. (1952). *Helv. Chim. Acta.* 45: 1939. (i) Neef, G., Eder, U., and Wiechert, R. (1978). *J. Org. Chem.* 43: 4679.

21 Paquette, L.A. and Han, Y.-K. (1981). *J. Am. Chem. Soc.* 103: 1831.

22 Yao, Q. (2001). *Org. Lett.* 3: 2069.

23 Kim, Y.J. and Lee, D. (2006). *Org. Lett.* 8: 5219.

24 (a) Tamao, K., Tanaka, T., Nakajima, T. et al. (1986). *Tetrahedron Lett.* 27: 3377. (b) Tamao, K., Nakajima, T., Sumiya, R. et al. (1986). *J. Am. Chem. Soc.* 108: 6090. (c) Tamao, K., Nakagawa, Y., and Ito, Y. (1990). *J. Org. Chem.* 55: 3438.

25 (a) O'Malley, S.J. and Leighton, J.L. (2001). *Angew. Chem. Int. Ed.* 40: 2915. (b) Spletstoser, J.T., Zacuto, M.J., and Leighton, J.L. (2008). *Org. Lett.* 10: 5593.

26 (a) Jones, G.R. and Landais, Y. (1996). *Tetrahedron* 52: 7599. (b) Polniaszek, R.P. and Dillard, L.W. (1992). *J. Org. Chem.* 57: 4103.

27 Ichihara, A., Kimura, R., Yamada, S., and Sakamura, S. (1980). *J. Am. Chem. Soc.* 102: 6355.

28 Büchi, G. and Wüest, H. (1966). *J. Org. Chem.* 31: 977.

29 (a) Schmid, G., Fukuyama, T., Akasaka, K., and Kishi, Y. (1979). *J. Am. Chem. Soc.* 101: 259. (b) Agtarap, A., Chamberlin, J.W., Pinkerton, M., and Steinrauf, L.K. (1967). *J. Am. Chem. Soc.* 89: 5737. (c) Pinkerton, M. and Steinrauf, L.K. (1970). *J. Mol. Biol.* 49: 533.

30 Kobayashi, Y., Kumar, G.B., Kurachi, T. et al. (2001). *J. Org. Chem.* 66: 2011.

31 Evans, D.A., Nagorny, P., Reynolds, D.J., and McRae, K.J. (2007). *Angew. Chem. Int. Ed.* 46: 541.

32 Waizumi, N., Stankovic, A.R., and Rawal, V.H. (2003). *J. Am. Chem. Soc.* 125: 13022.

33 Ireland, R.E., Dawson, M.I., Welch, S.C. et al. (1973). *J. Am. Chem. Soc.* 95: 7829.

34 Danishefsky, S., Cain, P., and Nagel, A. (1975). *J. Am. Chem. Soc.* 97: 380.

35 Wang, Z. and Deschênes, D. (1992). *J. Am. Chem. Soc.* 114: 1090.

36 Bonjoch, J., Solé, D., García-Rubio, S., and Bosch, J. (1997). *J. Am. Chem. Soc.* 119: 7230.

37 (a) Chaudhari, D.A., Pullaiah, K., and Fernandes, R.A. (2015). *RSC Adv.* 5: 42131.
(b) Li, J., Zhao, C., Liu, J., and Du, Y. (2015). *Tetrahedron* 71: 3885. (c) Fernandes, R.A. and Kattanguru, P. (2013). *Asian J. Org. Chem.* 2: 74.

38 (a) Curran, D.P. and Yu, H. (1992). *Synthesis* 123. (b) Nicolaou, K.C., He, Y., Fong, K.C. et al. (1999). *Org. Lett.* 1: 63.

39 Bartlett, M.F. and Taylor, W.I. (1960). *J. Am. Chem. Soc.* 82: 5941.

40 Qi, X., Bao, H., and Tambar, U.K. (2011). *J. Am. Chem. Soc.* 133: 10050.

41 Hoang, A., Popov, K., and Somfai, P. (2017). *J. Org. Chem.* 82: 2171.

42 Anderson, J.C., Denton, R.M., and Wilson, C. (2005). *Org. Lett.* 7: 123.

43 Griffith, D.R., Botta, L., St. Denis, T.G., and Snyder, S.A. (2014). *J. Org. Chem.* 79: 88.

44 Yu, Y.-M., Yang, J.-S., Peng, C.-Z. et al. (2009). *J. Nat. Prod.* 72: 921.

45 Thomson, M.I., Nichol, G.S., and Lawrence, A.L. (2017). *Org. Lett.* 19: 2199.

46 Croatt, M.P. and Carreira, E.M. (2011). *Org. Lett.* 13: 1390.

47 Kotammagari, T.K., Gonnade, R.G., and Bhattacharya, A.K. (2017). *Org. Lett.* 19: 3564.

Index

Protecting-Group-Free Organic Synthesis: Improving Economy and Efficiency, First Edition.
Edited by Rodney A. Fernandes.
© 2018 John Wiley & Sons Ltd. Published 2018 by John Wiley & Sons Ltd.